C0-BRL-304
WITHDRAWN

GEOLOGICAL INVESTIGATIONS OF THE NORTH PACIFIC

The Geological Society of America, Inc.
Memoir 126

Geological Investigations of the North Pacific

Edited by

JAMES D. HAYS

1970

Library of Congress Catalog Card Number 75-111442
S.B.N. 8137-1126-6

Published by
THE GEOLOGICAL SOCIETY OF AMERICA, INC.
Colorado Building, P. O. Box 1719
Boulder, Colorado 80302

Printed in The United States of America

The printing of this volume
has been made possible through generous
contributions to the Memoir Fund of
The Geological Society of America

Maurice Ewing

Dedication

Few have contributed as much to the exploration of the oceans as Professor Maurice Ewing. We owe most of the data contained in this volume to his foresight, energy and determination.

Maurice Ewing, known to many as "Doc," combines a variety of talents that has enabled him to achieve more than most men. His first love is science, and he possesses that keen scientific intuition that instinctively knows the important projects to follow and how to follow them. Through this searching he has made major contributions to a number of fields. Early in his career he contributed to our understanding of the structure and thickness of the continental margins and oceanic crust. Later, in a series of worldwide seismic reflection surveys, he delineated the pattern of sediment thickness in all the major ocean basins, and recently, he has been active in the study of lunar seismicity.

His interests, however, reached far beyond seismology; he has made major contributions to our knowledge of submarine topography, marine gravity and magnetics, as well as the sediments carpeting the ocean floor. In this latter area, he has added to our understanding of past climatic changes and proposed a theory to explain them.

Maurice Ewing is more than a researcher. He has excelled as an administrator and a professor. As a professor, his success is demonstrated by the prestige achieved by his students, many of whom are now leaders in their fields.

His life has been characterized by opportunities taken advantage of and challenges met face on. His greatness is marked by the size and importance of the problems he selected to study.

The year 1970 finds geology undergoing a profound revolution. Theories of the mobility of the earth's lithosphere and its shifting rigid plates are providing answers to questions geologists have been asking for generations, and new ones are being raised that will take years for geologists to answer. This revolution has occurred because of ocean basin exploration, to a great extent due to the work of Maurice Ewing.

Maurice Ewing was born in 1906 in Lockney, Texas, and his early life on the trackless open range of West Texas may have contributed to his later fascination for the rolling expanse of the open ocean. He studied physics at Rice Institute and while there published, at the age of 20, his first scientific paper. This paper, printed in *Science,* was entitled "Dewbows by Moonlight" and showed a keen scientific awareness even during his undergraduate nocturnal activities.

He received his Ph.D. in physics from Rice in 1931 and thereafter taught physics at both Pittsburgh and Lehigh Universities. During these years he began his exploration of the oceans with geophysical work on the eastern coastal plain and continental shelf. As his studies progressed toward deeper water, his interests expanded into other aspects of the sea. Instrumentation for exploring many aspects of the sea was, at that time, limited or nonexistent. A great deal of his effort was put into building equipment for exploration (bottom cameras, bathythermographs, corers, precision depth recorders, and so on). These early efforts in instrumentation were a prelude to the extent of Ewing's life's work. Later, when he came to Columbia (1944) and founded the Lamont Geological Observatory (1949), he sent Lamont's research vessels with their instruments (many of which he designed) to all corners of the world's ocean. His goals became the exploration and understanding of the floor of the entire ocean.

Few men have cut out for themselves such an ambitious task. Even fewer have accomplished so much once set on such a course. In order to do so, Maurice Ewing had to have a combination of talents that is rare even in great men. The exploration of the world's ocean, covering three-quarters of the Earth's surface, is not a task for a lonely investigator. It required marshalling much support in talent and funds, and inspiring those whom he gathered around him to achieve greatness themselves. While he strove to study the sea as he thought it should be studied, he built a large oceanographic institution that is now world famous, and has been its Director from its inception some 20 years ago. During these years, he has worked tirelessly to get the most from the people who work with him and from his ships. As a result, Lamont-Doherty ships spend little time in port, and when Maurice Ewing goes to sea, his tireless activity is an inspiration to all those who follow him. As a consequence, he has accomplished his goals of delineating the major features of the ocean's floor and understanding many of the processes that form and shape them. In spite of his many accomplishments and honors, Maurice Ewing continues to initiate new projects and generate new ideas. In recognition of his greatness, we wish to dedicate this volume to him.

J. D. H.

Contents

Preface

This volume is the natural outgrowth of the efforts of a group of scientists, primarily at the Lamont-Doherty Geological Observatory, who have a common interest—the North Pacific. In the summer of 1969, the work at Lamont-Doherty on the paleomagnetics of deep-sea cores was less than two years old, and the feeling of excitement and discovery that came with the application of this technique to each new area of the world's ocean was still running high. The North Pacific cores taken by R/V *Vema* on its twentieth cruise had yielded the best paleomagnetic results to date. These cores had much interest for a number of us; they contained high concentrations of diatoms, Radiolaria, and ice-rafted debris, as well as numerous discrete ash layers. They obviously provided an unprecedented opportunity to do a stratigraphic study of deep-sea cores.

The stratigraphers at Lamont-Doherty, however, were not the only ones interested in the North Pacific at this time. The sedimentologists were just embarking on a study of the granulometric properties of the sediments, and the marine seismologists were completing a study of the sediment thickness. The paleomagnetic and geomagnetic revolution was in full swing, and the specialists in geomagnetics were taking a second look at the magnetics of the now classic Juan de Fuca and Gorda ridges, as well as struggling with the problem of the magnetic bight in the Gulf of Alaska. In the meantime, new geomagnetic data and additional cores were pouring in from R/V *Conrad,* then operating on its tenth cruise in the western North Pacific.

With all this interest in the North Pacific, much of it interrelated, we decided we should do all we could to see that the papers that were a product of this work were published together. So was born, in the summer of 1967, the concept of our "North Pacific Volume." The idea was greeted with enthusiasm, and others volunteered to contribute. Although the final results of our efforts were at first not clear, nor did we know how the papers would relate to one another, we were all committed to producing a volume that would be comprehensive in scope.

The contributions group themselves into three categories. The first consists of three closely related papers dealing with the nature and distribution of bottom sediment, bottom water, and near-bottom suspended sediment. They contain new data on the size of the surface sediment particles, the flow of bottom water, and the interaction between bottom water and the sediments. They suggest sources for the sediment and the pattern of its dispersal across the sea floor, thereby providing a sound framework on which future investigators can build.

The second category, consisting of seven papers, is stratigraphic, and many use the excellent paleomagnetic data presented in the first of the series by Opdyke and Foster. These papers are closely interrelated and as a group make the following notable contributions. Paleomagnetic stratigraphies have been determined on the largest number of cores raised from a single ocean basin and reported in one paper. From these data, rates of sedimentation are ascertained and compared with total sediment thickness reported in previous studies. In general, slow rates of sedimentation correspond with thin sediments, and high rates with thick sediments. Exceptions to this occur south of the Aleutians and may be explained by anomalously high sedimentation rates in late Tertiary and Quaternary time. New information is presented on the geographic and paleogeographic distribution of diatoms and Radiolaria. Paleomagnetically correlated stratigraphies are established that define biostratigraphic zones based on these fossils back through the Upper Pliocene. The relationship between magnetic reversals and the extinction of diatoms and Radiolaria are examined and, in one case, the extinction is related to the evolutionary history of the species.

Paleoclimatic interpretations are made from the faunal and floral data gathered; pronounced climatic cycles are evident in the Brunhes magnetic epoch, while less pronounced cycles occur and generally milder conditions are indicated for the Matuyama reversed epoch below the Jaramillo event. The same climatic pattern is reflected in the ice-rafted detritus. Paleoclimatic evidence based on pollen grains is presented for the region bordering Mexico. The sequence of volcanic ash layers found in the far North and Northwestern Pacific sediments is correlated, dated, and the volcanic activity that produced them related to marginal underthrusting of the Pacific Plate.

The final paper of the volume presents much new geomagnetic data across the entire North Pacific and suggests ages for the basement rock. This paper also presents a model for the geological evolution of the Pacific floor through the Tertiary.

It is our hope that the broad scope of this volume and the diverse data it contains will make it a useful guide to future investigators studying the North Pacific.

We are very grateful to Dr. Tj. H. van Andel and the reviewers designated by him for their rapid and penetrating analyses of our manuscripts. Mrs. Rose Marie Cline's efforts through all phases of the production of this book are much appreciated.

J. D. H.

THE GEOLOGICAL SOCIETY OF AMERICA, INC.
MEMOIR 126, 1970

Sedimentary Provinces of the North Pacific

D. R. Horn
B. M. Horn
M. N. Delach

Lamont-Doherty Geological Observatory of Columbia University, Palisades, New York

ABSTRACT

A survey of 250 long piston cores from the North Pacific Ocean reveals that cores taken within definite areas of the sea floor have common sediment characteristics. These belong to distinct regions and are named sedimentary provinces.

Three orders of sedimentary provinces occur in the North Pacific Basin. First order provinces follow the general form of the basin and are of great areal extent. They are sites of pelagic deposition and are named the Central North Pacific, Japan-Kuril, and Aleutian-Alaskan Provinces. Because the North Pacific Basin is protected from large-scale basinward transfer of terrigenous sediment, the limits of first order provinces match the boundaries of major water masses.

Second order provinces are dominated by features of positive submarine relief, are restricted in size, and their boundaries follow the configuration of hills and ridges. They include the Hawaiian Ridge, Marcus-Necker Ridge, Shatsky Rise, Emperor Seamount Chain, and Ridge and Trough Provinces. Sediment accumulating on and around these features is a function of the depth of crestal portions of submarine highs.

Third order provinces constitute a narrow zone of terrigenous sediment around the rim of the North Pacific Basin. Included are the floors of the circum-Pacific trench system and a portion of the northeastern corner of the Pacific. Two are named: the Aleutian Trench and the Northeast Pacific Turbidite Provinces.

CONTENTS

INTRODUCTION

During the past five years, Lamont-Doherty Geological Observatory has conducted extensive oceanographic data gathering in the Pacific Ocean. Coring operations included in this program have resulted in the collection of 730 long piston cores from the North Pacific; 250 of these cores were taken north of a line that passes east-west through Hawaii (20° N. lat.) and constitute the basis of this report (Fig. 1).[1] The cores have been described and analyzed. Results indicate that there are large areas of the North Pacific within which cores have common characteristics. These regions have been delineated and are named sedimentary provinces.

The regional pattern of sedimentation within the North Pacific has been known for at least 35 years and this region has long been recognized as an area of very slow accumulation. Only the finest

[1]Latitudes and longitudes of coring stations are available from the Core Laboratory, Lamont-Doherty Geological Observatory, Palisades, New York 10964.

Figure 1. Index map of study area. Locations of the 250 cores that were analyzed are given on Plate 1.

sediment fractions derived from neighboring continents and the remains of planktonic plants and animals reach the floor of this deepest of ocean basins (W. Schott *in* G. Schott, 1935; Revelle, 1944; Sverdrup and others, 1942; Revelle and others, 1955; Goldberg and Arrhenius, 1958; Bramlette, 1961; Arrhenius, 1963; Menard, 1964; Ewing and others, 1968; Griffin and others, 1968; Windom, 1969; and many others).

Much of the early work was directed toward an understanding of the regional picture of pelagic sedimentation and associated geochemical processes. Recent articles tend to deal with problems of marine sedimentation within selected areas. For example, the sediments and marine geology of the Gulf of Alaska have received considerable attention during the past decade. Hurley (1959) provided an excellent survey of the submarine physiography and sediments of abyssal plains in the northeast corner of the Pacific; Nayudu (1958, 1959, 1964; *with* Enbysk, 1964) analyzed surface sediments of the Gulf; whereas Hamilton (1967) gave a complete picture of the marine geology of the region based on acoustic reflection profiles. Most recent is the work of investigators at Oregon State University who have added to the knowledge and understanding of deep-sea sedimentation off the northwest coast of the United States (Kulm and Nelson, 1967; Carlson, 1968; Duncan, 1968; Nelson, 1968; Griggs and others, 1969; Griggs and Kulm, 1969; and others).

The world-wide and deep-water coring activities of Lamont-Doherty Geological Observatory offer possibly the most complete collection of data on abyssal sediments of the world's ocean. In this article we have returned to the regional survey approach of early investigators in an attempt to better define major sedimentary provinces within the North Pacific Ocean. Results presented here are taken from a survey of sonic properties of bottom materials of the North Pacific (Horn and others, 1968).

The cores average 25 feet (7.6 m) in length and are 2.5 inches (6.4 cm) in diameter. Emphasis is placed on textures of the cores. Following the procedures outlined by Folk (1968), 1500 samples were processed. In short, sand and gravel fractions were sieved through 8-inch screens selected at ¼ phi intervals, and samples were agitated in a ROTAP shaker set at 15 minutes per sample. Sieves were calibrated prior to their use to determine accuracy of mesh sizes. Textures of silt, mud, and clay were determined by the pipette method, with aliquotes taken at ½ phi intervals. Cumulative curves were plotted on probability paper, and percentiles taken from the curves were inserted into formulas of graphic measures of grain size defined by Folk and Ward (1957). Textural classification and verbal limits of grain-size measures are those of Folk (1954, 1966, 1968) and Folk and Ward (1957).

Binocular microscope examination of all sand fractions was carried out. No attempt has been made to provide more than a cursory statement concerning biogenic materials in the cores because these have been the subject of earlier studies and are discussed in other contributions in this volume (Hays, 1970; Nigrini, 1970, and Donahue, 1970, this volume).

SEDIMENTARY PROVINCES

The North Pacific Basin is actually two depressions separated by a submarine high (Fig. 2). This high consists of a rugged line of mountains that constitute the Emperor Seamount Chain and the Hawaiian Ridge (Figs. 2 and 3). The basin is trapezoidal in outline with North America and Asia forming the eastern and western limits. The Aleutian Arc and Trench define the northern limit,

Emperor Seamount Chain, Hawaiian Ridge, and Ridge and Trough Province off the northwestern coast of the United States. Third order provinces include areas of relatively rapid terrigenous deposition and encompass the floors of the circum-Pacific trench system and the northeast corner of the Pacific Ocean. They include the Aleutian Trench Province and the Northeast Pacific Turbidite Province. No names have been given sedimentary provinces of the Kuril and Japan Trenches because no core data are available in these regions. Presumably, cores from these features will contain turbidite sequences and such regions will be third order sedimentary provinces.

First Order Provinces

Japan-Kuril Province and Aleutian-Alaskan Province. Western and northern regions of the North Pacific Basin are marked by a very broad zone of mud. The sediments are clay with admixed silt-sized biogenic material and volcanic detritus (Figs. 3 and 4; Pl. 1). The belt of sediment is at the periphery of the basin and extends at least 500 miles from the nearest land. Terrigenous volcanic silt occurs in some cores taken as much as 1000 miles from land. It is not to be confused with ice-rafted detritus. Presumably, the buoyancy and fine grain size of the particles result in their widespread dispersal by ocean currents.

We find that ubiquitous and volumetrically important fine-grained volcanic detritus occurs throughout the provinces and appears to reflect seaward dispersal of land-derived material by normal oceanic circulation. Some investigators have attributed much of the detrital phases of sediments of the North Pacific to wind transport (Rex and Goldberg, 1958; Griffin and others, 1968; Windom, 1969). However, when the results of measurements of sediment thicknesses (Ewing and others, 1968) are combined with new data on rates of deposition (Opdyke and Foster, 1970, this volume), the nepheloid layer (Ewing and Connary, 1970, this volume), and the location of major sedimentary provinces, it may well be that oceanic circulation and properties of major water masses control dispersal of finest detrital materials and pelagic sedimentation in the North Pacific.

The broad zone of abyssal sediment at the margin of the North Pacific Basin has been divided into two provinces. That portion west of the Emperor Seamount Chain is named the Japan-Kuril Province; that to the east is called the Aleutian-Alaskan Province (Fig. 3). Both are dominated by mud, which is intercalated with frequent, thin, silt-sized ash horizons (Fig. 4; Pl. 1). A major distinguishing feature, however, exists. The Japan-Kuril Province has white ash as a major constituent, whereas both brown and white ashes occur in cores from the Aleutian-Alaskan Province (Fig. 8). These provinces also are separated by a rugged topographic barrier (Emperor Sea-

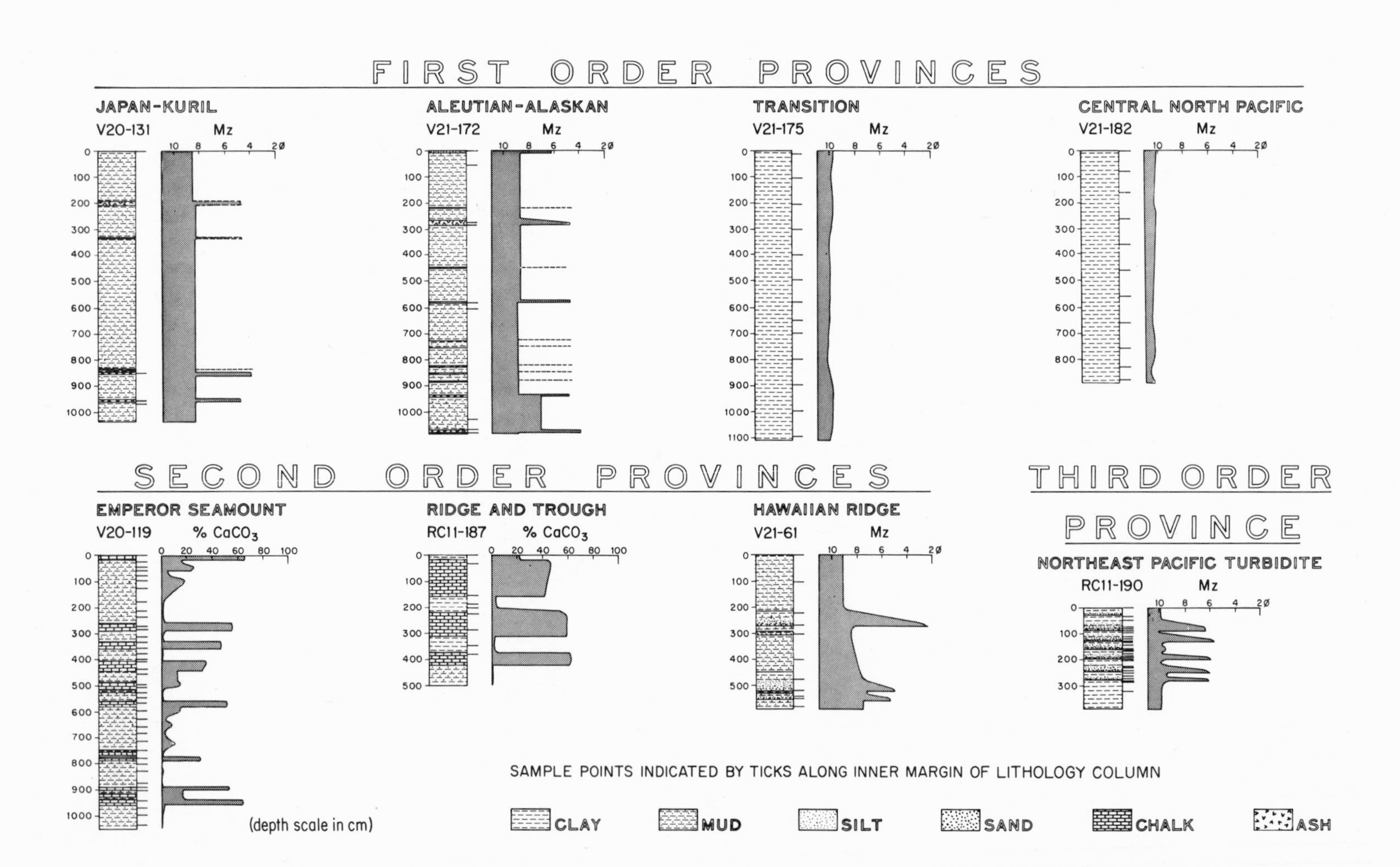
FIRST ORDER PROVINCES
JAPAN-KURIL
V20-131
Mz
ALEUTIAN-ALASKAN
V21-172
Mz
TRANSITION
V21-175
Mz
CENTRAL NORTH PACIFIC
V21-182
Mz
SECOND ORDER PROVINCES
EMPEROR SEAMOUNT
V20-119
% $CaCO_3$
RIDGE AND TROUGH
RC11-187
% $CaCO_3$
HAWAIIAN RIDGE
V21-61
Mz
THIRD ORDER PROVINCE
NORTHEAST PACIFIC TURBIDITE
RC11-190
Mz
(depth scale in cm)
SAMPLE POINTS INDICATED BY TICKS ALONG INNER MARGIN OF LITHOLOGY COLUMN
CLAY
MUD
SILT
SAND
CHALK
ASH

mount Chain) and represent peripheral areas of distinct sedimentary basins. Lithologies and textures of representative cores are given in Figure 4.

Biogenic silica in the form of diatom frustules and skeletons of Radiolaria is common throughout the province. However, it does not constitute 30 percent of the sediment and therefore these deposits cannot be considered diatom or radiolarian oozes. Only in rare instances are biogenic materials a dominant fraction of the sediment (*see* Plate 1, for example, Core RC10-202, the lower part of the core is pure diatomite). This corroborates Bramlette's finding (1961) that the diatomaceous deposits of the North Pacific are not diatom oozes, because they generally contain less than 10 percent diatom frustules. Similarly, the skeletons of Radiolaria rarely constitute more than a few percent of the total volume of the sediment. It is suggested that the Japan-Kuril and Aleutian-Alaskan Provinces are sites of slow accumulation of the finest sediment fractions derived from neighboring land. This material bypasses the major barriers and traps to seaward dispersal of sediment by being transported in suspension and finally settles along with biogenic material.

The southern limits of these first order provinces may be a function of two phenomena, which are in turn controlled by a feature common to both. The phenomena are the production of organic detritus and the oceanic transfer of inorganic particles in suspension by currents. Both may be directly related to the physical properties and circulation of major water masses within the North Pacific Basin.

Waters of maximum estimated photosynthetic productivity coincide with the Japan-Kuril Province and the Aleutian-Alaskan Province (*see* Fleming, 1957, Fig. 10). The southern limit of these provinces is also remarkably close to the *subarctic boundary* between

---

← Figure 4. Representative cores from the sedimentary provinces. Sediments of the Japan-Kuril and Aleutian-Alaskan Provinces are dominated by muds interlayered with thin ash horizons. There is an over-all decrease in Mz of sediments toward the center of the North Pacific Basin. Cores V20-131 and V21-172 have an average mean grain size of 8.5 phi, samples from the transition zone are just coarser than 10 phi, whereas those from the Central North Pacific Province characteristically are less than 10 phi. The seaward decrease in mean grain size is matched by a progressive reduction in rates of deposition (Opdyke and Foster, 1970, this volume), suggesting that proximity to land is reflected in the particle size of the sediments. Cores V20-119 and RC11-187 are from the crests of submarine topographic highs that lie above the compensation depth of calcium carbonate. The sediments are calcareous ooze interlayered with pelagic mud and clay. Cores taken in the vicinity of the Hawaiian Ridge contain turbidites (V21-61) indicating the importance of turbidity-current activity in the leveling of sea floor topography and the development of archipelagic aprons. Core RC11-190 is representative of turbidite sequences of the Northeast Pacific Turbidite Province. The textural profile reveals a history of repeated addition of relatively coarse terrigenous sediment by turbidity currents that flooded out across the southern Tufts Abyssal Plain.

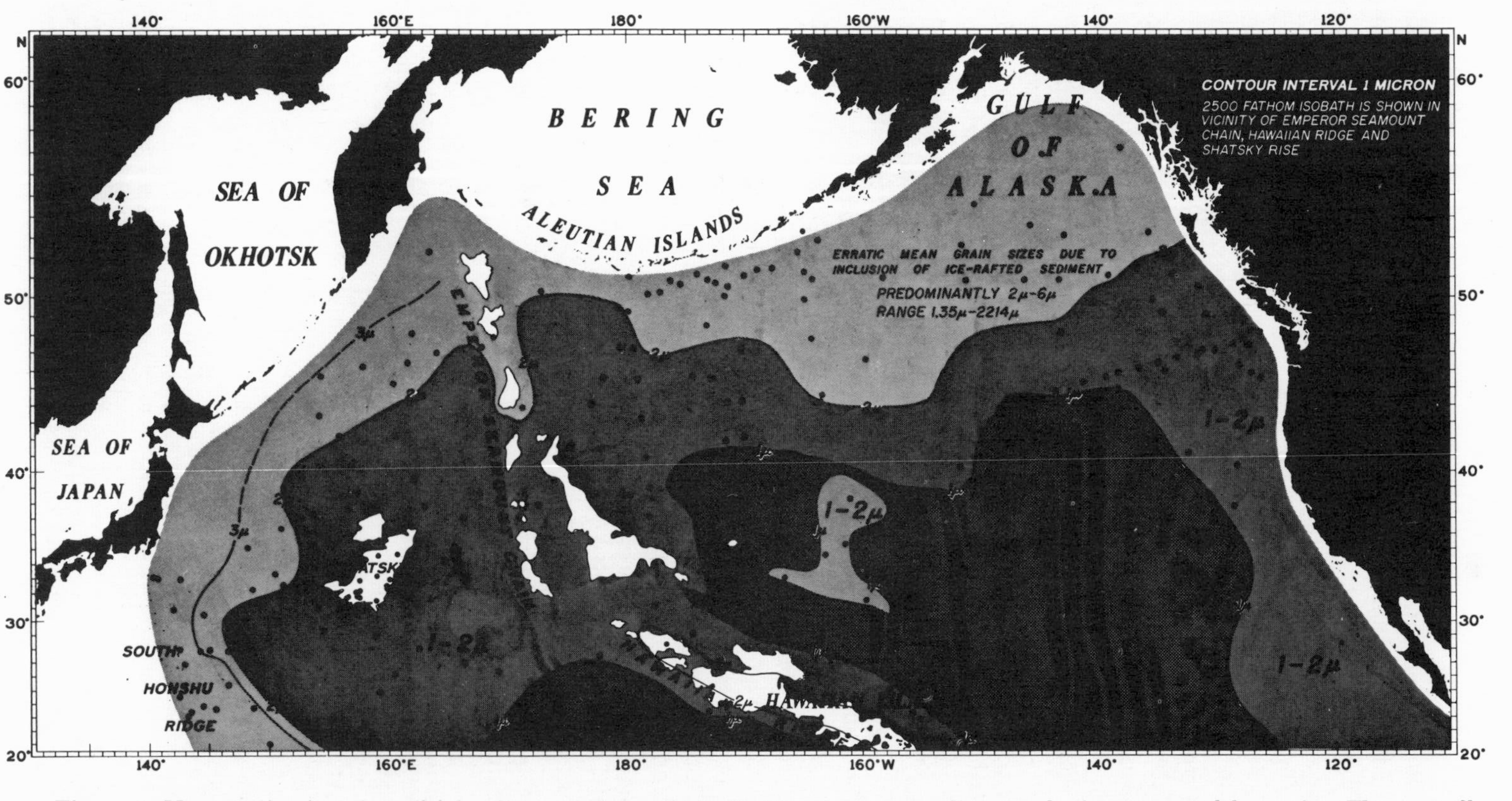

Figure 5. Mean grain size of surficial sediments of the North Pacific. Contours outline two basins separated by a rise. The over-all pattern of the grain size contours suggests that proximity of source, whether submarine or continental, determines the mean grain size of surficial sediments.

the West Wind Drift and Subarctic Current Systems, and the North Pacific Current (Dodimead and others, 1963, Fig. 109). It may be that sediments accumulating in the provinces are not truly pelagic (Bramlette, 1961) because of the addition of significant amounts of very fine-grained volcanic silt of low specific gravity (pumice shards) which are carried great distances by trans-Pacific, west- to east-moving current systems north of the *subarctic boundary*. The combination of high biogenic productivity in surface waters and oceanic transfer and deposition of terrigenous material may combine to produce the sediment characteristic of the Japan-Kuril and Aleutian-Alaskan Provinces.

A transition zone exists between the large peripheral provinces and the Central North Pacific Province (Fig. 3). It is approximately 400 miles wide in the northeast Pacific, and 800 miles wide in the northwest basin. Limits of the transition zone are not clearly defined due to a paucity of core data. It should be pointed out that cores from this zone are more closely related to the Central North Pacific Province than to areas lying to the north. Clays showing large-size burrow mottling and only traces of Radiolaria are characteristic. Enormous distances to the nearest land and extremely low rates of sediment accumulation (Opdyke and Foster, 1970, this volume) preclude abrupt boundaries between first order provinces.

Central North Pacific Province. This region encompasses two areas of "red clay" deposition separated by a submarine high—the Hawaiian Ridge. Cores from the Central North Pacific Province are very uniform light gray to light brownish-gray clay (Munsell soil color chart, 10YR 7/2 to 10YR 6/2). Colors commonly are darker at depth in the cores.

Between the Hawaiian Ridge and the North American continent, "red clays" extend from at least 20° N. (the southern limit of the area investigated) to approximately 40° N. West of the Hawaiian Ridge the clay belt is narrower and extends from the Marcus-Necker Ridge north to about 30° N. The restricted area of clay may be due to the smaller size of the Northwest Pacific Basin, shorter distances to the nearest land, and the presence of the Shatsky Rise in the middle of the basin (Fig. 3). Pure clays in the northwest Pacific clearly extend to 30° N. and might well have been the dominant sediment to 40° N. had it not been for the presence of the Shatsky Rise which represents a local source of sediment and blurs the northern limit of the transition zone in the northwest Pacific.

Only the finest sediment fractions have reached the Central North Pacific Province for an extremely long time. Mean grain sizes of surficial sediments are contoured in Figure 5. It is readily apparent that these deposits are well within the clay-size range throughout most of the North Pacific. The one micron contour (10ϕ) in

Figure 5 outlines the centers of the two basins. Distance from land (Schott, 1935, Fig. 12; Revelle and others, 1955, Fig. 3), solution of plankton tests, frustules and skeletons, and distribution of ice-rafted sediment influences the textures of the sediments. It is worth noting that, although there are variations of texture in cores from the Japan-Kuril and Aleutian-Alaskan Provinces, those taken within the Central North Pacific Province have uniform textures down the entire lengths of the cores (*see* Fig. 4, Core V21-182).

Many cores from the central North Pacific penetrate sediments that show either a progressive deepening of color with depth or an abrupt color break at some point down the core. The latter is always marked by a considerable darkening of color below the boundary. Upper portions of the cores are generally light gray to light brownish-gray (Munsell soil color chart, 10YR 7/2 to 10YR 6/2), whereas the underlying sediments range in color from dark grayish-brown to very dark grayish-brown (10YR 4/2 to 10YR 3/2). Examination of the cores reveals that much of the darker color at depth can be attributed to a major increase in the abundance of disseminated manganese micronodules.

The color change has been described by previous investigators (for example, Bramlette, 1961). It was interpreted by Bramlette as a reflection of the change in conditions in the Pacific Ocean during the transition from relatively warm waters of the Tertiary to lower temperatures of the Pleistocene.

Finally, there are some cores from the Central North Pacific Province that show evidence of local submarine volcanism. Core V21-184 taken approximately 200 miles north of the Hawaiian Islands, contains coarse gravel composed of fresh basalt fragments. Cores V20-67, RC11-196, and RC11-197 are from the Baja California Seamount Province (Pl. 1). All contain gravels which once were basalt fragments and now are almost completely altered to palagonite.

Second Order Provinces

Hawaiian Ridge, Marcus-Necker Ridge, Shatsky Rise, Emperor Seamount Chain, and Ridge and Trough Provinces. Specific areas of the North Pacific are dominated by elements of rugged relief, most of which are volcanic. They are characterized by sediments genetically related to these topographic highs and to the level the mountains and hills reach in the water column. Included are the Hawaiian Ridge, Marcus-Necker Ridge, Shatsky Rise, Emperor Seamount Chain and Ridge and Trough Provinces, named for the topographic elements with which they are associated (Fig. 3; Pl. 1).

Sediments within these provinces are to a great extent a function of the depth of crestal areas of elements of positive relief. The

Hawaiian Ridge breaks the surface of the North Pacific Ocean as the Hawaiian Islands and Midway Island. Subaerial weathering in conjunction with the development of carbonate reefs provides a source of considerable amounts of detritus. Coarse sediment moves downslope either by submarine slides (Moore, 1964; Hamilton, 1956) or through turbidity-current activity (Hamilton, 1956; Schreiber, 1968; Horn and others, 1969). As a result, there is a wide archipelagic apron surrounding the Hawaiian Ridge that received sediment derived from the islands and associated reefs.

Intermediate between seamounts that rise above sea level and the major groupings of seamounts that occur at considerable depth are isolated mountains of great relief. Because they are individual mountains it is impossible to give them province status. They must be described briefly, however, because they represent a class included in the spectrum of submarine mountains that constitute the framework of second order sedimentary provinces. Several occur within the Gulf of Alaska and have been cored by Lamont (for example, RC11-174, Parker Seamount; and RC11-175, Patton Seamount; Pl. 1). Coring sites were in depths of water less than 884 fathoms (1618 m), and in each instance the crests and upper flanks of the seamounts were covered by coarse sand and gravel. These are lag deposits resulting from current winnowing.

Most second order provinces are vast mountainous regions. Although they occur at considerable depths, their crests lie above the compensation depth of calcium carbonate (Bramlette, 1961; from 4500 to 5000 m or 2459 to 2732 fathoms). Included are the Marcus-Necker Ridge, Shatsky Rise, Emperor Seamount Chain, and Ridge and Trough Provinces (Fig. 3). Examples of cores from the Emperor Seamount Province (Fig. 4, V20-119) and the Ridge and Trough Province (Fig. 4, RC11-187) are considered representative. Crestal areas are characterized by alternating layers of foraminiferal ooze and carbonate rich clay. Each of these major areas of positive relief has surrounding it a "shadow" of fine-grained detritus derived from the mountains and dispersed by deep currents. The "shadow" extends to approximately the 2500 fathom (4575 m) isobath and follows this contour around the base of the mountains. For this reason the natural limit of second order provinces is defined as the 2500 fathom contour (Figs. 3 and 5). It is interesting to note that there is a general coincidence between the 2500 fathom depth to which sediment extends downslope from its source on oceanic mountains and the depth of the upper surface of the nepheloid layer in the North Pacific (*see* Ewing and Connary, 1970, this volume).

Third Order Provinces

Turbidity-current deposits of terrigenous silt and fine-grained

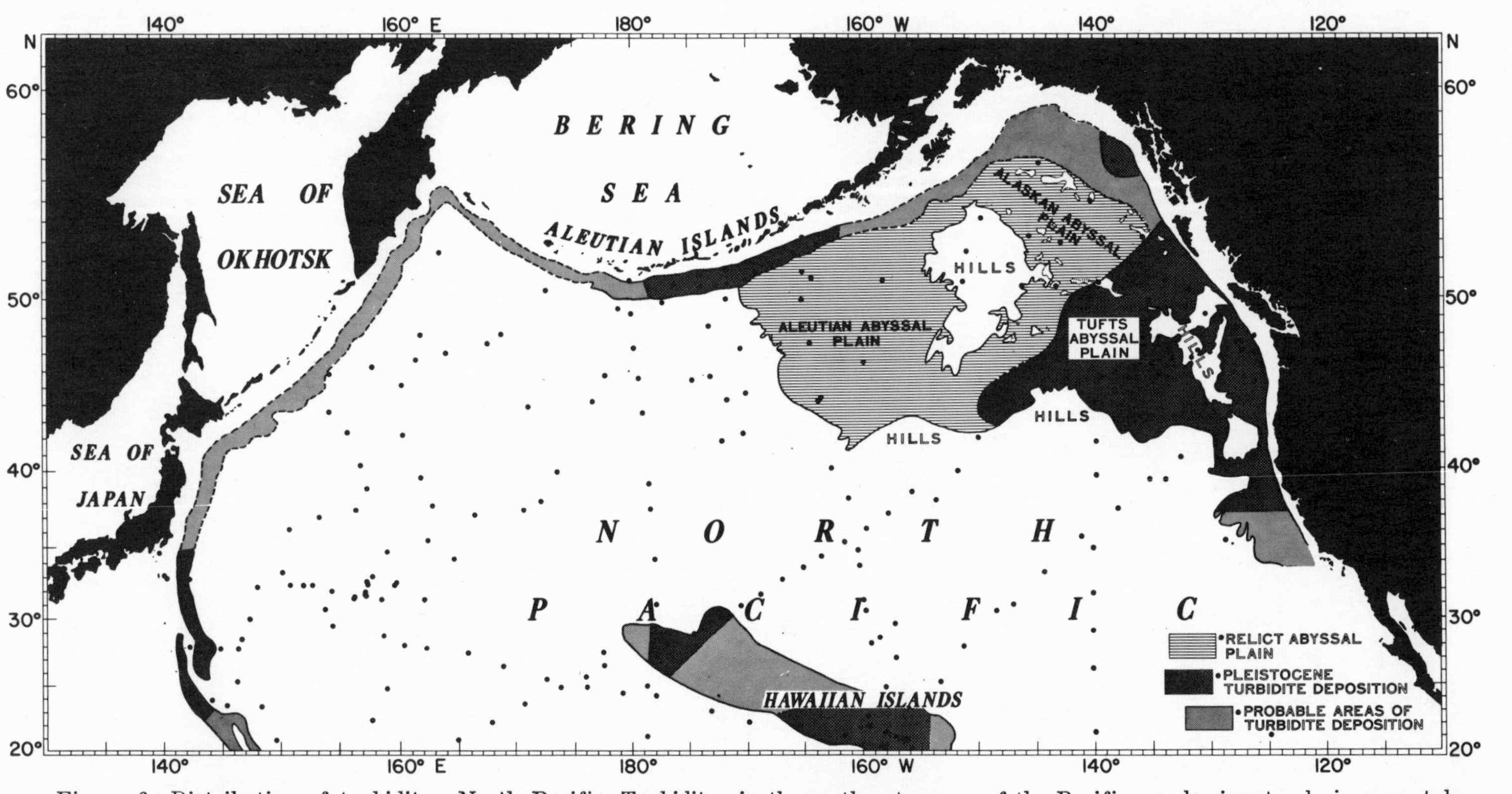

Figure 6. Distribution of turbidites, North Pacific. Turbidites in the northeast corner of the Pacific are dominant only in cores taken from the Tufts Abyssal Plain, the southeast part of the Alaskan Abyssal Plain, and the Cascadia Abyssal Plain (located immediately off the coast of Oregon). The turbidity currents that leveled the floors of the Aleutian Abyssal Plain and much of the Alaskan Abyssal Plain must have been active during an earlier sedimentary regime as cores from these features lack turbidites. The plains were considered fossil or relict by Hurley (1959) and Hamilton (1967).

sand are the dominant sediment of third order provinces. They occur as a narrow ribbon of sediment around the edge of the North Pacific Basin (Figs. 3 and 6; Pl. 1). The turbidites are of two mineralogical types, and this difference serves to distinguish provinces. The Aleutian Trench Province is the site of slow turbidite fill with brown volcanic silt and sand derived from the Aleutian Islands. The Northeast Pacific Turbidite Province was a region of rapid turbidite fill during the Pleistocene (Hurley, 1959; Hamilton, 1967). It is dominated by terrigenous, quartz-rich silt and sand derived from the North American continent.

Aleutian Trench Province. Cores are not available from the Japan and Kuril Trenches; therefore, the Aleutian Trench must serve as an index to processes of sedimentation within the circum-Pacific trench system. All level areas of the Aleutian Trench, whether terraces on insular slopes or the floor of the trench itself, contain turbidites (Horn and others, 1969). They occur as poorly defined layers of graded volcanic arenite and lutite. These thin layers are dominated by brown pumice and bubble wall shards presumably transported from contiguous volcanic islands. The deposits are products of turbidity currents moving down the steep insular walls of the trenches. Much of the sediment is trapped during transit on terraces or benches which appear to be characteristic of the steep landward slopes (Menard, 1964; Ross and Shor, 1965; Hamilton, 1967; and others). Additional core data are needed prior to further subdivision of the circum-Pacific trenches into sedimentary provinces.

Northeast Pacific Turbidite Province. Cores taken from the sea floor off the northwestern United States and British Columbia consistently penetrate turbidite sequences (Fig. 6). The position and thickness of these graded units were presented in a previous article (Horn and others, 1969). The distribution of turbidites with related mud and clay is shown on Plate 1. They have been grouped as a single sediment type to show the extent of terrigenous sediment influx into the northeastern Pacific at the close of the Pleistocene. This region is named the Northeast Pacific Turbidite Province.

Lamont data support the contention of previous workers that turbidity current activity in the northeast Pacific is far more restricted today than it was during the Pleistocene. It may well be that sediment dispersal by this process is confined to submarine channels and upper portions of submarine fans (Hurley, 1959; Hamilton, 1967; Griggs and Kulm, 1969; and others). Both Hurley and Hamilton provide considerable evidence in favor of the thesis that the abyssal plains of the northeast Pacific are relict or "fossil" features produced by pre-existing sedimentary regimes.

Most turbidites of the northeastern corner of the Pacific Ocean are silt and sandy silt grading upward to clay. We consider them deep-sea flysch deposits (*see* Kuenen, 1967) as they possess many of the primary structures attributed to deposition from turbidity currents. Figure 7 shows a textural profile of the upper portion of a core (RC11-190) taken from the southern and distal part of the Tufts Abyssal Plain. The textures are representative of many of the cores from the Northeast Pacific Turbidite Province. The textural profile suggests that at the close of the Pleistocene the southern Tufts Abyssal Plain was receiving frequent pulses of terrigenous sediment. The age of the turbidites has not been determined; however, Duncan (1968, p. 49) reported that in a core 1055 cm long from the southeast corner of the Tufts Abyssal Plain, the oldest sediments were approximately 35,000 to 40,000 years B.P. This information suggests that the distribution of turbidites in the northeast corner of the Pacific (Fig. 6) represents a sedimentary regime that existed at the close of the Pleistocene.

A complete account of the textural properties of the turbidites in the northeast Pacific is being prepared by the authors. Core RC11-190 is selected and described here as an example of turbidites which accumulate adjacent to but not within the main avenue followed by turbidity currents (Fig. 7). The profile of mean grain size

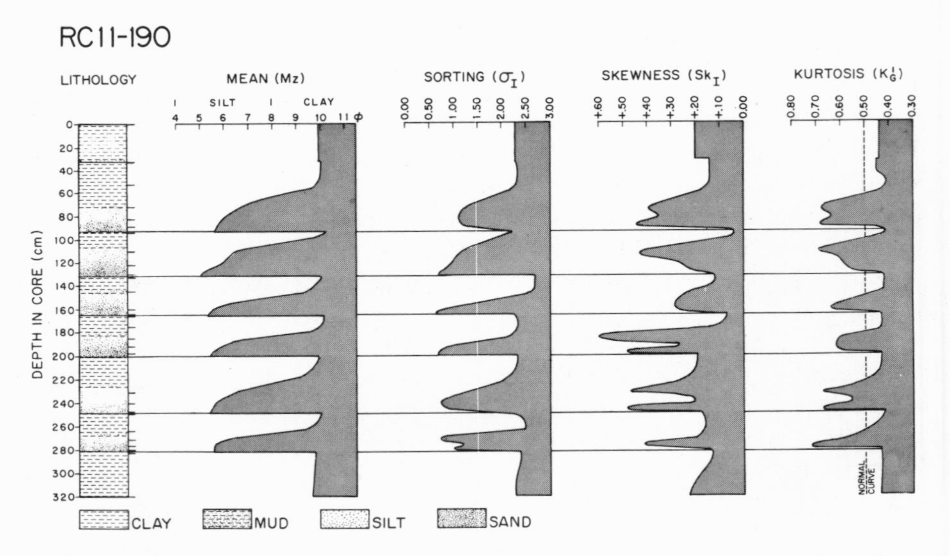

Figure 7. Textural profile of a turbidite core from the Tufts Abyssal Plain. Points where samples were taken for textural analysis are indicated by small ticks along the inner margin of the lithology column. The core contains several turbidite layers one on top of another. Rhythms of sediment that settled from turbidity currents have produced perfect grading of individual units. Note the progressive increase in mean grain size with depth within each turbidite. Sorting, skewness, and kurtosis, however, show considerable variation and reflect the amount of the secondary mode present in these bimodal sediments.

is regular and rhythmic, suggesting repeated addition of sediment of similar texture. It is important to note that, although mean grain sizes increase toward the base of each graded unit, the other textural parameters show considerable variation in their values. In core RC11-190, the uppermost graded unit becomes progressively coarser with depth, yet there is a reversal of sorting, skewness, and kurtosis within the thick basal section of silt. Fluctuations of sorting, skewness, and kurtosis define the degrees of mixing of principal size populations. Differences of values of these grain-size measures within a single turbidite are considered to be due, at least in part, to proximity to main avenues of transfer and to slight changes in the hydrodynamics of the transporting agent during deposition of the turbidite. It is variations in the amount of the secondary fine-sediment mode which produce the fluctuations in the values of sorting, skewness, and kurtosis.

VOLCANIC ASH HORIZONS

There is an important sediment type in the North Pacific whose distribution is not confined by provincial boundaries, but it is so widespread and significant that it must be mentioned to complete the picture of deep-sea sedimentation in the North Pacific. Volcanic ash occurs in cores taken oceanward of the island arcs and continents that form the rim of the North Pacific Basin (Fig. 8). The ashes lie within a broad zone 600 to 800 miles wide (970 to 1290 km). Their position and thickness in the cores is shown on Plate 1. The tephrochronology and genesis of these ash layers is the subject of an article by Hays and Ninkovich (1970, this volume).

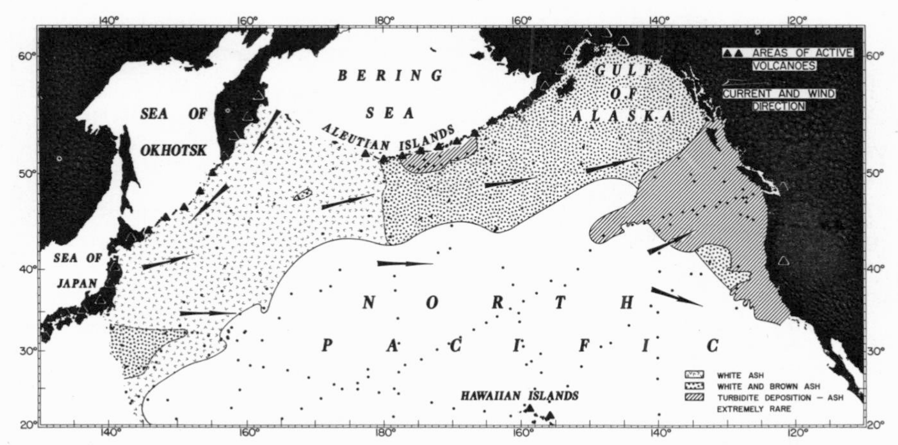

Figure 8. Distribution of volcanic ash layers, North Pacific. (*After* Horn and others, 1969.)

CONCLUSIONS

The location and extent of sedimentary provinces in the North Pacific Ocean are functions of the positions of major water masses, proximity of land, and the distribution and depth of submarine ridges. The North Pacific Basin, unlike the North Atlantic, is protected against major transfer of coarse terrigenous materials by an encircling system of sediment traps. As a result, first order sedimentary provinces have limits that are essentially those of major water masses within the North Pacific Ocean. The Japan-Kuril and Aleutian-Alaskan Provinces contain the finest fractions of silt and clay derived from neighboring continents and island arcs. This terrigenous sediment is apparently carried in suspension, traverses the Circum-Pacific traps, and is dispersed by the West Wind Drift and Subarctic Current Systems. The Central North Pacific Province lies beneath the North Pacific Current system, is most distant from a major continental source, and is the site of clay ("red clay") accumulation.

Second order provinces are located on and around regional elements of positive relief. The sediments that cap these features and accumulate on surrounding archipelagic aprons reflect the depth at which the crests occur. Those mountains, which break the ocean surface, are surrounded by locally derived reef debris and the products of subaerial and subsequent submarine dispersal. Peaks with summits at relatively shallow depths are capped by lag gravels and sands. Rises at greater depths, but above the compensation level of calcium carbonate, are covered by sequences of foraminiferal ooze interlayered with calcareous clay. The natural limit of archipelagic aprons around the ridges (those that do not break the surface) appears to be about the 2500 fathom isobath (4575 m).

Third order provinces include only peripheral areas of the North Pacific Basin. They are dominated by terrigenous silts. The Aleutian Trench Province contains turbidite sequences of brown volcanic arenite and lutite. The northeast Pacific has been an area of turbidite deposition. Turbidites occur in cores taken as much as 1100 miles west from Oregon. Presumably, during the late Pleistocene the Tufts and Cascadia Abyssal Plains received large quantities of coarse sediment from adjacent regions of the North American continent.

ACKNOWLEDGMENTS

We gratefully acknowledge the U. S. Naval Ship Systems Command for providing financial support for all analytical and technical phases of the study (Contract N00024-67-C-1186). Special thanks are due B. King Couper of NavShips and George M. Bryan of Lamont for their foresight and support throughout the investigation.

The cores were collected by scientists aboard the research vessels *Vema* and *Robert D. Conrad* under the direction of Professor Maurice Ewing. Maintenance of the Deep-Sea Core Library at Lamont is supported by the Office of Naval Research (Contract N00014-67-A-0108-0004) and the National Science Foundation (Grant NSF-GA-1193).

F. T. Ishibashi, G. P. Lamsfuss, M. Parsons, and R. C. Shipman provided research assistance. V. Rippon executed art work and drafting. Sincere appreciation is due J. D. Hays and R. R. Capo for aid in providing access to the large number of cores included in the study. The manuscript was critically reviewed by Tj. H. van Andel, R. L. Folk, P. D. Komar, J. Ewing and J. D. Hays.

REFERENCES CITED

Arrhenius, G., 1963, Pelagic sediments, p. 655-727 *in* Hill, M. N., *Editor,* The Sea: v. 3, New York, John Wiley & Sons, Inc.

Bramlette, M. N., 1961, Pelagic sediments: Am. Assoc. Adv. Sci. Pub. 67, p. 345-366.

Carlson, P. R., 1968, Marine geology of Astoria submarine canyon: Ph.D. dissert., Oregon State Univ., Corvallis, 259 p.

Dodimead, A. J., Favorite, F., and Hirano, T., 1963, Review of oceanography of the subarctic Pacific region: Internat. North Pacific Fisheries Comm. Bull. 13, 195 p.

Donahue, J. G., 1970, Pleistocene diatoms as climatic indicators in North Pacific sediments: Geol. Soc. America Mem. 126, p. 121-138.

Duncan, J. R., 1968, Late Pleistocene and postglacial sedimentation and stratigraphy of deep-sea environments off Oregon: Ph.D. dissert., Oregon State Univ., Corvallis, 222 p.

Ewing, J., Ewing, M., Aiken, T., and Ludwig, W. J., 1968, North Pacific sediment layers measured by seismic profiling, p. 147-173 *in* Knopoff, L., Drake, C. L., and Hart, P. J., *Editors,* The crust and upper mantle of the Pacific area: Am. Geophys. Union, Geophys. Mono. 12.

Ewing, M., and Connary, S. D., 1970, Nepheloid layer in the North Pacific: Geol. Soc. America Mem. 126, p. 41-82.

Fleming, R. H., 1957, General features of the oceans, p. 87-107 *in* Hedgpeth, J. W., *Editor,* Treatise on Marine Ecology and Paleoecology, Vol. 1: Ecology: Geol. Soc. America Mem. 67.

Folk, R. L., 1954, The distinction between grain size and mineral composition in sedimentary rock nomenclature: Jour. Geology, v. 62, p. 344-359.

—— 1966, A review of grain-size parameters: Sedimentology, v. 6, p. 73-93.

—— 1968, Petrology of sedimentary rocks: Austin, Texas, Hemphill's, 154 p.

Folk, R. L., and Ward, W. C., 1957, Brazos River Bar: A study in the significance of grain size parameters: Jour. Sed. Petrology, v. 27, p. 3-26.

Goldberg, E. D., and Arrhenius, G., 1958, Chemistry of Pacific pelagic sediments: Geochim. et. Cosmochim. Acta, v. 13, p. 153-212.

Griffin, J. J., Windom, H., and Goldberg, E. D., 1968. The distribution of clay minerals in the World Ocean: Deep-Sea Research, v. 15, p. 433-459.

Griggs, G. B., Carey, A. G., Jr., and Kulm, L. D., 1969, Deep-sea sedimentation and sediment-fauna interaction in Cascadia channel and on Cascadia Abyssal Plain: Deep-Sea Research, v. 16, p. 157-170.

Griggs, G. B., and Kulm, L. D., 1969, Glacial marine sediments from the northeast Pacific: Jour. Sed. Petrology (in press).

Hamilton, E. L., 1956, Sunken islands of the Mid-Pacific Mountains: Geol. Soc. America Mem. 64, 97 p.

—— 1967, Marine geology of abyssal plains in the Gulf of Alaska: Jour. Geophys. Research, v. 72, p. 4189-4213.

Hays, J. D., 1970, The stratigraphy and evolutionary trends of Radiolaria in North Pacific deep-sea sediments: Geol. Soc. America Mem. 126, p. 185-218.

Hays, J. D., and Ninkovich D., 1970, North Pacific deep-sea ash chronology and age of present Aleutian underthrusting: Geol. Soc. America Mem. 126, p. 263-290.

Horn, D. R., Delach, M. N., and Horn, B. M., 1969, Distribution of volcanic ash layers and turbidites in the North Pacific: Geol. Soc. America Bull., v. 80, p. 1715-1724.

Horn, D. R., Horn, B. M., and Delach, M. N., 1968, Sonic properties of deep-sea cores from the North Pacific Basin and their bearing on acoustic provinces of the North Pacific: Lamont Geol. Obs. Tech. Rept. no. 10, CU-10-68. Nav Ships N00024-67-C-1186, 357 p.

Hurley, R. J., 1959, The geomorphology of abyssal plains in the northeast Pacific Ocean: Ph.D. dissert., California Univ., Los Angeles, 173 p.

Kuenen, P. H., 1967, Emplacement of flysch-type sand beds: Sedimentology, v. 9, p. 203-243.

Kulm, L. D., and Nelson, C. H., 1967, Comparison of deep-sea channel and interchannel deposits off Oregon (abs.): Am. Assoc. Petroleum Geologists Bull., v. 51, p. 472.

Menard, H. W., 1964, Marine geology of the Pacific: New York, McGraw-Hill Book Co., 271 p.

Moore, J. G., 1964, Giant submarine landslides on the Hawaiian Ridge: U.S. Geol. Survey Prof. Paper 501-D, p. D95-D98.

Nayudu, Y. R., 1958, Recent sediments of the Gulf of Alaska (abs.): Geol. Soc. America Bull., v. 69, p. 1967.

—— 1959, Recent sediments of the northeast Pacific: Ph.D. dissert., Washington Univ., Seattle, 53 p.

—— 1964, Volcanic ash deposits in the Gulf of Alaska and problems of correlation of deep-sea ash deposits: Marine Geology, v. 1, p. 194-212.

Nayudu, Y. R., and Enbysk, B. J., 1964, Bio-lithology of northeast Pacific surface sediments: Marine Geology, v. 2, p. 310-342.

Nelson, C. H., 1968, Marine geology of Astoria deep-sea fan: Ph.D. dissert., Oregon State Univ., Corvallis, 287 p.

Nigrini, C., 1970, Radiolarian assemblages in the North Pacific and their application to a study of Quaternary sediments in core V20-130: Geol. Soc. America Mem. 126, p. 139-184.

Opdyke, N. D., and Foster, J. H., 1970, The paleomagnetism of cores from the North Pacific: Geol. Soc. America Mem. 126, p. 83-119.

Revelle, R., 1944, Marine bottom samples collected in the Pacific Ocean by the *Carnegie* on its seventh cruise: Carnegie Inst. Washington Pub. 556, Part 1, 180 p.

Revelle, R., Bramlette, M., Arrhenius, G., and Goldberg, E. D., 1955, Pelagic sediments of the Pacific: Geol. Soc. America Spec. Paper 62, p. 221-236.

Rex, R. W., and Goldberg, E. D., 1958, Quartz content of pelagic sediments of the Pacific Ocean: Tellus, v. 9, p. 18-27.

Ross, D. A., and Shor, G. G., Jr., 1965, Reflection profiles across the Middle America Trench: Jour. Geophys. Research, v. 70, p. 5551-5572.

Schott, W., 1935, *in* Schott, G., Geographie des Indischen und Stillen Ozeans: Hamburg, Boysen, 413 p.

Schreiber, B. C., 1968, Core, sound velocimeter, hydrographic, and bottom photographic stations-cores: Marine Geophys. Survey Prog. N62306-1688, area V, v. 8., 42 p.

Strakhov, N. M., 1967, Principles of lithogenesis: London, Oliver & Boyd, v. 1, 235 p.

Sverdrup, H. U., Johnson, M. W., and Fleming, R. H., 1942, The oceans: Their physics, chemistry and general biology: New York, Prentice-Hall, 1087 p.

Windom, H. L., 1969, Atmospheric dust records in permanent snowfields: Implications to marine sedimentation: Geol. Soc. America Bull., v. 80, p. 761-782.

LAMONT-DOHERTY GEOLOGICAL OBSERVATORY CONTRIBUTION NO. 1483
MANUSCRIPT RECEIVED APRIL 10, 1969
REVISED MANUSCRIPT RECEIVED JULY 18, 1969

THE GEOLOGICAL SOCIETY OF AMERICA, INC.
MEMOIR 126, 1970

North Pacific Bottom Potential Temperature

ARNOLD L. GORDON
AND
ROBERT D. GERARD

Lamont-Doherty Geological Observatory of Columbia University, Palisades, New York

ABSTRACT

The bottom potential temperature distribution for the Pacific Ocean north of 11° S. is presented, based upon 495 data points derived from hydrographic stations, the Lamont thermograd, and an *in-situ* salinity-temperature-depth recorder. Although the entire range of potential temperature is only 1.2° C, an oceanwide pattern is apparent. The coldest bottom water (< 0.8° C potential temperature) occurs in the equatorial region between 160° and 180° W. This feature can be traced into the area north of Hawaii by paths both east and west of these islands with a warming of only 0.5° C.

In order to separate the effect of the decrease of potential temperature with depth (which occurs in the upper 4000 to 5000 m) from the bottom variation due to influx of cold water, the anomaly of bottom potential temperature is plotted. This parameter is defined as the difference between the observed bottom potential temperature for a particular station and the average bottom potential temperature for the North Pacific at the depth of the observation. The water between 160° and 180° W. has the largest negative anomaly, 0.3° C colder than the average t_p found at corresponding depths. The water north of Hawaii has a positive anomaly, indicating that in this region the deeper water is relatively warm compared to that in the equatorial region between 160° and 180° W., even after the depth effect is removed. The bottom water over and east of the crest of the mid-ocean ridge at 120° W. has large positive anomalies, suggesting the possibility that this water is warmed by geothermal heating.

From the bottom potential temperature distribution and the anomalies of these values, together with recently obtained Lamont bottom current measurements, it is possible to trace the basic pattern of bottom circulation (which should also reflect the circulation for the water below 2000 to 2500 m). The sole influx of water is the Antarctic Bottom Water flowing northward along 175° W. from 20° S. to 16° N. This current is no doubt a continuation of the deep western boundary current along the Tonga Trench described recently for the South Pacific. From this point, the flow divides into two zonal branches. The eastern branch passes between the Johnston Island and Christmas Island Ridges and the western branch passes between the Marshall Islands and the Marcus-Necker Ridge. After transversing these respective passages, both flows turn northward, eventually converging in the North Pacific Basin between the Hawaiian and the Aleutian Islands, where deep upward transport is expected. It is estimated that the time necessary for the water to flow from 175° W., 10° S. to the center of the convergence area is 750 years.

CONTENTS

INTRODUCTION

Though the Pacific Ocean comprises approximately 46 percent of the total volume of the world ocean, the range of temperature and salinity encountered in its deep and bottom layers is surprisingly small. The large areal extent and the homogeneous nature of these layers make it particularly difficult to deduce information about the circulation of the water from standard hydrographic measurements. Not only are many hydrographic stations necessary, but their absolute accuracy must be high in order to successfully study the small variations of the water characteristics. An early attempt at constructing a bottom potential temperature map was made by Wüst (1937). With the data presently available, a greatly improved map can be constructed from which the basic bottom circulation and spreading can be learned. This will also aid in the planning of future oceanographic work.

The deep and bottom water of the Pacific Ocean, in contrast to that of other oceans, is derived from a single source, the lower layer of the Antarctic Circumpolar Current, which is composed to varying degrees of Antarctic Bottom Water and Circumpolar Deep Water. The influx of this homogeneous water and the lack of other source regions in the Pacific Ocean leave only the normal vertical diffusive processes to alter it along the path of flow. Since these processes are slow, it follows that the deep and bottom water throughout the Pacific will not vary substantially from its properties at the entrance point. (For further information *see* Knauss, 1962.)

Recent cruises of the USNS *Eltanin* in the South Pacific support this view in that a deep northward flowing current of Antarctic Bottom Water was found along the western boundary of the deep South Pacific Ocean (Reid and others, 1968; Warren, 1968). The western boundary nature of this current is consistent with the deductions of Stommel and Arons (1960a, 1960b) and the earlier findings of Knauss (1962).

The deep western boundary current of Antarctic Bottom Water must continue into the North Pacific Ocean, where it is the sole source of the water below 2000 to 2500 m. The attempts of Wooster and Volkmann (1960) and Knauss (1962) to determine the path of this flow led to only very general results and the likely conclusion that the deep and bottom water converge and are transported upward in the North Pacific south of the Aleutian Islands. Deep vertical transport in this vicinity is an important process by which the surface water sinking to the deep ocean in the North Atlantic, Antarctic waters, and a few marginal seas returns to the surface layers. The other basic method for such a return flow is the slow upward diffusion that must occur across the main thermocline to counter the downward flux of heat.

The purpose of this study is to present more detailed maps of the distribution of bottom potential temperature in the North Pacific and, in particular, in that region between 20° N. and 20° S., where new measurements of bottom current and bottom t_p have been obtained on a recent cruise of the Lamont-Doherty Research Vessel *Robert Conrad*. From these maps, information on the bottom circulation is deduced, and areas are noted where future oceanographic field work should be carried out. Since there is no other source of baroclinicity in the deep and bottom North Pacific Ocean, this deduced bottom circulation should represent the flow below 2000 to 2500 m.

THE DATA

The bottom temperature data used in this study are all deeper than 2000 m and within 550 m of the sonic depth. There are three kinds of data used: hydrographic data (reversing thermometers), thermograd (T-grad) measurements (thermistors), and the sensors of the *in-situ* salinity-temperature-depth recorder (STD). The hydrographic data were obtained from the Scripps Institution of Oceanography data compilations for the Pacific Ocean, the National Oceanographic Data Center's Marsden square files for the regions south of 20° N., the IGY data reports (Capurro, 1961), and recent Lamont-Doherty Geological Observatory measurements.

There are a total of 345 hydrographic observations with reversing thermometers. The estimated limits of error of good hydrographic station data are ± 0.02° C; however, somewhat larger variations in the accuracy are expected among the different ships. The T-grad data are obtained by thermister probes, which are attached to the Ewing bottom-sediment corer immediately before entrance into the bottom sediments to determine geothermal heat flow (Langseth, 1965). They, therefore, represent the immediate bottom conditions. There are usually 3 to 5 thermistors on the core pipe. The absolute accuracy of these thermistors compared to reversing thermometers is poor, being approximately ± 0.1° C; however, the averaging of the data and "weeding out" the data of thermistors that have altered characteristics at each station lead to more reliable information. In addition, the importance of random errors is decreased because of the large quantity of T-grad stations (140) taken aboard the R/V *Vema* and R/V *Conrad*.

The STD data were collected aboard the R/V *Conrad* on Cruise 12 during July, 1968. There are 10 STD stations aligned north of the Hawaiian Islands (159° W.). The accuracy of these temperature measurements is 0.02° C, using the calibration method described by Gerard and Amos (1968).

The positions of the bottom temperature data points used in this study are shown in Figure 1. The locations of bottom observations from R/V *Conrad* Cruise 12 are shown in Plate 2. The hydrographic stations are mostly in the eastern and north Pacific and in the equatorial region. The T-grad data fall in the broad area of the deep northwestern North Pacific and central Pacific.

The basic bottom topography of the North Pacific Ocean is shown in Plate 1 (from the U.S. Naval Oceanographic Office, 1966). The areas deeper than 3000 fm (5480 m) are confined to the regions north and northwest of the Hawaiian Island Ridge with some isolated patches to the southwest. The mid-ocean ridge occurs in the southeast where depths of less than 2000 fm (3655 m) are common.

BOTTOM POTENTIAL TEMPERATURE DISTRIBUTION NORTH OF 11° S. LAT.

The *in-situ* bottom temperature values were converted to potential temperature, t_p, using the tables of Wüst (1961). The t_p distribution is shown as Figure 2. A 0.2° C contour interval is used with the 1.1° C isotherm inserted in the extremely homogeneous area north of 20° N. The bottom t_p ranges from 0.7° C near 10° S. and 170° to 180° W. up to 1.9° C in the West Carolina Basin. Therefore, the total t_p range for the bottom water below 2000 m is only 1.2° C for the entire North Pacific.

The coldest bottom water (< 0.8° C) is found in the equatorial area between 160° and 180° W. and extends as a narrow feature 16° to 17° N. along 178° W. From this position, a cold water area (0.8° to 1.0° C) extends zonally between 15° to 20°, passing south of the Marcus-Necker and Hawaiian Island Ridges. The eastern branch spreads to the east of Hawaii; the western branch spreads into the Mariana Basin and to the north.

North of 20° N., the bottom water of the deep North Pacific varies only between 1.0° and 1.2° C. The 1.1° C isotherm extends far north, both east and west of Hawaii, but not between 155° to 170° W. The Aleutian Trench is colder than 1.1° C (as low as 1.03° C) but this cold water apparently does not have a direct connection with the bottom water of less than 1.1° C farther south. The source of the cold water is not defined by the data distribution available, and remains an interesting problem to be solved. Perhaps the small difference in the t_p is due to inaccuracies in the adiabatic correction table.

The bottom t_p along the eastern boundary of the North Pacific shows relatively large gradients. The coastal area is, for the most part, between 1.6° to 1.8° C, with the exception of the warmer water immediately to the southwest of the Panama Canal. The influence

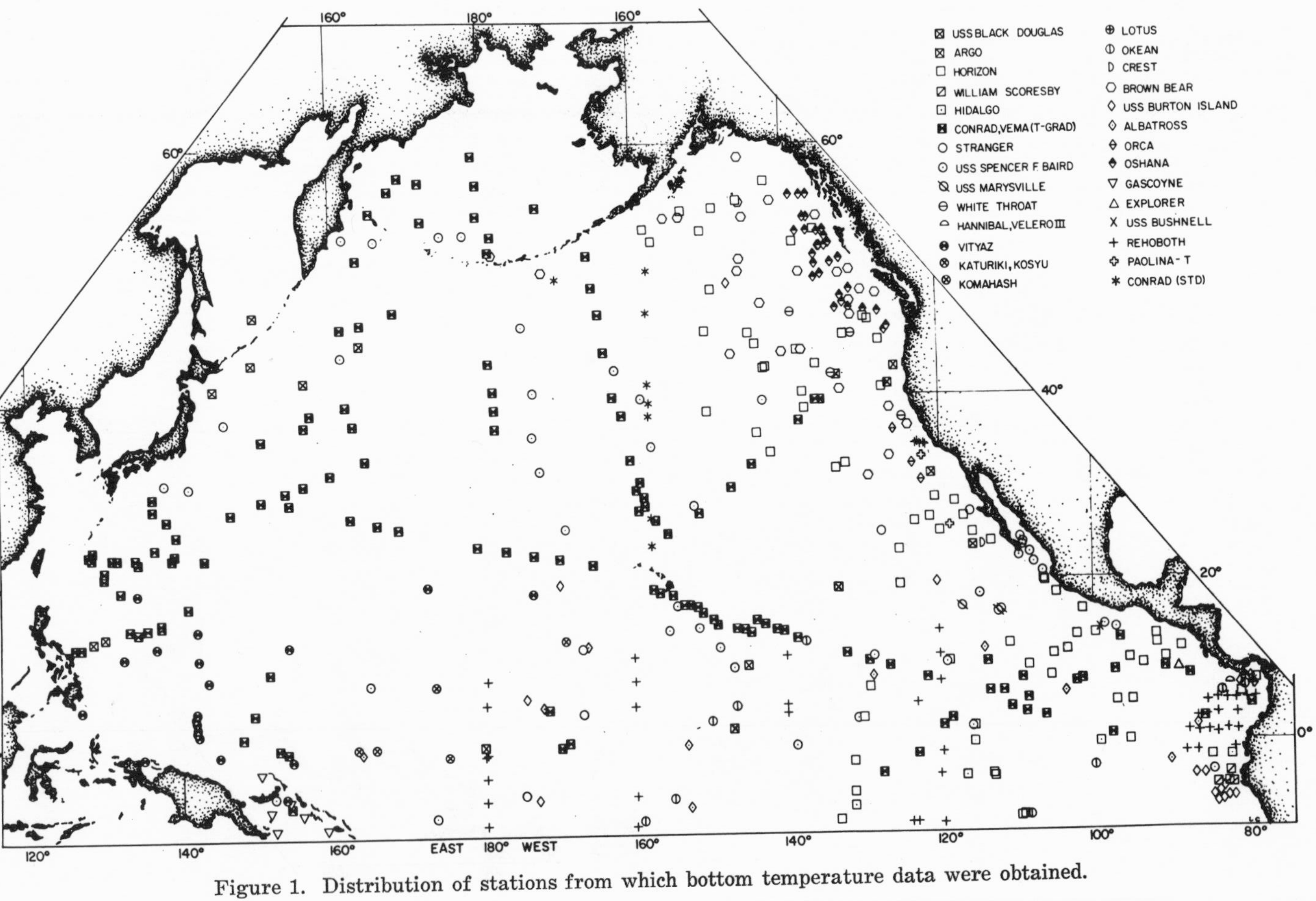

Figure 1. Distribution of stations from which bottom temperature data were obtained.

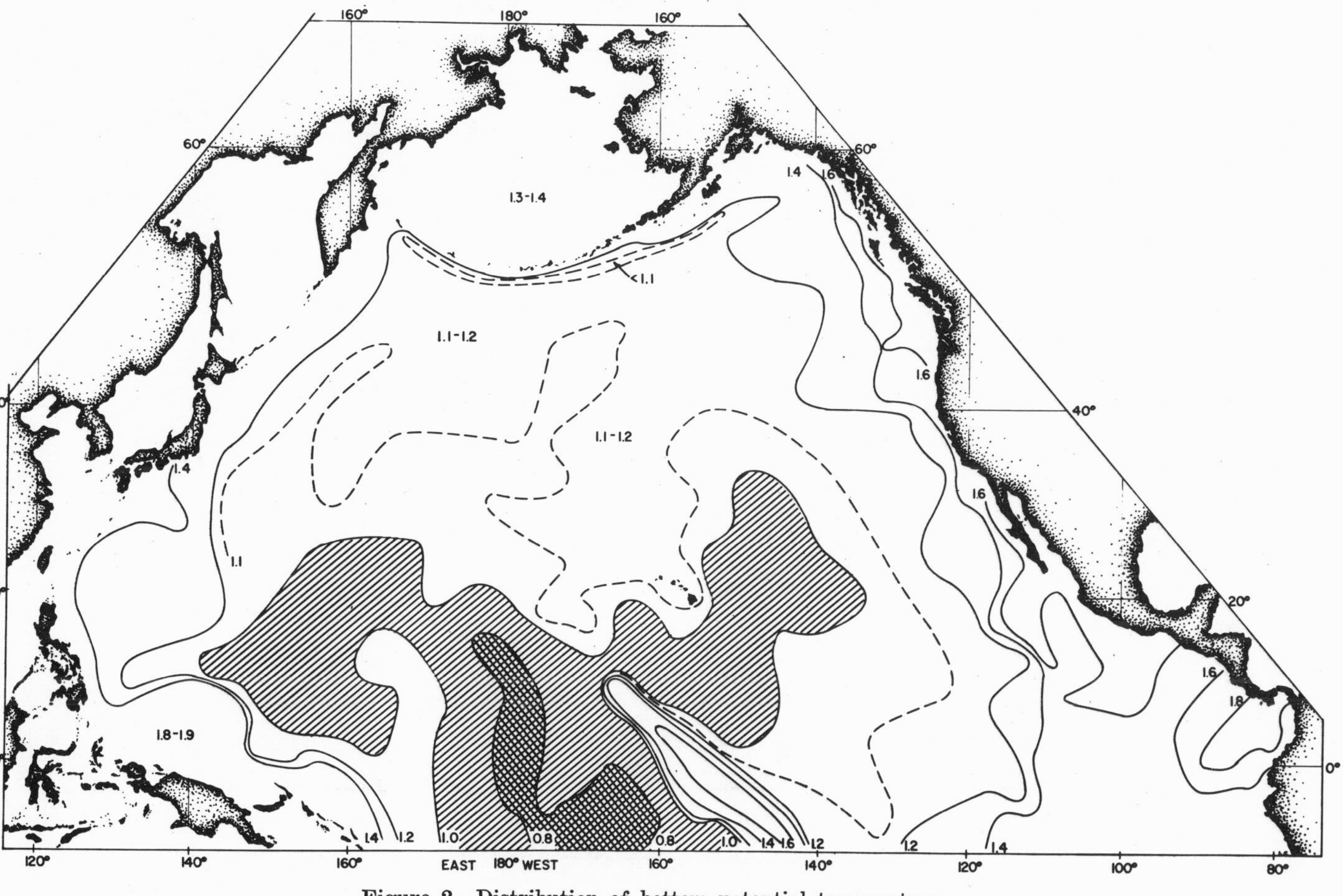

Figure 2. Distribution of bottom potential temperature.

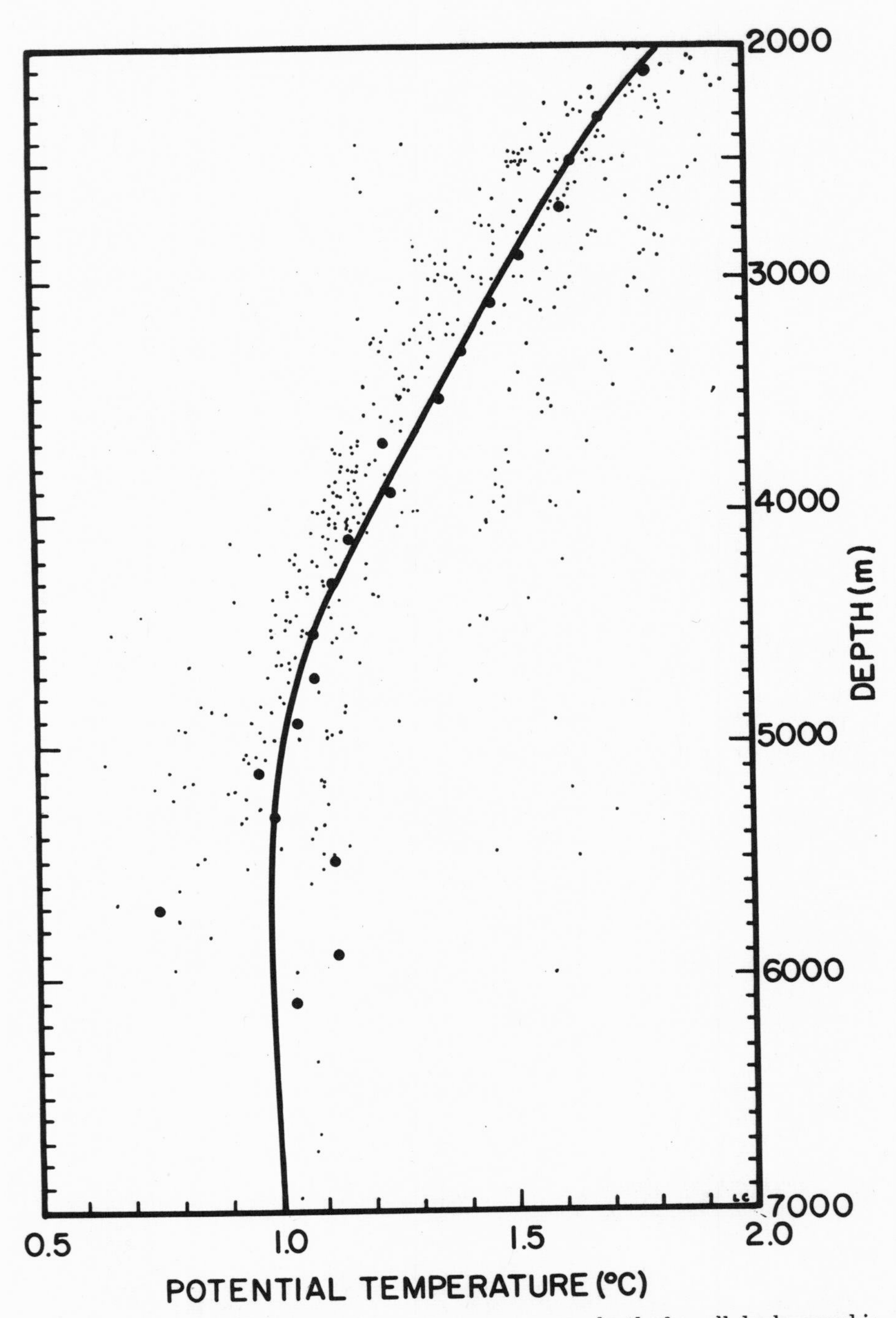

Figure 3. Bottom potential temperature versus depth for all hydrographic station observations (small dots), average for each 200-m interval (large dots), and best fit (by eye) average potential temperature versus depth (curve).

of the Mendocino Escarpment is apparent in the trend of the isotherms at 40° N. The t_p distribution in the eastern Pacific, as well as the close packing of the isotherms on the Christmas Island Ridge, is no doubt due to bottom topographic slopes. The influence of the bottom topography and the natural decrease of t_p with depth are expected to be larger in these regions because their depths are less than 4500 m; above this depth the variation of t_p with depth is large (*see* Fig. 3). For the deeper Pacific to the west, this influence is not as great since at those deeper depths t_p varies little with depth.

BOTTOM WATER TEMPERATURE ANOMALIES

The problem of separating the effect of depth and t_p variation with depth from the bottom t_p variation due to cold water influx (representing real currents) has led some researchers to abandon the bottom t_p method and plot deep temperatures on standard levels. The bottom t_p distributions have in the past led to important oceanographic conclusions which have been subsequently confirmed by other methods. The best example is the confirmation of Wüst's 1933 work on the bottom t_p of the Atlantic Ocean by recent work done at Lamont (Heezen and Hollister, 1964; Le Pichon and others, 1969a, 1969b). In order to answer the recent criticism, the following method was devised. The natural bottom t_p variation with depth and the variation due to current influences may be separated by determining the anomaly of bottom potential temperature. This is done by finding the average bottom t_p as a function of bottom depth and calculating the difference of each observed bottom t_p from the average. The bottom t_p (small dots) against depth and the average bottom t_p (large dots) for the North Pacific hydrographic station data are shown in Figure 3. The smooth best fit curve (plotted by eye) is compared with each bottom t_p observation for calculation of the anomaly of bottom potential temperature, $\delta\ t_p$. The distribution of $\delta\ t_p$ is shown on Figure 4. They are calculated only from the hydrographic station data, and averaged for 5° squares. Naturally, the non-uniform distribution of these stations causes the "average" bottom t_p to be heavily weighted by the eastern Pacific values. This is a drawback to the method; however, it would become less important as the data become more uniformly distributed.

The largest negative anomaly occurs in the cold water feature of the equatorial region between 160° to 180° W. Moderate negative or near zero anomalies extend from this region to the east and north, generally within the 1.1° C isotherm, with the exception of an extension of negative anomalies to the coastlines of the United States and Canada. Large positive anomalies are found west of Mexico and the Isthmus of Panama, and in the region from the Aleutian Islands

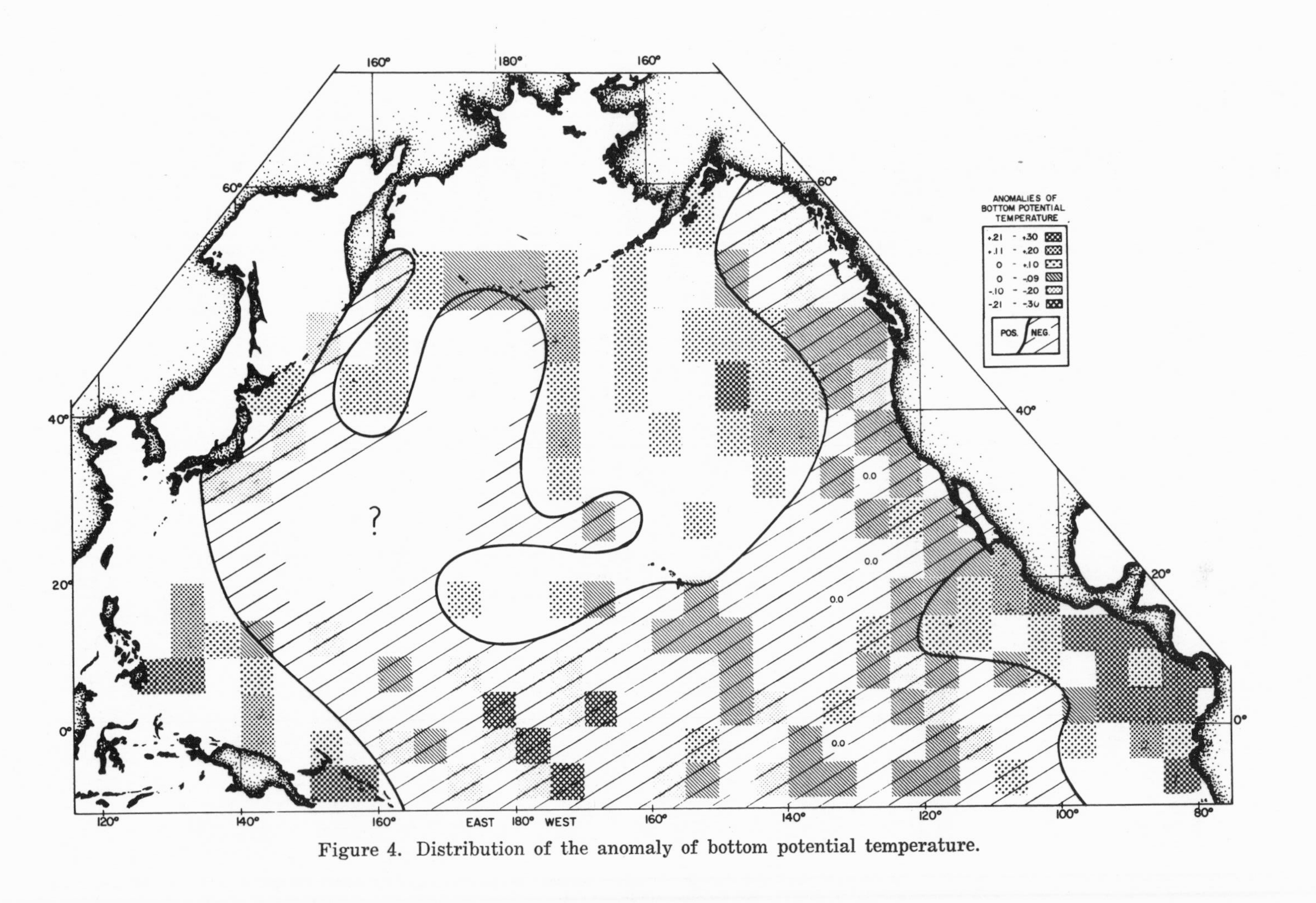

Figure 4. Distribution of the anomaly of bottom potential temperature.

to Hawaii. The warm water over the Christmas Island Ridge does not show up as large anomalies and is no doubt only the influence of bottom topographic slopes.

GENERALIZED BOTTOM CIRCULATION PATTERN

The bottom t_p and δ t_p distribution may be used to obtain information on the bottom circulation. An influx of cold water warms along its path of flow by geothermal heating and the downward flux of heat. When a deep temperature minimum exists, which is the case for most of the North Pacific (Olsen, 1968), the bottom water is warmed primarily by the first method, since the diffusive heat transport in the water column is upward at depths below the temperature minimum layer. Therefore, the flow is from regions of negative anomalies to those of positive anomalies. The angle made by the streamlines and isotherms would depend on the ratio of the mixing coefficient to the bottom current; however, some generalities can be made (Gordon, 1966):

(1) The flow is mostly parallel to the isotherms where the isotherms are closely packed, but in the main section of the North Pacific, the extremely small bottom t_p gradients do not permit determination of the flow direction, and the only method is the transfer from negative to positive anomalies.

(2) Water warmer than 1.2° C is only found adjacent to the coastal regions, and it is expected that the general paths of flow are parallel to the isotherms.

The circulation of the water colder than 1.2° C is as follows:

(1) The influx is no doubt in the equatorial region between 160° to 180° W. This current is a continuation of the deep western boundary current described by Warren (1968). The thickness of the equatorial sediment thins considerably in this region (Ewing and others, 1968), which may result from the eroding capabilities of the northward-flowing current. The low temperature of the water in this current and its relatively high oxygen value (4.5 ml/l) indicate that it is composed of Antarctic Bottom Water (Gordon, 1966).

(2) On reaching the latitudes of 15° to 16° N., the current divides into an eastward- and a westward-flowing branch. Both currents are constricted by narrow passages. To the east, the current flows between the Johnston Island and Christmas Island Ridges; the western branch enters the passage between the Marshall Islands and the Marcus-Necker Ridge. It is expected that the bottom current should reach a maximum velocity at these passages. The west current flows into the Mariana Basin and northward into the deep northwest North Pacific. The east current turns northeast to flow into the northeastern North Pacific area.

(3) The two branches of bottom flow enter the Pacific Ocean

north of 20° N. where they converge in the region of positive anomaly of potential temperature between the Aleutian Islands and Hawaii. Within this convergence area, upwelling must occur. The influx of bottom water across the entire North Pacific Ocean and the subsequent upwelling in the north-central areas show the Pacific to be similar to a large estuary (Tully and Barber, 1960).

The large positive anomalies of the eastern tropical Pacific Ocean may be the result of the high heat flow over the mid-ocean ridge. Since this area is far from the main bottom flow pattern, it is probable that the bottom current is very small, and the long resident time permits substantial warming by geothermal heating.

From the variation of bottom t_p and oxygen values, it is possible to calculate the time necessary for water to flow from 10° S., 175° W. to the center of the convergence area north of Hawaii. The temperature increases by 0.5° C and the oxygen decreases by 1.5 ml/l. Using an oxygen consumption rate of 2×10^{-3} ml/l yr (Arons and Stommel, 1967), it is found that 750 years are needed to accomplish the flow.

Assuming all the warming is due to geothermal heating and that this heat is distributed uniformly for the lower 1000 m (roughly the layer below the deep temperature minimum layer; *see* Olsen, 1968) it is found that the heat flow necessary to accomplish the necessary temperature increase in 750 years is 64 cal/cm^2/yr. Such a value is not far from the average heat flow value for the North Pacific (Langseth, 1968, oral commun.). Therefore, a 750-year period for the movement of water across the North Pacific Ocean seems reasonable. It is possible that the slow introduction of heat from the sea floor eventually leads to an unstable lower water column. The ensuing convective overturning may be the driving mechanism of the deep North Pacific upward transport.

The total length of the above path is 9500 km, both east and west of the Hawaiian Islands. The calculated mean velocity is 0.04 cm/sec, not far from the 0.05 cm/sec calculated by Knauss (1962) on the basis of C^{14} abundance.

Bottom currents for the entire North Pacific are sluggish except where the bottom t_p's are below 0.8° C. The velocity of the current will depend strongly on the local topography, being faster in constricting passages than over broad basins. The flow at the areas with bottom t_p above 1.2° C must be exceedingly small. Perhaps the small mean currents result in the tidal currents being the dominant feature of the bottom flow.

The bottom circulation pattern probably extends upward 2000 to 2500 m above the sea floor. Therefore, even though the velocities are very small, the total transports of the deep and bottom water of the Pacific Ocean may be large.

NEW BOTTOM WATER MEASUREMENTS IN THE CENTRAL PACIFIC

The foregoing discussion has dealt with the broad-scale features of North Pacific bottom circulation. On a recent cruise of the Lamont-Doherty Research Vessel *Robert D. Conrad* in October, 1968, new bottom temperature and current measurements were obtained in the central Pacific between 15° N. and 15° S. lat.

These new measurements, shown in Plate 2, permit a more detailed view of the complex bottom circulation in the region between the northern Tonga Trench and the Marcus-Necker Ridge-Hawaii area. Table 1 summarizes the new bottom temperature observations at 19 stations plotted in Plate 2. These measurements are of two kinds: reversing thermometer bottom temperatures, and STD profiles. The former (listed under Camera Stations, Table 1) were obtained by attaching reversing bottles to the bottom camera-current meter tripod. The thermometers are actuated by bottom contact at exactly 2 m above the ocean bottom. The STD observations listed in Table 1 represent temperatures at the deepest level at each station. These measurements combine results of electronic sensors and reversing thermometers in a manner described by Gerard and Amos (1968). The maximum limits of error for the temperature measurements are estimated to be ± 0.02° C.

Table 2 summarizes results of eleven bottom current measurements taken with the Thorndike photographic bottom current meter (Bruce and Thorndike, 1967). This instrument, housed in a tripod frame, is lowered from a drifting ship using the hydrographic wire and winch. It measures current during several 5 to 10 minute intervals at each station while the tripod rests on the bottom in a stable position. The instrument is capable of measuring current speeds as low as 2 cm/sec and is linear in the range of 2 to 15 cm/sec.

In Plate 2, a larger scale and more detailed review of the bottom potential temperature distribution for the central Pacific is presented. This construction is based upon data points already shown (Fig. 1) and 14 additional data points from the same sources for the region below 11° S., in addition to the recent *Conrad* measurements.

The coldest bottom water (t_p 0.65° C) of all the measurements available is that observed at *Conrad* Station 238 in the northern Tonga Trench at a depth of 6642 m. This minimum value, and two additional low values (0.67° C, t_p at Stations 236 and 237) along long. 169° W., northward to 10° S., confirm the belief that the Tonga Trench provides the chief avenue for influx of Antarctic Bottom Water into the central Pacific. North of 10° S., the flow divides, the eastern part moving northward along the axis of the north Tokelau Trough. The western part of the current moves from the

TABLE 1. *Conrad*-12 BOTTOM TEMPERATURE MEASUREMENTS, HONOLULU TO SUVA

Ship Sta.	Date	Position[1] Lat.	Long.	Temp.[2] Obs.	*In Situ* Temp. °C Corr.	Depth Temp. (m)	Bottom[3] Depth (m)	Potential[4] Temp. °C
217	4×68	14°44′N.	167°32.7′W.	STD 20	1.35	4900	5170	0.90
218	5×68	13°48.8′N.	167°16.2′W.	Camera 155	1.24	5028	5028	0.80
219	6×68	12°46.2′N.	167°04.3′W.	STD 21	1.37	5271	5310	0.89
220	6×68	11°54.5′N.	166°40′W.	Camera 156	1.28	4594	4594	0.89
222	9×68	9°46′N.	174°18′W.	STD 22	1.31	5053	5800	0.86
223	10×68	9°45.6′N.	175°36′W.	Camera 157	1.39	5697	5697	0.85
224	11×68	9°45.1′N.	176°50.7′W.	STD 23	1.32	5041	5660	0.87
225	11×68	9°42.4′N.	177°59′W.	Camera 158	1.41	5964	5964	0.85
226	12×68	9°39.7′N.	179°19′W.	STD 24	1.33	5048	6100	0.88
227	12×68	9°43′N.	179°22′E.	Camera 159	1.40	5822	5822	0.86
228	13×68	8°51′N.	178°52′E.	STD 25	1.33	5006	6100	0.88
229	13×68	7°35′N.	178°25′E.	Camera 160	1.37	5547	5547	0.85
231	17-18×68	4°16.5′S.	176°11.5′E.	STD 26	1.24	4824	5030	0.82
232	20×68	6°34′S.	177°14.8′W.	Camera 162	1.38	5951	5951	0.80
233	21×68	7°05′S.	175°36.3′W.	STD 27	1.26	5007	5935	0.82
235	22-23×68	8°28′S.	171°31.2′W.	Core 205	1.24	4850	4850	0.82
				STD 28	1.23	4722	4850	0.81
236	23-24×68	9°25.2′S.	169°W.	Camera 163	1.14	5318	5318	0.67
				STD 29	1.14	5246	5275	0.68
237	25×68	14°37′S.	168°30′W.	Camera 164	1.12	5058	5058	0.67
238	27×68	16°44.5′S.	172°22′W.	Core 208	1.31	6642	6642	0.65

[1]Positions determined by satellite navigation method (Talwani and others, 1966).

[2]STD indicates measurements obtained with the Bissett-Berman Model 9006 Instrument, together with reversing thermometer measurements obtained simultaneously using the Lamont command sampler (Gerard and Amos, 1968).

Camera refers to measurements obtained on the ocean bottom with reversing thermometers used on the Lamont tripod bottom camera (Thorndike, 1963).

Core refers to measurements obtained on the ocean bottom with reversing thermometers attached to the Ewing piston corer.

[3]Depth measured by precision depth recorder and corrected for sound velocity (using Matthews, 1939).

[4]Temperatures corrected for adiabatic increase using table compiled by Wüst (1961).

10° S. area northward along long. 178° W. At 5° N., it turns northeastward, dividing into an eastern and western branch upon reaching 13° N. Bottom currents measured in this area range from 2 cm/sec to 10 cm/sec, the strongest yet measured in the deep North Pacific (M. Ewing, 1968, oral commun.). Current speeds of 2 cm/sec are too close to the low threshold limit of the current meter to be of use in studying large-scale circulation. However, current readings of 4 cm/sec and greater are believed significant in this region.

East of the Samoa Islands at about 15° S., a moderate eastward flow was measured (*see* Table 2). This is in agreement with the pattern of flow deduced from potential temperature. The measurement at 9° 25′ S. is located in a narrow topographic gap, causing it to be one of the principal passages for the bottom current, as evidenced by the remarkably swift northward current.

At 1° 28′ N. and 7° 35′ N., west of the assumed axis of bottom flow, currents of 5 and 4 cm/sec, respectively, are shown flowing eastward. South of Johnston Island there is support for plotting an eastward branch of bottom current, based upon a measurement which shows a moderate southeastward flow.

The new work confirms and gives detail to the main features of central Pacific deep circulation discussed earlier in this paper. It is hoped these results will stimulate further work on the many problems remaining in our understanding of the deep circulation of the Pacific Ocean.

TABLE 2. *Conrad*-12 BOTTOM CURRENT METER MEASUREMENTS, HONOLULU TO SUVA

Ship Sta.	Date	Current Meter Obs.	Position[1] Lat.	Long.	Measurements Bottom Depth[2] (m)	Current Dir.[3]	Current Speed
218	10/5/68	113	13°49′N	167°16′W	5028	118°	Moderate[4]
220	10/6/68	114	11°55′N	166°40′W	4594	204°	Weak[5]
223	10/10/68	115	09°46′N	175°36′W	5697	209°	Weak
225	10/11/68	116	09°42′N	177°57′W	5964	95°	Weak
227	10/12/68	117	09°43′N	179°22′E	5822	84°	Weak
229	10/13/68	118	07°35′N	178°25′E	5547	96°	Moderate
230	10/16/68	119	01°28′N	174°25′E	4693	98°	Moderate
232	10/20/68	120	06°34′S	177°15′W	5951	252°	Weak
236	10/23-24/68	121	09°25′S	169°00′W	5318	12°	Strong[6]
237	10/25/68	122	14°37′S	168°30′W	5058	103°	Moderate
239	10/28/68	123	17°19′S	175°34′W	2000	145°	Moderate

[1]Positions determined by satellite navigation method (Talwani and others, 1966).
[2]Depth measured by precision depth recorder and corrected for sound velocity, using Matthews (1939).
[3]Direction (corrected for magnetic variation) toward which the current flows.
[4]Between 4 and 8 cm/sec.
[5]Less than 3 cm/sec.
[6]Above 9 cm/sec.

ACKNOWLEDGMENTS

The authors are grateful to Dr. M. Langseth, who provided thermograd measurements of water temperature, and to Dr. Maurice Ewing for the bottom current meter data. The drafting was done by L. Childs and the secretarial duties by J. Stolz. The work was supported by Office of Naval Research Contract N00014-67-A-0108-0004 and by Atomic Energy Commission Contract AT(30-1) 2663.

REFERENCES CITED

Arons, A. B., and Stommel, H., 1967, On the abyssal circulation of the World Ocean-III. An advection-lateral mixing model of the distribution of a tracer property in an ocean basin: Deep-Sea Research, v. 14, no. 4, p. 441-458.

Bruce, J. G., and Thorndike, E. M., 1967, Photographic measurements of bottom currents, *in* Hersey, J. B., *Editor,* Deep Sea Photography: Johns Hopkins Press, Baltimore, Maryland, p. 107-110.

Capurro, L. R. A., 1961, Oceanographic observations in the intertropical regions of the World Ocean during IGY and IGC: Part IIb, Pacific Ocean, IGY Oceanography Rept. 3, p. 1-298.

Ewing, J., Ewing, M., Aitken, T., and Ludwig, W., 1968, North Pacific sediment layer measured by seismic profiling: Am. Geophys. Union Mono. 12.

Gerard, R., and Amos, A. F., 1968, A surface-actuated multiple sampler: Marine Sciences Instrumentation, v. 4, Plenum Press, N. Y., p. 682-686.

Gordon, A. L., 1966, Potential temperature, oxygen and circulation of bottom water in the Southern Ocean: Deep-Sea Research, v. 13, p. 1125-1138.

Heezen, B. C., and Hollister, C., 1964, Deep-sea current evidence from abyssal sediments: Marine Geology, v. 1, p. 141-174.

Knauss, J. A., 1962, On some aspects of the deep circulation of the Pacific: Jour. Geophys. Research, v. 67, no. 10, p. 3943-3954.

Langseth, M. G., 1965, Techniques of measuring heat flow through the ocean floor: Am. Geophys. Union Mono. 8, Terrestrial Heat Flow.

Le Pichon, X., Eittreim, S., and Ludwig, W. S., 1969a, Sediment transport and distribution in the Argentine Basin (1) Entrance of Antarctic Bottom Current through the Falkland Fracture Zone: Physics and chemistry of the Earth, v. 4 (in press).

LePichon, X., Ewing, M., and Truchan, M., 1969b, Sediment transport and distribution in the Argentine Basin (2) Antarctic Bottom Current passage into the Brazil Basin: Physics and chemistry of the Earth, v. 4 (in press).

Matthews, D. J., 1939, Tables of the velocity of sound in pure water and sea water for use in echo-sounding and sound-ranging: Hydrographic Dept., Admiralty, London, 52 p.

Olsen, B. E., 1968, On the abyssal temperatures of the world ocean: Ph.D. dissert., Oregon State Univ., Corvallis, Oregon.

Reid, J., Stommel, H., Stroup, E. D., and Warren, B. A., 1968, Detection of a deep boundary current in the western South Pacific: Nature, v. 217, p. 937.

Scripps Institution of Oceanography, University of California. Oceanographic observations of the Pacific: Eleven sections, 1960-1965.

Stommel, H., and Arons, A. B., 1960a, On the abyssal circulation of the World Ocean (1) Stationary planetary flow patterns on a sphere: Deep-Sea Research, v. 6, no. 2, p. 140-154.

—— 1960b, On the abyssal circulation of the World Ocean (II) An idealized model of the circulation pattern and amplitude in oceanic basins: Deep-Sea Research, v. 6, no. 3, p. 217-233.

Talwani, M., Dorman, J., Worzel, J. L., and Bryan, G. M., 1966, Navigation at sea by satellite: Jour. Geophys. Research, v. 71, no. 24, p. 5891-5902.

Thorndike, E. M., 1963, A suspended-drop current meter: Deep-Sea Research, v. 10, p. 263-267.

Tully, J. P., and Barber, F. G., 1960, An estuarine analogy in the subarctic Pacific Ocean: Jour. Fisheries, Research Board Canada, no. 17, p. 91-112.

U.S. Naval Oceanographic Office, 1966, Chart of the World: H.O., 1262A.

Warren, B. A., 1968, General comments on the circulation of the South Pacific (symp.): Scientific Exploration of the South Pacific (in press).

Wooster, W. S., and Volkmann, G. H., 1960, Indications of deep Pacific circulation from the distribution of properties at five kilometers: Jour. Geophys. Research, v. 65, no. 4, p. 1239-1249.

Wüst, G., 1933, Das Bodenwasser und die Gliederung der Atlantischen Tiefsee: Wiss. Erg. Deutsch. Atlant. Exp. *Meteor,* **1925-27, Bd. VI,** Teil 1, Lief 1.

—— 1937, Bodentemperatur und Bodenstrom in der pazifischen Tiefsee: Veroff. Inst. Meeresk. Univ. Berl., Neue Folge, A. Geogr. naturwiss., v. 35, p. 56.

—— 1961, Tables for rapid computation of potential temperature: Tech. Rept. CU-9-61-AT(30-1)1808 Geol. (unpub.)*.

LAMONT-DOHERTY GEOLOGICAL OBSERVATORY CONTRIBUTION NO. 1484
MANUSCRIPT RECEIVED APRIL 10, 1969
REVISED MANUSCRIPT RECEIVED JUNE 30, 1969

**Available on request from Lamont-Doherty Geological Observatory*

THE GEOLOGICAL SOCIETY OF AMERICA, INC.
MEMOIR 126, 1970

Nepheloid Layer in the North Pacific

MAURICE EWING
AND
STEPHEN D. CONNARY

Lamont-Doherty Geological Observatory of Columbia University, Palisades, New York

ABSTRACT

A distinct nepheloid layer with a vertical gradient of light scattering is observed on the continental rise of western North America. Throughout the deep North Pacific Basin the nepheloid layer extends from the bottom to the temperature minimum. The intensity of light scattering in the nepheloid layer of the North Pacific is substantially less than that of the Argentine Basin, except on the Alaskan and Tufts Abyssal Plains where it is comparable. Light scattering measured in this nepheloid layer with a photographic nephelometer is due to particulate matter; a model is proposed according to which this matter is derived principally from continents bordering the North Pacific and transported to the sea largely by rivers. Suspended particles are carried from the continental margin to the deep basin by lateral mixing and transport within the nepheloid layer. They are distributed throughout the large basins of the North Pacific by counterclockwise gyres and replace particles lost by sedimentation. Marginal trenches contain a strong nepheloid layer of trapped particles which ultimately settle to the trench floor.

The particulate matter in the nepheloid layer enters the water in the North Pacific because the homogeneous bottom water moving northward through narrow equatorial channels along about 175° W. has only a weak nepheloid layer. Rapid increase in temperature and maintenance of an adiabatic gradient during transit of the channels from about 17° S. to 15° N. are attributed to strong vertical turbulence associated with vigorous flow and admixture of warmer water from above the deep temperature minimum.

The vertical gradient of light scattering in the nepheloid layer is the result of a balance between particle settling and turbulent mixing in the homogeneous water below the temperature minimum. When data on settling rate of the particulate matter become available, this gradient may be used to estimate the intensity of vertical turbulence in the homogeneous water. Evidence supports the hypothesis that the nepheloid layer is a steady-state phenomenon.

CONTENTS

INTRODUCTION

The following is an interim report on *in situ* measurement, by means of a nephelometer, of the intensity of light scattering as a function of depth at several hundred stations in the North and Equatorial Pacific. A nepheloid layer, whose strength shows considerable regional variations and whose thickness ranges from 100 to 1500 fm, has been found to characterize all areas where soundings exceed 2500 fm. A similar layer also occurs in many areas of lesser depth, particularly on the continental rises. Major difficulties have been encountered in the attempt to explain the nepheloid layer and related phenomena in terms of the accepted model of deep circulation in this basin.

The long-standing uncertainties about the pattern of circulation of the deep and bottom water in the North Pacific Basin arise from the control of this circulation by complex, incompletely known topography, and from the absence of strong variations in the characteristics of the deep water. This water comes from a single source, the western boundary current of Antarctic Bottom Water (AABW) which enters the basin through a gap in its southern boundary at about 10° N. 175° W. The conclusion of earlier workers, summarized and extended by Gordon and Gerard (1970, this volume), is that this flow divides at about 10° N. into two zonal branches—the eastern one passing between the Johnston Island and Christmas Island Ridges (sill depth about 2800 fm), and the western one passing between the Marshall Islands and the Marcus-Necker Ridge (sill depth uncertain, assumed to be about 2800 fm). Gordon and Gerard interpret anomalies of bottom potential temperature as confirmation of the opinions of earlier workers (Knauss, 1962; Wooster and Volkmann, 1960) that the two branches of flow turn northward and converge in a region somewhere between the Hawaiian and the Aleutian Islands where they suppose this water is transferred to the upper layers by upwelling as well as mixing.

Gordon and Gerard note the absence of an explanation for the low potential bottom temperature (as low as 1.03° C) of the water in the Aleutian Trench, since no connection between it and either of the two areas of equally cold water shown in their Figure 3 has been found. They state that (except in constricted passages) bottom currents are sluggish (perhaps small compared with tidal currents) for the entire North Pacific, but that even so, the transport of the currents is large because they extend several km above the bottom of the basin. A deep temperature minimum is observed over much of the North Pacific (Olson, 1968) at a depth of about 2200 fm, below which salinity is practically constant and the thermal gradient is approximately adiabatic. Lynn and Reid (1968) ". . . have examined the abyssal properties in the world ocean with the newer data and found them consonant with the conventional notions of deep flow." They state that in the northern North Pacific the vertical and horizontal gradients of salinity, potential temperature and potential density near the bottom are relatively weak, and that the values given for these quantities are representative of large areas. Evidently they assign no important role to boundary currents here.

The light scattering in the nepheloid layer results from particles so small that they may remain in suspension for years. This layer is of interest in connection with provenance and distribution of the lutite component of deep-sea sediments sampled in cores and measured by acoustic reflections. It is also of interest because the particles may serve to identify the water mass in which they are suspended and to supplement other data for study of the circulation of the deep water.

The observed properties of the nepheloid layer are most easily explained by assuming that the main feature of the deep circulation in the North Pacific is a system of counterclockwise gyres whose associated deep boundary currents follow the topographic boundaries of the basin. Strong boundary currents have been described by Warren (1968) and Reid and others (1968) as following the Tonga-Kermadec Ridge in the South Pacific. It is highly probable that such a boundary current in the North Pacific acquires a sediment load while skirting the North American borderland and supplies sediment-laden bottom water to form the nepheloid layer in the main basin and in the marginal trenches.

NEPHELOMETER TECHNIQUE

At each station the nephelometer designed by Thorndike and Ewing (1967) was lowered to the ocean floor. This instrument contains an underwater camera which records continuously the intensity of light scattered at small angles on a single strip of slowly advancing photographic film. Time and depth marks are simultaneously recorded during the ascent and descent of the nephelometer. It also monitors the speed of film transport and the intensity of light that has passed directly from the source to the camera. This technique serves to identify spurious variations in film density attributable to instrumental effects or to film processing, and facilitates correlation of true light scattering between various stations. After processing, the nephelometer film is measured with a photodensitometer which constructs a graph of intensity of scattered light versus depth and a graph of direct light versus depth for the descent and ascent of the nephelometer. The graph for scattered light gives a qualitative picture of the variation of light scattering with depth at that station. For quantitative measurements, the graph for the direct beam is used to correct for variations in the light source and in the film transport speed. Sensitometric patches recorded before lowering on the film from sources of known relative intensity are used to obtain "logarithm of light intensity" from optical density calculated from the graphs.

In our present analysis we have not made quantitative corrections for variations of instrument light intensity or film developing, but instead have rejected data for which these effects are significant. A qualitative judgment of gross differences between films may be easily and rapidly made by examination of the film itself and of the densitometer record of direct light transmission.

Two quantities are measured at each station—the intensity of scattering and the depth to the top of the nepheloid layer. In the simplest case the depth of the clearest water is the obvious choice for the top of the layer. In more difficult cases we choose the depth of clear water, not necessarily the clearest, from which light scattering increases significantly and uniformly to the ocean floor. In this analysis we express light scattering intensity in terms of photodensitometer deflection. The deflection values are read for the top and bottom of the nepheloid layer, and the corresponding ratio R of these values serves as a measure of the strength of the layer. A station with no nepheloid layer thus has an R of 1.0; R increases with increasing intensity of the layer. For a given value of R, it can then be said that the densitometer deflection at the bottom of the nepheloid layer is R times that at the top, which is the point of clearest water. This then provides a measure of the intensity of that part of the nepheloid layer in which there is the highest concentration of particulate matter. For stations in the North Pacific, R varies between 1.0 and about 15, with the bulk of values in the range 1.0 to 6.0. Over most of the North Pacific, with the exception of the Gulf of Alaska and the Aleutian and Tufts Abyssal Plains, the intensity of light scattering in the nepheloid layer is substantially less than that in the Argentine Basin where the average value of R is about 6.

We acknowledge that some details and high accuracy may be

lost by this approach, but at present we seek only to describe the observations in a manner which stresses the main features and is sufficiently accurate to provide a meaningful description of the main nepheloid layer over a large geographic area. For this purpose we have assumed that the clearest water recorded is essentially the same at all stations and used it as a reference to which light scattering intensity at other depths may be compared. This assumption is justified on the grounds that the measured variations in light scattering are at least one order of magnitude and that particulate matter samples taken in the clear water above the nepheloid layer show low particle concentration (Eittreim and others, 1969).

At a typical station, light scattering is large near the surface and decreases (monotonically in the simplest case) to some depth at which it reaches an absolute minimum; this is taken as the reference level. From this depth, light scattering increases to the ocean floor. The zone of increased light scattering adjacent to the bottom has been called the main nepheloid layer (Ewing and Thorndike, 1965). Frequently, smaller variations in scattering higher in the water column define zones which are probably related to the interfingering of water masses. A detailed explanation of the generation and maintenance of such maxima and minima of light scattering over a large region will undoubtedly lead to increased knowledge of oceanic circulation and the physical properties of water masses at these depths. In this study we focus attention on the bottom, or main nepheloid layer, a feature most remarkable in its persistence in time and place.

A quantitative relation between nepheloid intensity and mass of suspended matter per unit volume has not yet been determined, but a rough proportionality has been established through continued studies by Marian Jacobs and by Stephen Eittreim on samples extracted by means of a centrifuge or filters. In view of this uncertainty, we have given a qualitative designation to R values as follows: 1.0 to 1.5 is weak; 1.5 to 2.5 is moderate; greater than 2.5 is strong.

HYPOTHESIS ON DEEP CIRCULATION AND THE NEPHELOID LAYER

A model is presented for the deep circulation in the North Pacific. With it we attempt to explain our data on the nepheloid layer and those on temperature, salinity, and oxygen in deep and bottom water. Several other lines of evidence support the hypothesis.

For the purposes of this model, the North Pacific Basin is considered to be bounded on the west by the ridges associated with the Mariana-Japan-Kuril Trench system, and on the north by the Aleutian Ridge. The eastern boundary is the continental margin of North America and the northern end of the East Pacific Rise. A broad rise in the vicinity of the equator joins the Christmas Island Ridge to form the eastern part of the southern boundary. The western part is the Caroline-Marshall-Gilbert Islands group.

In each basin of the central and North Pacific there is a deep *in situ* temperature minimum. The depth to this minimum and the corresponding temperatures have been studied by Olson (1968), particularly for the areas deeper than 2500 fm. His results are summarized in Table 1. From approximately the depth of the temperature minimum to the bottom, the water is typically remarkably homogeneous. Salinity and dissolved oxygen in it are nearly constant and the temperature increases at essentially the adiabatic rate.

The nepheloid layer in each of these basins is characterized by a decrease of light scattering from a relative maximum at the seafloor to a minimum at the top of the layer, which is at approximately the level of the temperature minimum. Within the rather rough limits of accuracy with which scattering intensity at one station may be compared with that at another (particularly if the two stations were occupied on different cruises), the strength of the nepheloid layer appears to be nearly constant throughout each major basin. The light scattering is attributed to particles whose tendency to settle to the bottom is counteracted by vertical turbulence in the homogeneous water.

These facts imply that the vertical mixing within the homogeneous layer is rather uniform over a basin and that horizontal circulation (presumably in the form of a large counterclockwise gyre) is sufficiently vigorous to maintain an approximately uniform distribution of the nepheloid layer despite the addition of particulate matter along some parts of its boundary and depletion by sedimentation over most of the basin.

The water injected into this system is AABW, which is channeled through relatively narrow passages south and west of the Hawaiian Islands. This water is characterized by a thick but very weak nepheloid layer. The opinion that most of the particulate matter in the North Pacific gyres is supplied from the continental borders and not from the surface water is based on the clarity of the intermediate water over the entire basin.

TABLE 1. SUMMARY OF DEPTH AND TEMPERATURE *in situ* AT THE TEMPERATURE MINIMUM IN THE NORTH PACIFIC

	Depth (fm)	T (°C)
Northeast Pacific Basin	2130	1.45 - 1.50
Northwest Pacific Basin	2185 - 2460	1.45 - 1.55
Central Pacific Basin	2460 - 2620	1.20 - 1.30

(*From* Olson, 1968)

The nepheloid layer on the continental rise has a vertical gradient of scattering intensity and tends to maintain a constant thickness well up on the rise. This water appears to have a slight vertical gradient of density on the basis of hydrographic stations examined (Fig. 1), but good observations extending near the bottom are rare. The existence of vertical turbulence capable of maintaining the nepheloid layer here is attributed to interaction between the boundary current and the sea floor.

Loss of water from the homogeneous layer is ascribed to mixing with the overlying water mass throughout the entire area of the basin. The water so lost carries with it practically no particulate matter.

In the model chosen as compatible with nephelometer data, the injection of AABW into the Northeast Pacific Basin is restricted to the passage between the Hawaiian Islands and the Christmas Island Ridge. Most of the bottom water in the Northwest Pacific Basin is assumed to be received in the vicinity of the north end of the Emperor Seamount Chain. The AABW, which flows westward between the Marshall Islands and the Marcus-Necker Ridge, apparently is constrained by the Marcus-Necker Ridge so that most of

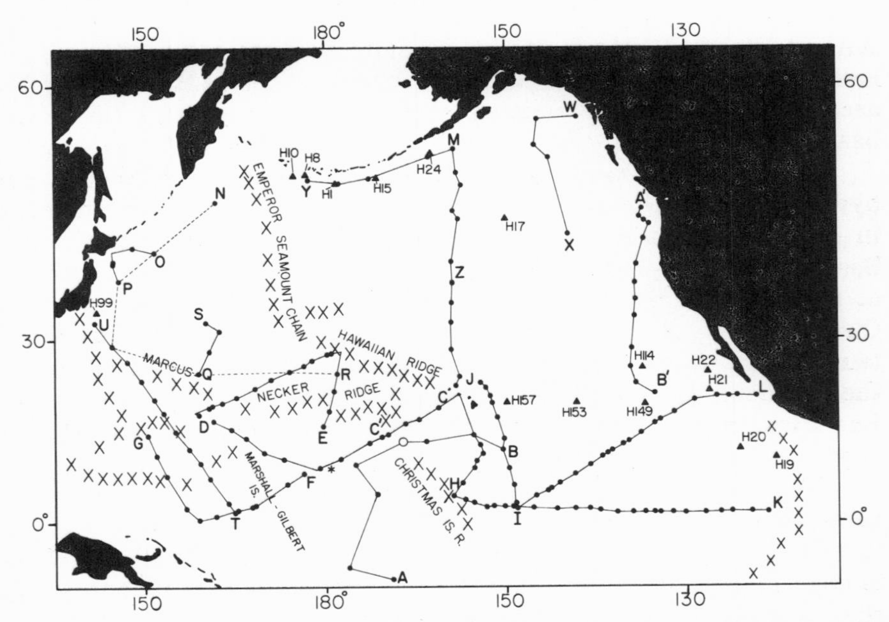

Figure 1. Position chart of nephelometer stations (solid circles) and hydrographic stations (solid triangles) in the North Pacific. Hydrographic stations are listed in Table 3. The station C12—N154 whose vertical light scattering profile is illustrated in Figure 3a is indicated by an asterisk in the Central Pacific Basin. The open circle in section A-B marks the postion of C12-N151 illustrated in Figure 3b. C12 refers to *Robert D. Conrad* Cruise 12. Major topographic highs are indicated by X's.

it flows into the Mariana and neighboring basins. Some of this water does appear to penetrate the Ridge and enter the Northwest Pacific Basin (Gordon and Gerard, 1970, this volume).

When more information about sizes and settling rates for the particles as well as velocities in the gyre can be obtained, calculations of sediment budget similar to those for the Argentine Basin (Ewing and others, 1970) may be made. At that time it should also be possible to estimate, from the vertical gradient of light scattering, the intensity of vertical turbulence below the temperature minimum.

CENTRAL PACIFIC BASIN

The Central Pacific Basin is bounded on the north by the Marcus-Necker Ridge, on the east by the Christmas Island Ridge, and by the Marshall-Gilbert Islands on the west (Fig. 1). Available bathymetric data indicate a deep passage in the northern Christmas Island Ridge and one between the Marcus-Necker Ridge and the Marshall Islands. On the south the basin has no major topographic boundary.

Gordon and Gerard (1970, this volume, Fig. 6) show AABW flowing north as a western boundary current along the axis of the Tonga Trench and passing into the Central Basin. During its transit from the Samoa Islands at about 15° S., to the region in which AABW branches at about 13° N., potential temperature increases nearly 0.25C°. This temperature increase cannot be reasonably ascribed to geothermal heating of the bottom water in this small basin.

In Figure 2b, temperature *in situ* versus depth is plotted for hydrographic stations occupied by *Brown Bear* and *Spencer F. Baird* in the Japan and Aleutian Trenches and in the Northeast Pacific Basin (Table 3). A temperature minimum of about 1.46° C occurs at about 2200 fm; this value agrees well with the value given by Olson (1968) for the Northeast Pacific Basin (Table 1). Below the temperature minimum, the temperature increases adiabatically to the sea floor, indicating that throughout the deep Northeast Pacific Basin and in the marginal trenches, the water below about 2200 fm is strikingly homogeneous.

For the Central Basin, bottom temperatures *in situ* are plotted versus depth (Fig. 2a). The latitudinal increase of bottom temperature here differs markedly with the regional uniformity of bottom temperature in the North Pacific. A remarkable fact illustrated by these curves is that, despite the rapid northward warming of the AABW, all bottom temperature values at closely spaced stations (grouped according to latitude) lie on a line whose slope is nearly that for the homogeneous water in the Aleutian and Japan Trenches and in the Northeast Pacific Basin (Fig. 2b). This evidence strongly suggests that in the Central Basin there is a rapid northward in-

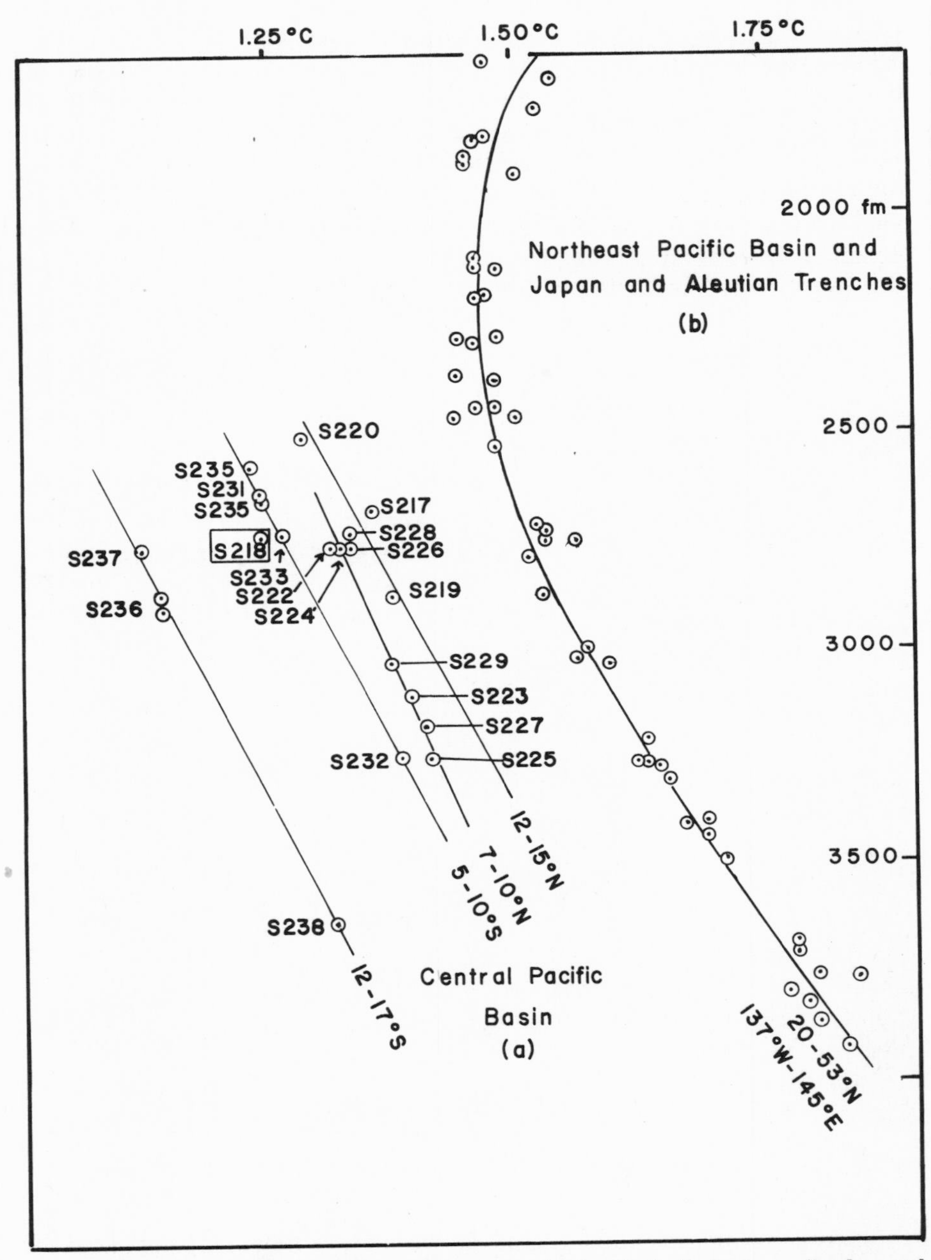

Figure 2. (a) Temperature *in situ* versus depth (fm). Data *from* Gordon and Gerard (this volume, Table 1) for the Central Pacific Basin. The temperature at S218 appears to be a reading error as S217-S220 are very closely spaced (*see* Gordon and Gerard, this volume, Figure 6, for positions). (b) Composite curve for hydrographic stations (H) in the Aleutian Trench, in the Japan Trench, and stations, including H153 and H157, in the Northeast Basin. *See* Figure 1 for positions. Note that below about 3000 fm temperature increases at approximately the adiabatic lapse rate.

crease of bottom temperature, which may be attributed to dilution of the AABW by warmer water above the temperature minimum due to strong vertical turbulence in the underlying homogeneous water. Olson (1968) has stated that the depth of the temperature minimum in this basin is 2460 to 2620 fm. (At one station, C12-S218, an apparently low value of bottom temperature is reported. All other values are consistent; this one value is possibly a reading error as it would lie on the proper curve if the value were larger by 0.1 C°.)

The difference in bottom temperature in the Christmas Island Ridge Gap and the Northeast Pacific Basin suggests that the last appreciable mixing between the homogeneous water and that overlying temperature minimum takes place in the channel between the Central Basin and the main Northeast Basin.

Associated with the AABW in the Central Basin is a distinct, weak nepheloid layer. In Table 2 are listed nephelometer stations which are typical for the Central Basin, and in Figure 3a is shown a representative scattering profile. This and all subsequent scattering profiles are tracings of the densitometer records drawn with zero deflection on the extreme right and maximum deflection on the extreme left. Light scattering intensity (I) increases to the right along the abscissa of each record. At C12-N154 (Fig. 3a), light scattering decreases from the surface to about 2300 fm, the depth of clearest water which is chosen as the depth of the top of the nepheloid layer. From 2300 fm to the bottom at 3073 fm, light scattering increases; the intensity of light scattering at this station is weak.

In Figures 4 and 5, north-south and east-west profiles in the Central Pacific Basin and bordering areas are presented. These profiles show that a generally weak nepheloid layer persists throughout this basin. The depth of the top of the layer is about 2300 fm; from there the light scattering increases steadily to the bottom. Throughout the basin, the top of the nepheloid layer occurs at

TABLE 2. NEPHELOMETER STATIONS IN THE CENTRAL BASIN

Station	Sounding	H	Thickness	Intensity R	Date
C12-N151	2715	1750	965	1.4	5 October 1968
C12-N152	3017	2320	700	1.2	10 October 1968
C12-N153	3152	2200	950	1.8	11 October 1968
C12-N154	3073	2300	775	1.3	12 October 1968
V24-N75	2810	2250	590	1.3	2 April 1967
C10-N105	2860	2355	505	1.3	7 April 1966
C12-N158	2893	2340	555	1.2	21 October 1968
C12-N159	2820	2275	545	1.3	23 October 1968

All depths are in fathoms. H is depth to top of the nepheloid layer.

nearly this same level, which is, within the limit of error, the depth given by Olson (1968) for the temperature minimum. This evidence supports the hypothesis of considerable turbulent mixing in the AABW of the Central Basin; it is consistent with the uniform distribution of temperature, salinity, and oxygen observed below the temperature minimum. These facts suggest that the temperature minimum is the upper level to which turbulence associated with the flow of bottom water can extend and the limit to which particulate matter is lifted.

In Figure 3b is shown the scattering profile for C12-N151 located in the Christmas Island Ridge gap (Figs. 1 and 4). The intensity of light scattering here is about the same as at other stations in the Central Basin. Light scattering is uniform below 2300 fm, and the clearest water is at a depth of 1750 fm. These features are probably related to turbulence in the strong deep currents through this relatively constricted passage. It is interesting that at this station the top of the layer of uniform scattering coincides approximately with the average depth of the temperature minimum and the clearest water in the Central Basin.

Nephelometer data have been obtained in the Central Basin during a period of about three years. Data for Figure 4 were ob-

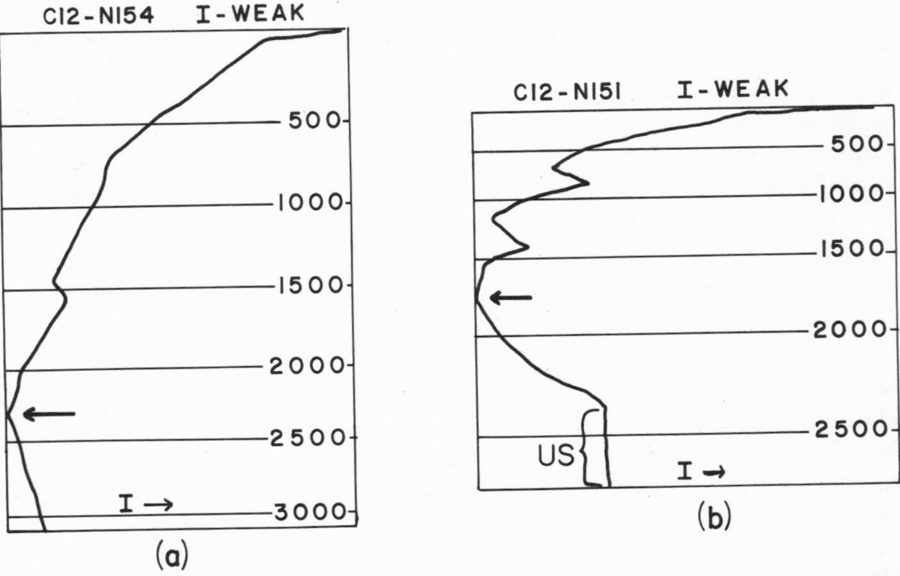

Figure 3. Scattering profiles in the Central Pacific Basin. *See* Figure 1 for positions. I is intensity of light scattering measured to the right from the reference of clearest water. The arrow indicates the top of the nepheloid layer. Depths are in fathoms. Figure 3a is a typical profile in the Central Basin. Figure 3b is located in the Christmas Island Ridge gap. Uniform scattering (US) is observed from about 2300 fm to the bottom.

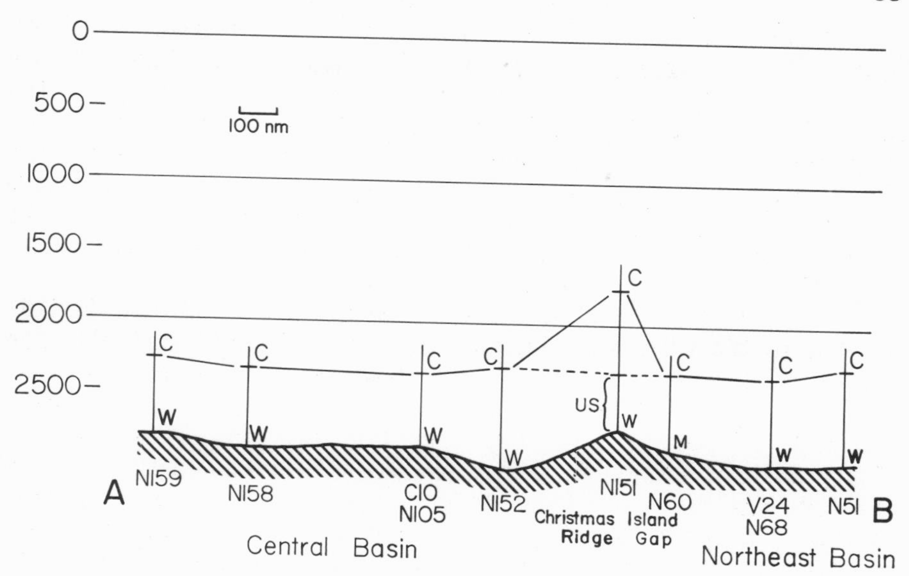

Figure 4. Profile A-B of the nepheloid layer in the Central and Northeast Basin (Fig. 1). The level (C) of clearest water is chosen as the top of the nepheloid layer. W indicates weak intensity, M moderate intensity of light scattering. US indicates the region of uniform light scattering shown in Figure 3b. Depth is in fathoms.

tained in October, 1968, and that for Figure 5 during April, 1967. The consistency over this time interval of the intensity of scattering and the depth to the top of the nepheloid layer suggest that the nepheloid layer is a steady-state phenomenon. The weakness of the nepheloid layer in this region suggests that the AABW does not contribute a significant quantity of particulate matter to the nepheloid layer in the North Pacific Basin.

NEPHELOID LAYER IN THE NORTHEAST PACIFIC BASIN

Gordon and Gerard (1970, this volume) conclude that at about 13° N., 178° W., the northward-flowing AABW separates into two branches, each of which is constrained to follow moderately narrow channels for about 900 nm before entering the broad basins of the North Pacific. Here the Coriolis force deflects the flow to the right to form a boundary current limited on its right-hand side by suitable isobaths (about 2500 fm in this case). Our interpretation of the distribution and intensity of the nepheloid layer in the North Pacific suggests that the main basin is divided into two parts by a somewhat permeable barrier, probably the Emperor Seamount Chain-Hawaiian Ridge. A counterclockwise boundary current of AABW circulates in each basin with appreciable interchange of AABW between the two systems.

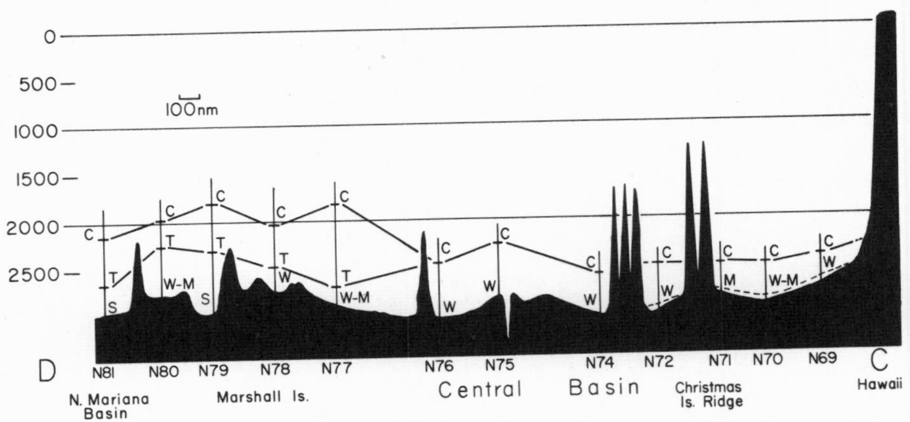

Figure 5. Profile of the nepheloid layer along C-D in the Central Basin and adjoining areas (Fig. 1). W is weak intensity of light scattering; W-M is weak to moderate; M is moderate; S is strong; C is depth of clearest water; T is depth at which higher gradient of light scattering begins. *See* text for explanation. Depths are in fm.

Bottom water in the eastern branch of the AABW must pass through a narrow gap in the northern Christmas Island Ridge. Bottom scour, bare rocks, nodules, and cores of Eocene sediments confirm the existence of strong currents through the constricted passage between the Hawaiian Platform and the Christmas Island Ridge at about 155° W. (Fig. 6). Particulate matter from islands

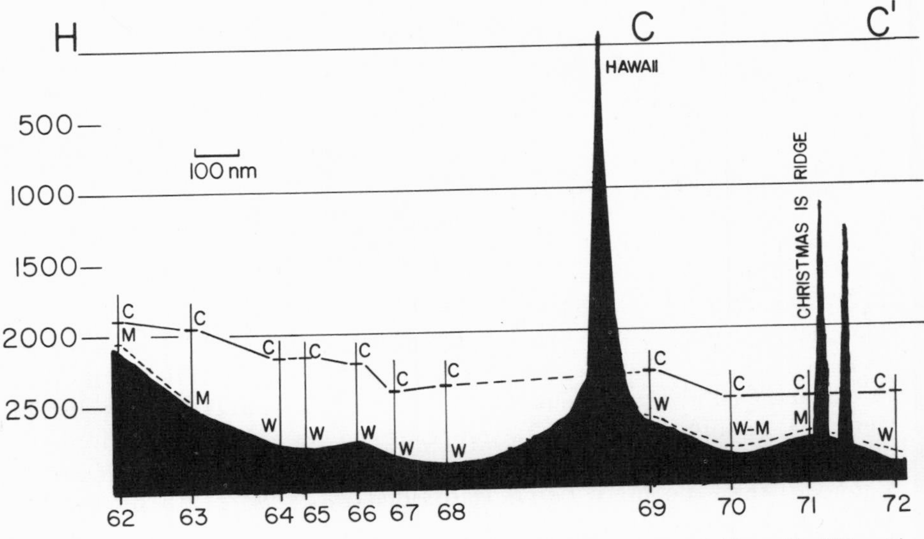

Figure 6. Profile of the nepheloid layer H-C′ south of Hawaii (Fig. 1). Symbols as in previous figures. Strong scour is observed in photographs at N62, 63, 64 and 68. At N65, 100% manganese nodule cover and at N66 many rocks were photographed.

in this region may contribute to the somewhat stronger nepheloid layer seen in the region immediately east of this gap. In the region of deep water close to the southeastern side of the island of Hawaii, a strong layer is observed to begin at about 2330 fm (Fig. 7). This is attributed to winnowing by the AABW of the finer fractions of the volcanic and carbonate sediments which are being deposited on the archipelagic apron (Horn and others, 1970, this volume).

In Figures 6 and 7 are presented north-south sections across the flow of AABW. As in the Central Pacific Basin, a generally weak nepheloid layer is observed in this region of the Northeast Basin. The layer is traced from the southern end of the Central Basin to C12-N51 in Figure 4. The regional uniformity of weak light scattering intensity and of the depth at which the main nepheloid layer begins are notable.

An interesting feature is observed at C12-N52 (Fig. 7). Light scattering increases from the depth (C) of clearest water to about 2730 fm, as indicated in Figure 7. Below this level, scattering decreases to the bottom. This pattern is generally not observed throughout the North Pacific and suggests that it is a phenomenon related to the local topography.

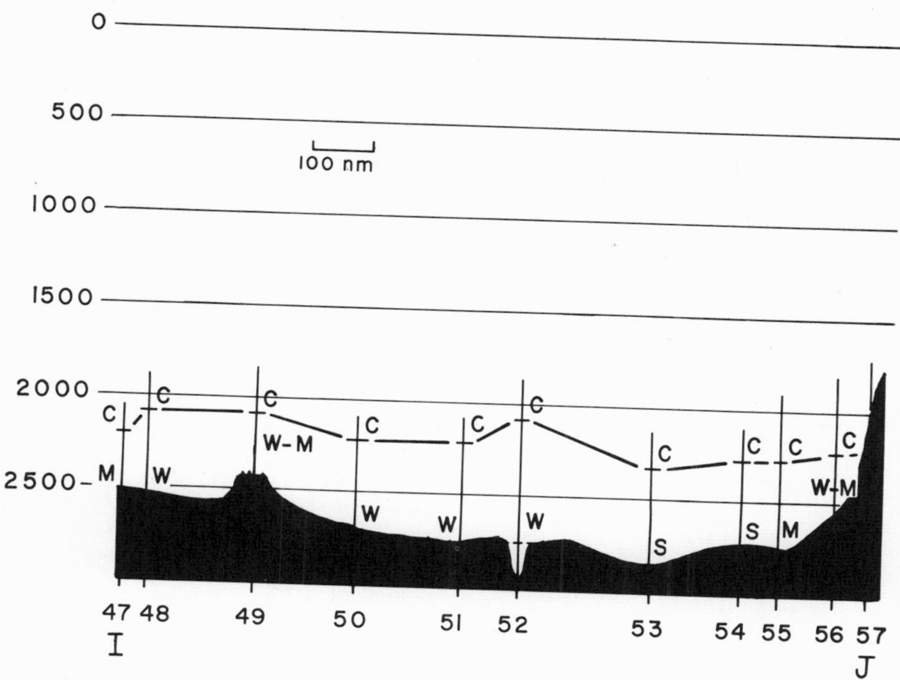

Figure 7. Profile of the nepheloid layer along I-J southeast of Hawaii (Fig. 1). Symbols are same as in Figure 5. At C12-N52, the depth of maximum scattering is at 2730 fm. See text for explanation. Depths are in fm.

In V24-N69 to N72 (Fig. 6), and V24-N62, N63 (Fig. 8), there exists at the bottom of the main nepheloid layer a zone 50 to 100 fm thick in which the gradient of light scattering is notably higher than that in the water above this zone. This zone is also present in the Marshall Islands, Mariana Basin and in the basin between the Hawaiian and Marcus-Necker Ridges. Its occurrence seems to be limited to regions where very weak bottom currents would be expected. Our interpretation of this feature is that the clearest water indicates the top of the main nepheloid layer, and that the lower zone, with its higher gradient of light scattering, forms because turbulent mixing in these semi-isolated regions is too weak to be effective very far above the bottom.

Along the equatorial rise (Fig. 8), the depth of clearest water is generally about 2000 fm, and a moderate intensity of light scattering persists along the equator to the Christmas Island Ridge. A profile from the equator at about 148° W. to the northern Middle America Trench is shown in Figure 9. The top of the main nepheloid layer is at a depth of about 2200 fm in water deeper than 2500 fm. The top of the layer here is at about the same depth as the weak layer south of Hawaii (Figs. 4, 6, and 7), but the light scattering intensity is uniformly moderate in the southeastern corner of the basin.

Opdyke and Foster (1970, this volume, Fig. 23) show the latitudinal variation of sedimentation rate north of the equator along 140° W. A relatively high value of over 15 mm/1000 yrs on the equator contrasts sharply with a value of about 2 mm/1000 yrs from about 8° N. to 36° N. Light scattering intensity of the nepheloid layer in this region (Figs. 8 and 9) is generally moderate at best. This suggests that, at least in this area, the rain of particulate matter from the surface water contributes little to the light scattering measured near the bottom and emphasizes the fact that areas of known high sedimentation rate and thick sediments (Ewing and others, 1968) may have only a weak or moderate nepheloid layer.

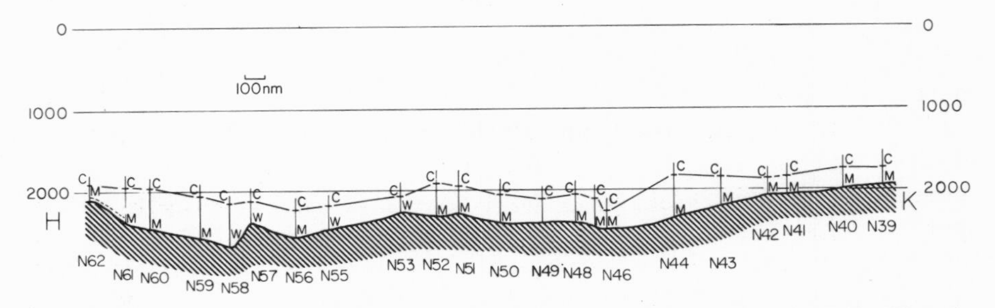

Figure 8. Profile of the nepheloid layer along the equator (H-K, Fig. 1). Symbols are as in Figure 5. The top of the nepheloid layer is the depth (C) of clearest water. Depths are in fm.

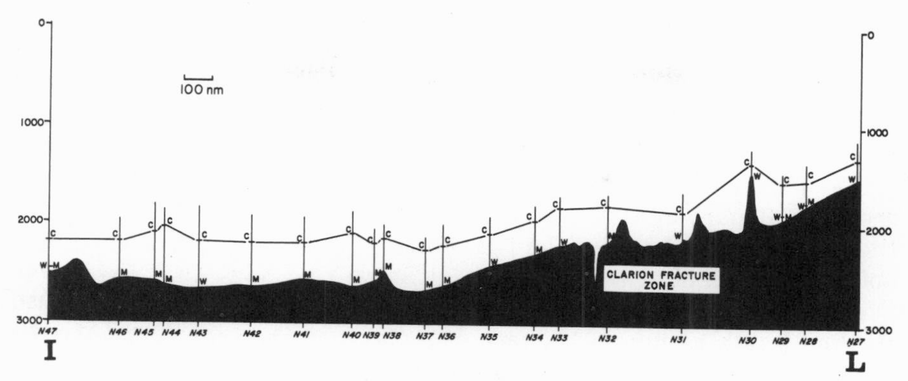

Figure 9. Profile of the nepheloid layer along I-L in the southeast portion of the NE Basin (Fig. 1). Symbols are as in Fig. 5. The top of the nepheloid is the depth of clearest water. Depths are in fm.

According to the Mie theory of light scattering for a dilute suspension of independent non-absorbing spheres, the intensity of scattering in sea water by particles greater than about 2 microns is directly proportional to particle surface area (Jerlov, 1968). This implies that large particles would contribute significantly to sediment volume without proportional addition to the intensity of light scattering. Those particles settling directly to the bottom would have an approximately uniform vertical distribution and hence would not affect the shape of a vertical profile of light scattering. Microscopic examination of water samples obtained from the nepheloid layer in the Aleutian Trench shows infrequent microfossils (Jacobs, 1970, personal commun.).

This is not an unexpected result, as larger particles are probably not suspended and concentrated within the nepheloid layer by vertical turbulence. This observation suggests that in regions of high productivity, the number of organic particles per volume within the nepheloid layer may be small compared to the lutite component. This supports the claim that such biogenic material has a negligible effect on light scattering in the nepheloid layer.

At present, we have no nephelometer measurements in the tongue of AABW defined by the 1.0°C contour of potential temperature (Gordon and Gerard, 1970, this volume, Fig. 3) in the central part of the Northeast Pacific Basin. The Hawaii-Aleutian Trench section (Fig. 10) is located about 13° to the west of the supposed axis of flow and is taken to be representative of the central part of this basin. Among the significant characteristic features of these curves is a sharply defined, frequently thin nepheloid layer at each of the 12 stations shown. It is striking that throughout the deep basin the main nepheloid layer begins at about 2600 fm; this is in contrast with a depth of about 2300 fm for the nepheloid layer associated with the AABW in the southern part of this basin and in the Central Basin.

Immediately north of Hawaii (C12-N148 and N149, Fig. 10), the nepheloid layer begins at about 1800 fm, 800 fm shallower than in the deep basin. In and near the Aleutian Trench (C12-N137 and N138, Fig. 10), clearest water is found 500 fm shallower than in the central part of the basin. It is suggested that at both margins the elevation of the nepheloid layer may indicate the presence of a boundary current.

Gordon and Gerard (1970, this volume, Fig. 3) show the AABW flowing north along the eastern boundary of the Northeast Pacific Basin. We have only two nephelometer stations (not shown) in the western part of this basin. The presence of a moderate to strong nepheloid layer at these stations, at each of the 12 stations between the Aleutian Ridge and Hawaii (Fig. 10), and in the southeastern part of this basin (Fig. 9), suggests that this layer is to be found throughout the basin. At least part of the AABW in the northern part of the basin probably flows south along the eastern side of the Emperor Seamount Chain-Hawaiian Ridge. Transport of the suspended particulate matter from the probable region of input in the Gulf of Alaska and western coast of Canada and the United States to the area north and southeast of Hawaii, is most reasonably done by means of a gyre in this basin. Lateral mixing in the

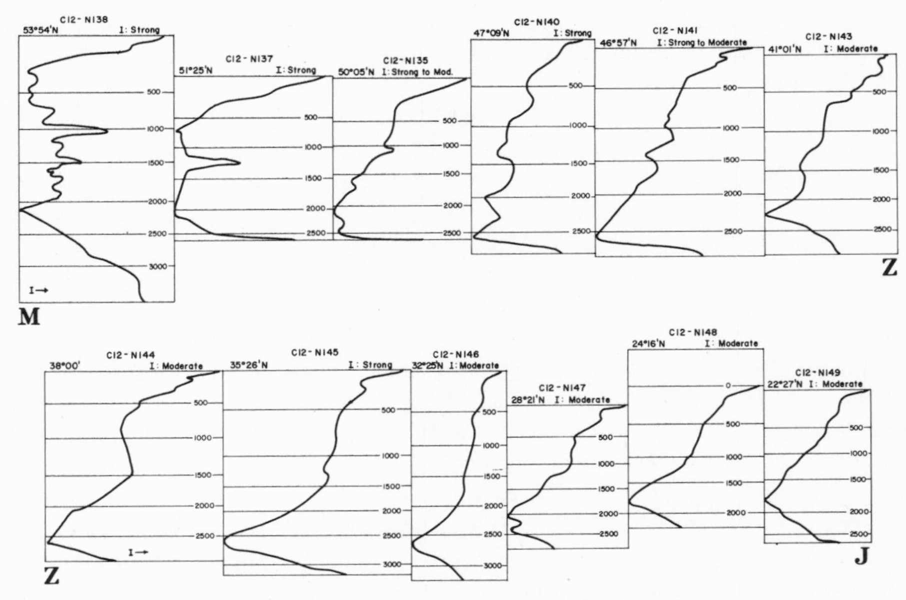

Figure 10. Scattering profiles from the Aleutian Ridge to the Hawaiian Islands (M-Z-J, Fig. 1). These are tracings of densitometer records. I is intensity of light scattering. Depths are in fm.

homogeneous water throughout this basin must occur at a sufficient rate to maintain the uniformity of the nepheloid layer by replacing the particulate matter lost by sedimentation.

This proposed pattern of circulation would receive additional support if the 1.1° C contour of bottom potential temperature (Gordon and Gerard, 1970, this volume, Fig. 3) could be relocated. This suggestion is not in conflict with data on bottom potential temperature (A. Gordon, 1969, oral commun.). We propose that the two tongues of water with bottom potential of temperature $< 1.1°$ C be joined together in the northeast part of the basin and closed at the Emperor Seamount Chain. There are some gaps in the Chain and it is by no means impossible that some bottom water may pass from the western into the eastern basin, as Gordon and Gerard indicate. Additional data on potential bottom temperature may also indicate a connection between the 1.1° C contour of the AABW in this basin and the 1.1° C contour outlining the Aleutian Trench. All other indications are strong that AABW reaches and fills the Aleutian Trench.

The zones of coarser sediments mapped by Horn and others (1970, this volume) along the margins of the basin probably coincide with boundary currents relatively stronger than those in the central part of the basin. From C12-N143 south to C12-N147 (Fig. 10), the intensity of light scattering is generally moderate, that is, comparable with that at the south end and less than that at the north end of the traverse. These stations lie in a region of very fine sediments ($< 1\mu$) where Opdyke and Foster (1970, this volume) and Horn and others (1970, this volume) have concluded that accumulation of sediment is less rapid than toward the margin of the basin.

In the Northeast Pacific Basin, the depth to the temperature minimum, which is the top of the homogeneous water, is about 2150 fm (Olson, 1968). At the northern end of the profile (Fig. 10), this level is in good agreement with the top of the nepheloid layer. In the deeper part of the basin to the south, corresponding to the area bounded by the 1μ contour of Horn and others (1970, this volume, Fig. 6), the top of the nepheloid layer is generally at about 2600 fm. We attribute this regional difference in the depth of the clearest water to a difference in turbulent conditions between the homogeneous water along the boundary and that in the central part of the gyre.

A noteworthy feature of Figure 9 is the continuity of the nepheloid layer between the deep basin and the continental rise. Our interpretation of the nepheloid layer on the rise is that particulate matter is in transit from the adjacent continent to the deep basin. The particles are transported in a nepheloid layer which is main-tained by the vertical component of turbulence produced by the interaction of topography and bottom currents. Data from hydrographic stations (Fig. 1 and Table 3) obtained in the vicinity of section I - L (Fig. 9) suggest that, although a temperature minimum is not clearly observed on the rise because adiabatic warming at these depths approaches the limit of accuracy of the temperature measurements, temperature *in situ* is nearly uniform below about 1800 fm. This corresponds to the top of the nepheloid layer at C12-N31, N32, and N33 in Figure 9 and to the depth of clearest water observed along the rise south of the Mendocino Escarpment (Fig. 11.) The shallower nepheloid layer north of the escarpment is attributed to lesser water depth. The nepheloid layer is conspicuously weak south of the escarpment. However, in the vicinity of and north of the escarpment, the layer increases to a moderate or strong intensity, possibly due to the influx of fine particulate matter from the Columbia River.

At two stations adjacent to the Aleutian Trench (Fig. 10), the depth is less than elsewhere in the northern part of the section, and the main nepheloid layer is thinner and more intense than to the south. The continuity of the top of the nepheloid layer between the Aleutian Abyssal Plain and the trench is impressive, and if one advocates flow from the plain over a sill to the trench, the observed intensity of light scattering along the trench axis is explained. AABW, which has acquired particulate matter from the Gulf of Alaska, flows over the trench, exchanges oxygen, salt, and heat, and contributes particles.

TABLE 3. HYDROGRAPHIC STATIONS ON THE CONTINENTAL RISE OFF WESTERN CENTRAL AMERICA

H19,	*Spencer F. Baird,* Dec. 9, 1954, sounding 1800 fm
H20,	*Spencer F. Baird,* Dec. 13, 1954, sounding 2170 fm
H21,	*Spencer F. Baird,* Dec. 17-18, 1954, sounding 2125 fm
H22,	*Spencer F. Baird,* Dec. 18, 1954, sounding 2120 fm
H149,	*Spencer F. Baird,* Sept. 10, 1955, sounding 2400 fm
H114,	*Stranger,* August 15, 1955, sounding 2300 fm
	HYDROGRAPHIC STATIONS IN ALEUTIAN AND JAPAN TRENCHES AND IN NORTHEAST PACIFIC BASIN
H1,	*Brown Bear,* August 9, 1957, sounding 7185 m
H8,	*Brown Bear,* August 14, 1957, sounding 7095 m
H10,	*Brown Bear,* August 17, 1957, sounding not given
H15,	*Brown Bear,* August 27, 1957, sounding 7095 m
H24,	*Spencer F. Baird,* August 10, 1953, sounding 3390 fm
H99,	*Spencer F. Baird,* October 26, 1953, sounding 4665 fm
H17,	*Spencer F. Baird,* August 5, 1953, sounding 2740 fm
H153,	*Spencer F. Baird,* Sept. 6, 1955, sounding 2800 fm
H157,	*Spencer F. Baird,* Sept. 3, 1955, sounding 2850 fm

Data *from* Norpac Committee and NODC, Washington, D. C.

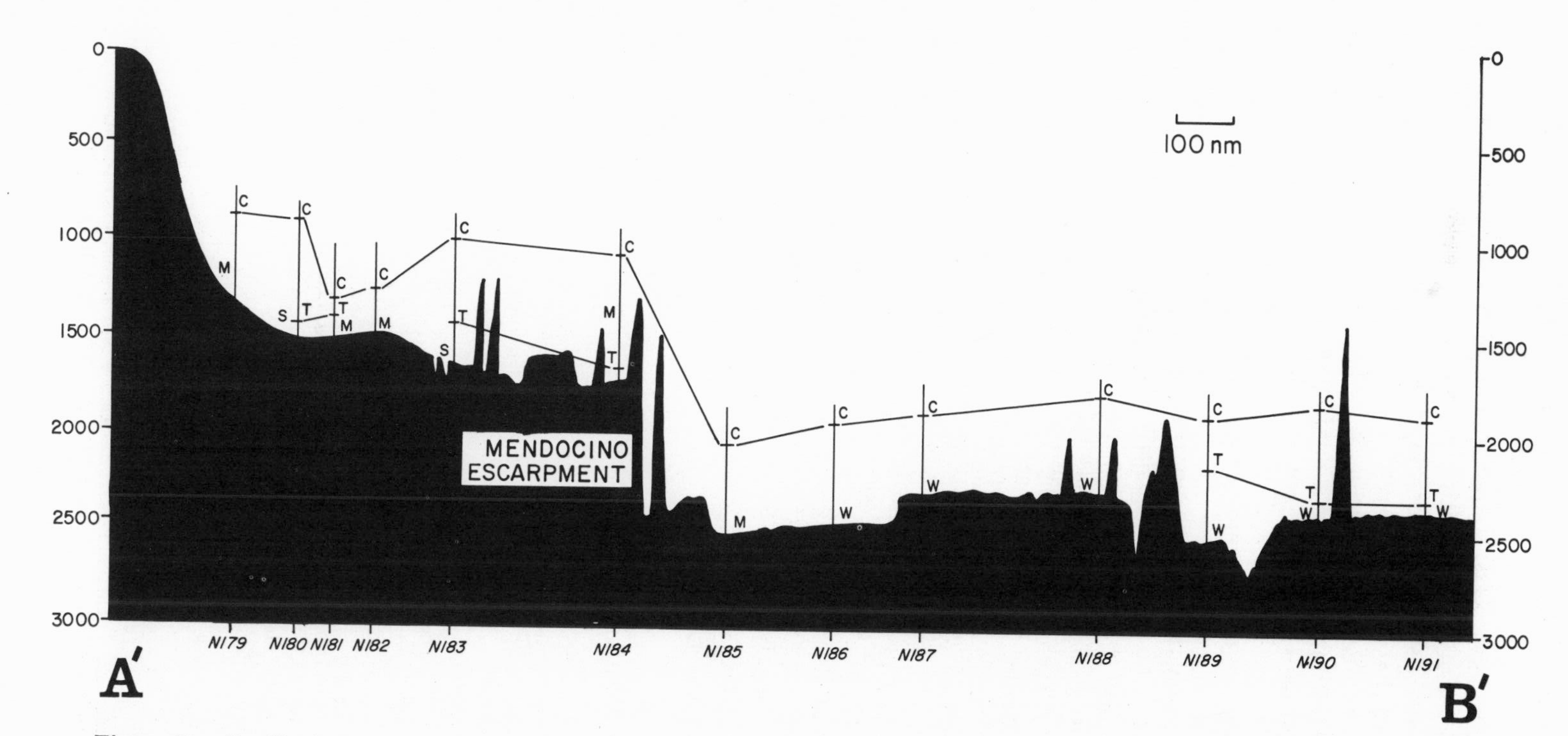

Figure 11. Profile A′-B′ along about 130°W on the continental rise (Fig. 1). South of the Mendocino Escarpment the nepheloid layer is generally weak; a trend of increasing strength toward the northern end of the profile is noted. This is the region of discharge of the Columbia River.

It has been stated that the deep and bottom water converge south of the Aleutians with a consequent massive upwelling (Gordon and Gerard, 1970, this volume; Reid, 1965; Knauss, 1962). We consider that the vertical or regional distribution of the nepheloid layer supports the concept that water is lost from the homogeneous layers to the overlying water by a mixing process occurring over the entire basin, rather than by massive upwelling in a restricted area.

NORTHWEST PACIFIC BASIN

At present we have only a few nephelometer stations of suitable quality in the deep Northwest Pacific Basin. In Figure 12 these stations are presented as a section which follows the margin of the basin in a counterclockwise manner from the northern end to the southwest corner between the Hawaiian and Marcus-Necker Ridges (Fig. 1). At all stations in the deeper part of the basin (for example, V21-N56, C10-N132, V24-N89, and V24-N94, Fig. 12) light scattering in the nepheloid layer is moderate to strong, and the depth of clearest water is about 2100 to 2500 fm. The depth of the temperature minimum in the Northwest Basin is stated by Olson (1968) to be 2190 to 2460 fm. This corresponds rather well with

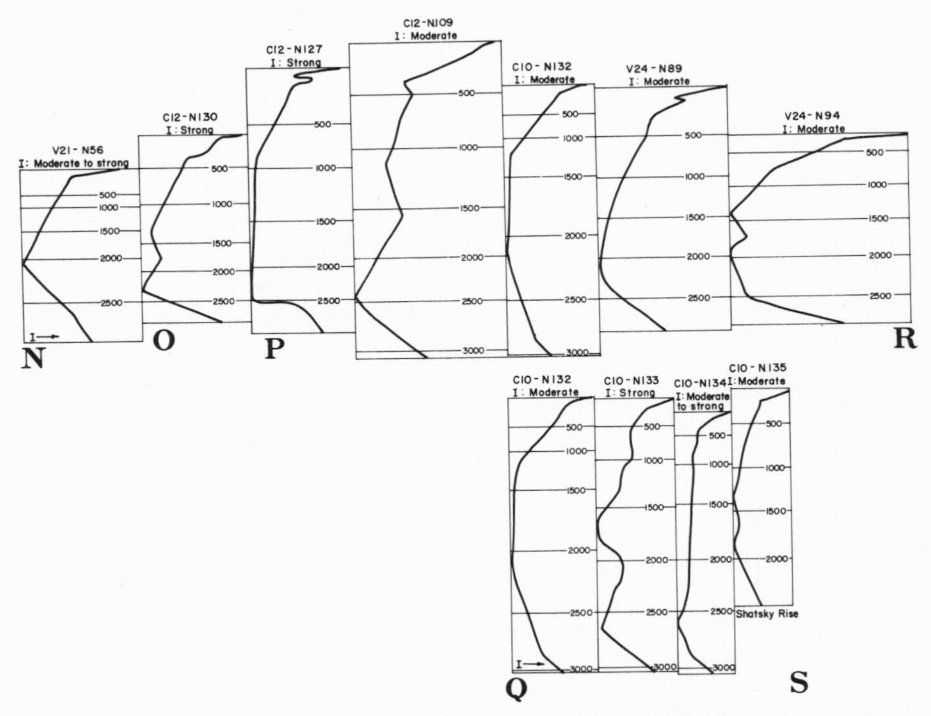

Figure 12. Scattering profiles in the Northwest Pacific Basin along N-O-P-R and Q-S (Fig. 1). Depths are in fm. I is intensity of light scattering, as in Figure 3.

the top of the nepheloid layer in this basin. As in the other basins of the North Pacific, the nepheloid layer characterizes the homogeneous water, and its gradient of scattering is attributed to a balance between the vertical component of turbulence and the gravitational settling of the light scattering particulate matter.

As in the Northeast Basin, at all nephelometer stations a distinct nepheloid layer is seen. We interpret this fact to mean that in the Northwest Basin there exists, probably as a counterclockwise gyre, a closed circulation which maintains the nepheloid layer from which particles are lost by sedimentation. Consistent with such a gyre would be the elevation of the level of clearest water along the margin of the basin. This is suggested in section Q-S (Fig. 12) in which the depth of clearest water is about 500 fm shallower adjacent to the northern side of the Marcus-Necker Ridge than toward the center of the basin, although the depth remains about the same. Several lines of evidence indicate that there is a flow from the Northeast to the Northwest Pacific Basin. Because of lack of nephelometer data, we have not yet been able to determine whether the flow is through a gap in the Emperor Seamount Chain or around the north end of it.

The strikingly asymmetrical distribution of chlorite (Fig. 15) shows high concentrations relative to the deep basin in a band along the northern and western margin of the North Pacific Basin, indicating in a general manner the existence of an interconnection in this region and of a counterclockwise gyre in the Northwest Basin. The chlorite appears to enter the North Pacific Basin principally in the region of the Gulf of Alaska. This observation is supported by Jacobs (personal commun.) who characterizes the bottom water and sediments of the Gulf of Alaska, Aleutian Trench, and North Pacific east of the Emperor Seamount Chain as chlorite-rich, and the deep North Pacific water west of the Emperor Seamount Chain as having notably lesser chlorite concentrations.

Additional support for a flow into the Northwest Basin at the north end of the Emperor Seamount Chain is given by Olson (1968, Pl. 2). Olson shows that at the deep temperature minimum, the temperature *in situ* is less than 1.50° C over much of the Northeast Basin, whereas in the Northwest Basin water of similar temperature is found only along the margins. The 1.50° C isotherm suggests that the homogeneous water of the Northeast Basin enters the Northwest Basin at the northern end of the Emperor Seamount Chain and flows as a western boundary current.

The map of Gordon and Gerard (1970, this volume, Fig. 3) suggests the uniformity of bottom potential temperature throughout the Northwest Basin. The influence of the Shatsky Rise is evident in their Figure 3. A long narrow tongue of water is shown reaching to about 44° N. on the west side of the rise. We suggest the possibility that the 1.1° C contour is broken by the Emperor Seamount Chain and that this contour could be joined to those north of the Rise.

Horn and others (1970, this volume) identify the Shatsky Rise as a local sediment source. To what extent it contributes to the nepheloid layer is difficult to determine. A nepheloid layer of moderate intensity is observed to begin at about 1850 fm on the rise (C10-N135, Fig. 11). The relation between an elevated nepheloid layer such as this and the main nepheloid layer may be the same as in the Northeast Basin where a nepheloid layer associated with the homogeneous water is observed on the continental rise.

Profiles in the western part of the Central Basin, the Mariana Basin, and the southern part of the Northwest Basin are presented in Figures 5, 13, 14, and 16. In Figure 5 from N75 to N80, the homogeneous water is characterized by a generally weak nepheloid layer as was observed throughout the Central Basin. This flow of the western branch of AABW (Gordon and Gerard, 1970, this volume) is constrained by the Marcus-Necker and Mariana Ridges to flow into the Mariana Basin. Some AABW may escape through the Marcus-Necker Ridge into the Northwest Basin (Gordon and Gerard, 1970, this volume; Olson, 1968).

The general continuity of the top of the nepheloid layer throughout this complex of islands, seamounts, and narrow passages is striking. An interesting feature of the scattering profiles is illustrated by V24-N94 in Figure 16. The clearest water, indicated by C, in Figures 5, 12, 13, and 14, is at about 2000 fm; a zone with a higher

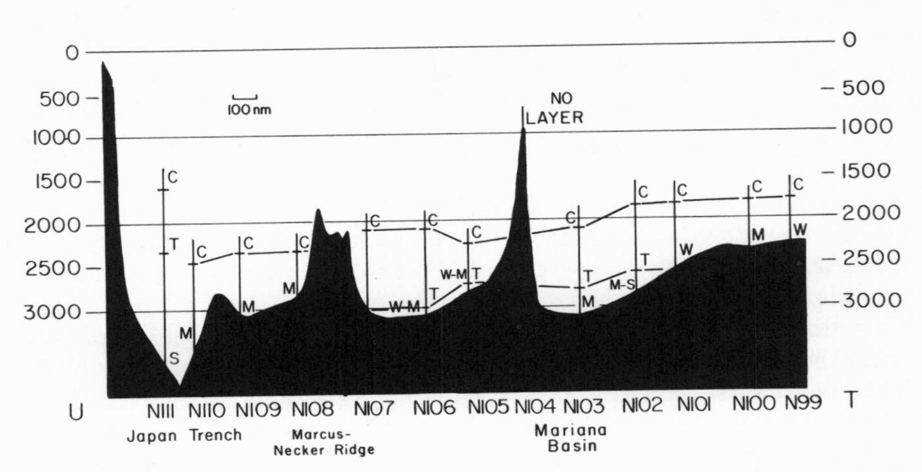

Figure 13. Profile of the nepheloid layer along U-T in the Mariana Basin (Fig. 1). Symbols are as in Figure 5. The Marcus-Necker Ridge appears to be an effective boundary separating water masses with distinctly different scattering profiles. Depths are in fm.

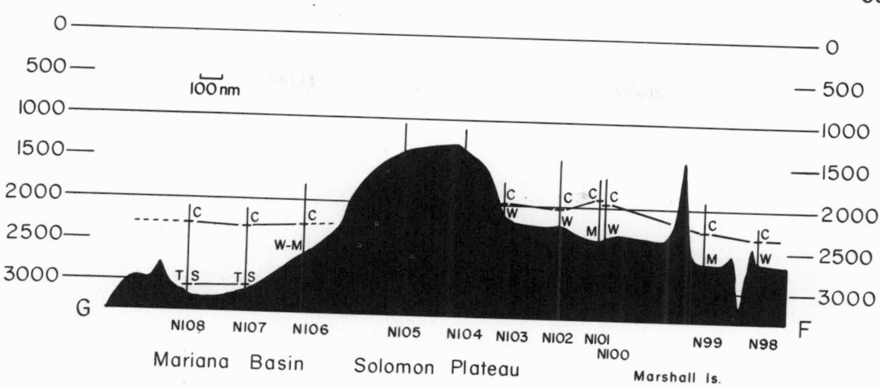

Figure 14. Profile of the nepheloid layer along G-F (Fig. 1). Symbols are as in Figure 5. Depths are in fm.

gradient of light scattering extends from about 2500 fm to the bottom and is indicated by T. This feature is persistent at all stations in the Marshall Islands and the Mariana Basin. Our interpretation of this feature is that in these regions suspended particulate matter is being concentrated in the lower part of the main nepheloid layer. This would be expected in areas of weak currents where the stirring effect of turbulence no longer balances gravitational settling. These regions then are settling basins for the suspended particles. The concentration of particulate matter in this lower zone of the nepheloid layer explains the moderate to strong light scattering intensities measured in the Mariana Basin.

In Figure 16 at N94 and N95, this lower zone in the nepheloid layer is observed in the semi-isolated basin between the Hawaiian and Marcus-Necker Ridges. Horn and others (1970, this volume, Fig. 6) show in this basin an area of sediments with mean diameters of less than 1 micron. It would not be unreasonable to claim that the deposition of such fine material would be expected in an area of very weak bottom currents and low turbulence that might be characteristic of the central part of the gyre.

At N86, N87, and N88 in Figure 16, the nepheloid layer is generally weak to moderate and at N108 and N109 in Figure 13 it is moderate. The vertical gradient of light scattering is approximately constant at these and other stations in the deep Northwest Basin. This would be expected in a basin in which the circulation is sufficiently active to continuously maintain a moderate to strong nepheloid layer in the homogeneous bottom water. In the Mariana Basin (Fig. 13) the nepheloid layer has a typical vertical profile of light scattering illustrated by V24-N94 (Fig. 16). The lower part of the layer is marked by a gradient of intensity larger than in the upper part of the layer. This we attribute to less turbulent conditions and a resultant concentration of particles to the lower part of the layer.

The Marcus-Necker Ridge appears to be a topographic barrier separating the deep water of the Northwest Basin, which exhibits a uniform vertical gradient of intensity, and the deep water of the Mariana Basin which does not.

SOURCES OF SUSPENDED SEDIMENT IN THE DEEP WATER

A study by Ewing and others (1970) indicates that continents are probably the principal contributors of particulate matter to the nepheloid layer. The dominant process by which terrigenous matter reaches the nepheloid layer appears to be river transport to the coast and current action across the continental margin. The intense nepheloid layers in the Argentine Basin and in the North American Basin appear to be satisfactorily explained by transport of fine particles down the slope and rise and distribution of the particles throughout the deep water of the basins by strong western boundary currents and their associated eddies.

The regional pattern of light scattering intensity in the North Pacific suggests that western North America is the most important source of suspended particulate matter in the deep water. Although Asia is bordered by an effective sediment barrier which prevents influx of detritus, particularly the coarser fractions, Griffin and others (1968) consider that the distribution of illite in sediments suggests that Asia is also an important source of some of the major components of deep-sea sediments in the North Pacific.

On the basis of nephelometer data we cannot support or refute this interpretation. At present, the best we can say is that we do not observe any strengthening of the intensity of light scattering in the nepheloid layer which might be attributed to the input of fine particulate matter from the Asian continent. If an appreciable quantity of light scattering particles of Asian origin are reaching the nepheloid layer, they are doing so by a presently undetected path.

The map of distribution of chlorite in "pelagic" sediments (Griffin and others, 1968) is presented in Figure 15. Contours of chlorite concentration in the size fraction < 2 microns of sediments indicate > 50 percent chlorite in sediments over a small region in the northern Gulf of Alaska. C11-N135, showing an extremely strong nepheloid layer (Fig. 17), is located in the proximity of the 50 percent chlorite contour. Jacobs (personal commun.) has noted the high concentrations of chlorite in the water and sediment of the Gulf of Alaska and Aleutian Trench and has suggested that chlorite may be supplied from the gulf of the trench. Griffin and others (1968) conclude that the chlorite distribution along the Alaskan coast is most likely the result of glacier and river input. The chlorite is derived from the predominantly high-latitude physical weathering of chlorite-bearing sedimentary and low-grade metamorphic rocks of

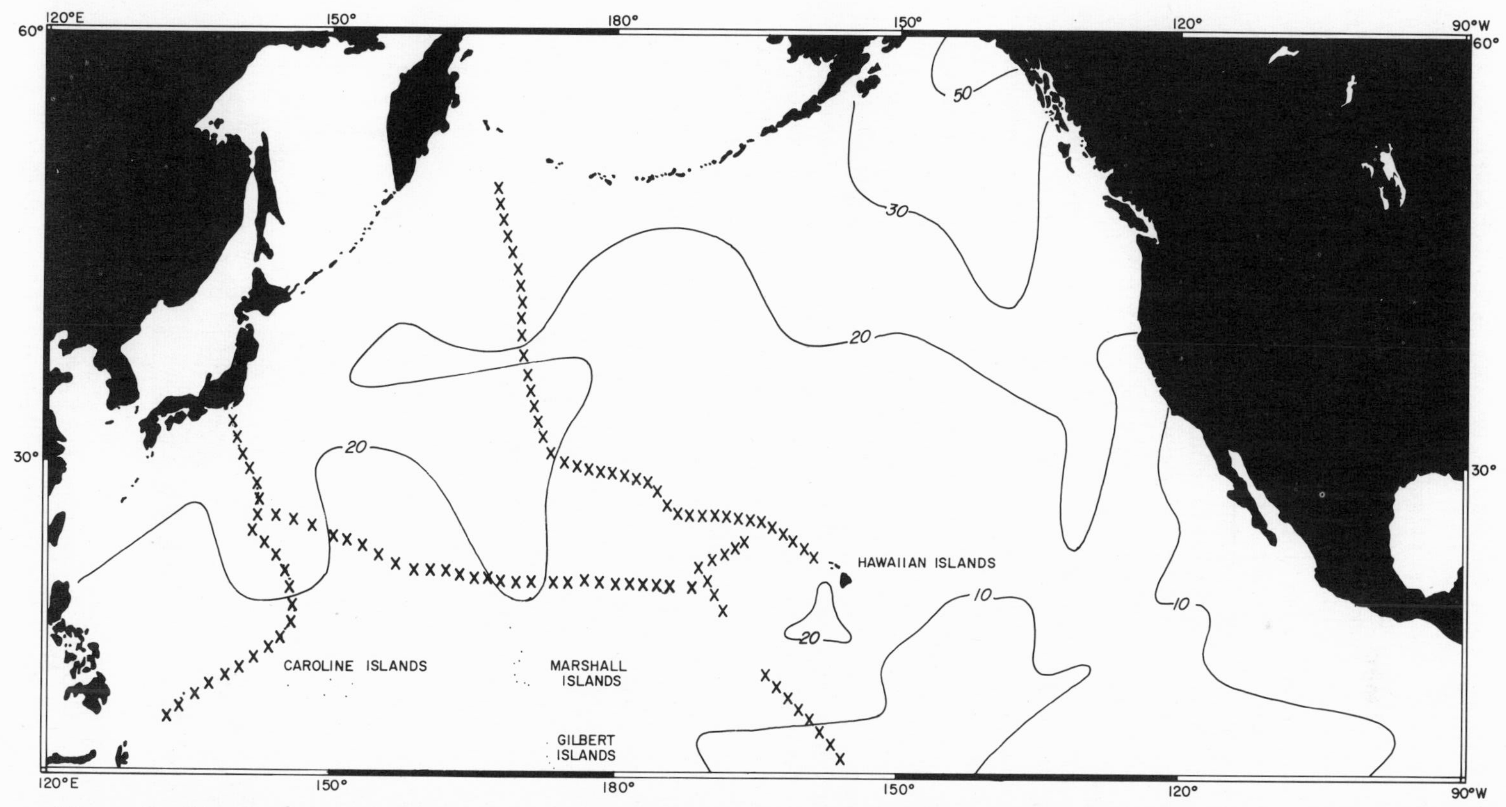

Figure 15. Distribution of chlorite in the < 2 micron size fraction of sediments in the North Pacific. Values are in percent. Adapted *from* Griffin and others (1968, Fig. 2). The high concentrations observed in the northeast Pacific reflect river input. The pattern may be interpreted to result from an east to west transport, presumably via the nepheloid layer, and distribution over the basins by counterclockwise gyres.

the Alaskan region. This primary mineral provides an excellent tracer for the movement of water masses once its sources are identified. The Copper River, which empties into the Gulf of Alaska has abundant chlorite and illite (Table 4).

The pattern of circulation as described by Neumann and Pierson (1966) and Reid (1965) for the Gulf of Alaska is counterclockwise in the surface and intermediate waters to about 1100 fm. Once chlorite is introduced into the surface water, it will be distributed by the counterclockwise surface gyre. Loss of sediment from this gyre will occur by mixing with the southeastward-flowing California Current and by precipitation to the deeper waters. Fine chlorite particles settling through the water column are suspended above the ocean floor in the nepheloid layer. In the Gulf it is evident from the distribution of high chlorite sediments (Fig. 15) that some grains (< 2 microns) settle to the bottom before they can be removed as a suspended load. In the region enclosed by the 30 percent chlorite contour, which coincides rather well with the 2500 fm isobath, we consider that deep currents are sufficiently weak so that many larger particles and clusters of fine particles settle directly to the bottom when they drop below the surface layer. Because of their size these particles probably do not contribute appreciably to the gradient of light scattering. The chlorite tongue extending south to about 23° N., 130° W. lies in water that is generally shallower than 2500 fm; this tongue is attributed to a large chlorite input into the northeast Pacific (Table 4), the California Current, and weak bottom currents on the continental rise.

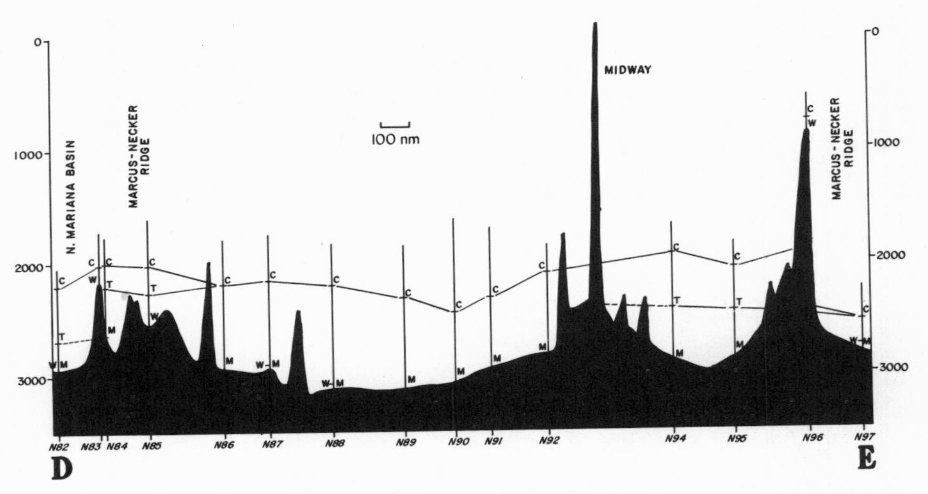

Figure 16. Profile of the nepheloid layer along D-E (Fig. 1). Symbols are as in Figure 5. The Marcus-Necker Ridge appears to be an effective barrier between the water masses in the Northwest Basin and Mariana Basin.

TABLE 4. CLAY MINERAL DATA FOR RIVERS ALONG THE WEST COAST OF NORTH AMERICA

River	Sample type	% Mont.	% Illite	% Chlorite	% Kaolinite
Matthole R., California	Bed Material	10	44	46	?
Eel R., Calif.	Bed Material	18	38	35	9
Mad R., Calif.	Bed Material	—	50	50	?
Redwood Ck., California	Bed Material	—	46	54	—
Klamath R., California	Bed Material	6	44	50	?
Smith R., Calif.	Bed Material	Trace	52	48	?
Rogue R., Oregon	Bed Material	High	Present	?	?
South Umpqua R., Oregon	Bed Material	>80	Trace	?	Present
N. Umpqua R., Oregon	Bed Material	>80	Present	Present	Present
Columbia R., Washington	Suspended	28	42	(Kaolinite + chlorite, 30%)	
Deschutes R., Oregon	Bed Material	>90	Trace	Trace?	Trace?
John Day R., Oregon	Bed Material	>90	Trace	?	Trace?
Snake R., Washington	Bed Material	68	20	?	12
Fraser R., British Columbia	Suspended	32	34	34	?
Copper R., Alaska	Suspended	7	58	36	?

(*from* Griffin and others, 1968)

Another line of evidence in agreement with the identification of continents as the source of the suspended particulate matter in the nepheloid layer is the distribution of primary terrigenous montmorillonite in sediments (Griffin and others, 1968). This mineral, an alteration product of the weathering of igneous and pyroclastic rocks, is found in abundance in many rivers along the Pacific coast of North America (Table 4). The volcanic ash soils of Japan, Kamchatka, and the Aleutians (Dudal, 1964, *in* Griffin and others, 1968) may contribute montmorillonite to the North Pacific. In contrast, Windom (1969) has shown that atmospherically transported dusts obtained from glaciers are essentially devoid of montmorillonite. This suggests that nearly all the montmorillonite component of North Pacific sediments is transported by surface and deep currents. Submarine alteration of volcanic material to montmorillonite is not assigned special importance in the North Pacific by Griffin and others (1968). Rather they suggest that areas of high montmorillonite concentration along the borders of the North Pacific reflect continental input of this mineral. Montmorillonite in moderate amounts was observed by Jacobs (personal commun.) in water samples obtained from the nepheloid layer in the Gulf of Alaska, Aleutian Trench, and northeast Pacific. The general uniformity of montmorillonite concentration in sediments of the central North Pacific suggests horizontal mixing and transport within and between the Northeast and Northwest Basins. The montmorillonite distribution presented by Griffin and others (1968, Fig. 3) is notably unaffected by such topographic highs as the Emperor Seamount Chain. This may be partly due to an insufficient number of closely spaced samples in this region.

In our regional analysis of the main nepheloid layer, we have observed that as a general rule no strong nepheloid layer is found in the North Pacific in water shallower than about 2500 fm. There is one exception, however, and that is the Gulf of Alaska, a large area whose depth gradually increases from the foot of the continental slope to the 2500 fm contour.

The stations listed in Table 5 were occupied between July 31 and August 24, 1967, in the Gulf of Alaska during the eleventh cruise of *Robert D. Conrad*. Their positions are shown in Figure 1 and the light scattering profiles in Figure 17. The main nepheloid layer here is only 100 to 350 fm thick, but the scattering is remarkably strong as represented by the densitometer deflection ratios which range from 4 to about 15. These R values may be compared with those of about 4 to 6 which are well above average for the North Pacific Basin and are typical for the trenches and the Aleutian Abyssal Plain and Argentine Basin. At C11-N136, located about 200 nm west of Queen Charlotte Islands (Fig. 1), the complicated

TABLE 5. NEPHELOMETER STATIONS IN THE GULF OF ALASKA

Station	Sounding	H	Thickness	Intensity R	Date (1967)
C11-N135	2050	1700	350	15.0	31 July
C11-N136	1600	1460	140	4.9	1-2 August
C11-N137	2185	1950	235	5.0	4 August
C11-N138	2070	1850	220	4.0	5 August
C11-N145	2345	2100	245	7.3	24 August

H is the depth of the top of the main nepheloid layer.
Depths are in fathoms.

nature of the scattering (Fig. 17) may be attributed to its proximity to a terrigenous sediment source. The presence of a strong nepheloid layer close to the continental border in this region and to the west of the Columbia River (Fig. 12) suggests that many of the light scattering particles transported to the sea in these areas reach the nepheloid layer fairly rapidly. To investigate in greater detail the process by which this fine particulate matter is carried across the shelf and slope to deep water, more near-shore nephelometer stations are required in the region of a known sediment source. The association of a strong nepheloid layer and a region of known par-

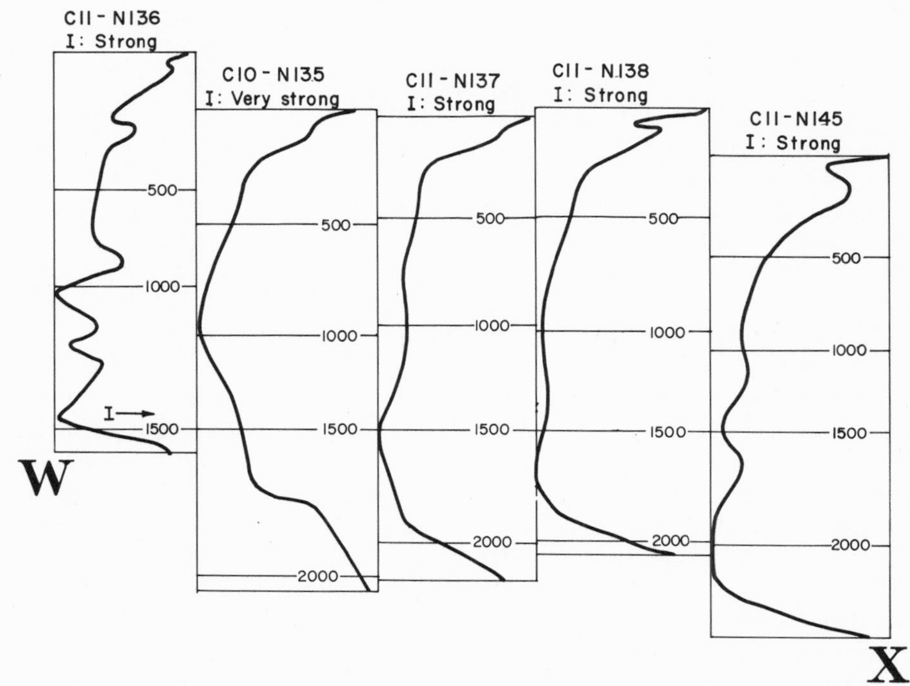

Figure 17. Scattering profiles in the Gulf of Alaska along W-X (Fig. 1). I is intensity of light scattering increasing to the right from the reference of clearest water. Depths are in fm.

ticulate matter input suggests that the suspended particles are in transit from a continental source to the nepheloid layer in the deep basin. Transport to the deep basin results from the action of the various types of deep currents that are present on the rise.

To what extent re-suspended sediments contribute to light scattering in the nepheloid layer is difficult to deduce from nephelometer data alone. It is generally claimed (for example, Gordon and Gerard, 1970, this volume) that bottom velocities are very weak throughout the deep North Pacific.

It has also been assumed, without much observational support, that deep currents are steady. Studies on the continental rise off California (Nowroozi and others, 1968; Schwartzlose and Isaacs, 1969) demonstrate the existence of relatively high-speed, transient deep currents, the origin of which is uncertain. Speeds of 16 to 18 cm/sec were measured during the transient events in areas where a bottom current of less than 1 to about 4 cm/sec was typical. At present we cannot rule out the possibility that bottom sediments may be re-suspended by such transient currents and become light scattering particles in the nepheloid layer. The significance of this process depends on the presently unknown frequency, areal extent, and intensity of this phenomenon.

It is our belief, on the basis of the clear intermediate water above the nepheloid layer, and of the regional uniformity of this layer and the nepheloid layer, that volcanic ash and other windborne minerals do not contribute significantly to the gradient of light scattering in the nepheloid layer. Horn and others (1970, this volume, Fig. 9) have mapped the distribution of volcanic ash horizons in the North Pacific. That these layers are only found in a zone about 1000 km wide bordering the line of active volcanoes of the North Pacific suggests that this is the greatest distance to which volcanic materials can be ejected to form ash layers on the ocean floor. Once the ash falls into the ocean, it sinks relatively rapidly to the bottom, as is proven by the abrupt initiation and termination of ash deposition. The ash grains are very coarse (about 5ϕ; Horn and others, 1970, this volume, Fig. 4) relative to the clay (nearly 9ϕ). (By definition, $\phi = -\log_2$ [diameter in mm], that is, $5\phi = 2^{-5}$ mm $= 31\mu$.) Hence, they would settle rapidly under any regime of currents which would permit clay deposition. That ash horizons are not observed over the entire North Pacific Basin also implies that surface and bottom currents are not significantly effective in the basin-wide dispersal of volcanic ash. The pattern of ash layers shows no clear relation to wind except possibly a general displacement from west to east.

Very fine volcanic dust and other terrigenous materials are transported in significant quantities by wind and water, as shown by Windom's (1969) analysis of the inorganic component of deep-sea sediments. The slow settling rate in air and water would allow wide dispersal in the ocean; no readily identifiable volcanic dust horizons would be expected to form in deep-sea sediments from the very fine particles as they do for the coarser ash. Only the finest fraction of volcanic dust could ultimately be suspended and become a component in the nepheloid layer. However, the very slow settling rates of the most efficient light scattering particles—about 200 to 800 years for quartz spheres 1 to .5 micron in diameter, settling 2700 fm in distilled water (20° C) according to Stokes Law (Sverdrup and others, 1942)—severely limits the contribution that fine ash or other particles of small dimension might make to the gradient of light scattering in the nepheloid layer.

Windom's (1969) work effectively supplements that of Goldberg and his associates in showing that wind-borne material is a significant component of the non-biogenic fraction of North Pacific sediments. A model which would permit the material finer than 2 microns to contribute to sediment accumulation without proportionate contribution to the light scattering would involve the combination of the small particles into larger units or clumps which would sink rapidly without appreciable disturbance by currents, especially near the bottom. Preparation of samples by dispersion with an ultrasonic generator, as used by Windom (1969), or by other means (Griffin and others, 1968), might destroy the evidence of this flocculation process. Such floccules might have a diameter of about 5 to 15 microns (Sverdrup and others, 1942) and settle in quiet water 25 to 225 times faster than particles about 1 micron in diameter.

A second model would depend on the slow sinking of particles in the surface and bottom light scattering layers but much faster sinking in the clear intermediate layer attributed to the absence of vertical turbulence. Dispersion of the floccules into their finer grained components would take place within the nepheloid layer and provide fine, light scattering particulate matter. We have no explanation for this redispersion process and consider this second model very unlikely, as well as the appreciable contribution of very fine particulate settling from the surface to the nepheloid layer.

Windom (1969) estimates that no more than about 50 percent on the average of the less than 2 micron dispersed fraction can be attributed to wind transport. At present we are not able to attribute all the remaining fraction to bottom current transport via the nepheloid layer because of the uncertainty in the original particle sizes and in the degree to which surface currents may distribute large agglomerates of minerals, such as montmorillonite, which are notably lacking (Windom, 1969) in air-borne dusts.

It is interesting to note that, according to Windom's calculations, the non-aeolian contribution to North Pacific sediments finer than 2 microns has negligible amounts of chlorite. This is in apparent disagreement with river data (Table 4) and chlorite distribution (Fig. 15) presented by Griffin and others (1968) who identify many of the rivers of western North America as important chlorite contributors to North Pacific sediments whose mean diameter is less than 2 microns.

TRENCHES

A number of stations covering a range of 27° of longitude (Table 6), have been occupied in the Aleutian Trench; several are in the Kuril and Japan Trenches (Figs. 18 and 19). The curves for light scattering versus depth (Fig. 19) are remarkably similar below the 2000 fm level throughout the trench system. This is an observation of some significance for it suggests that the explanation of the nepheloid layer is the same for the three trenches.

The main nepheloid layer in the Aleutian Trench extends from the bottom to about 2400 fm. At the easternmost station, C12-N138 (Fig. 19), the depth to the top of the nepheloid layer is 2100 fm, which is the same depth at which it begins at C12-N137, 2° to the south, and at C11-N145 on the Tufts Abyssal Plain (Figs. 1, 10, and 17). The depth of 2100 fm is considerably above the trench sill depth of about 2500 fm in this region. In the Kuril and Japan trenches a similar continuity of the top of the nepheloid layer between the plain and the trench is observed (Figs. 1 and 18). To the west of C12-N138 (Fig. 19), the top of the layer is generally about 300 fm deeper, and at C11-N127 at 177° 31′ E., the top of the layer occurs at 2550 fm, which is below the sill depth of about 2450 fm for this area. We take 2350 fm to be an average depth to the top of the nepheloid layer in the Aleutian Trench, and about 2400 fm in the Kuril Trench.

TABLE 6. NEPHELOMETER STATIONS IN THE ALEUTIAN TRENCH

Station	Sounding (fm)	H (fm)	Nepheloid thickness	Intensity R	Date
C12-N138	3374	2100	1275	4.6	2 Sept. 1968
C11-N134	3161	2440	1275	4.1	26 July 1967
C10-N166	3833	2400	1435	2.9	27 July 1966
C12-N132	3852	2300	1550	6.9	18 July 1968
C11-N127	3625	2550	1075	5.8	11 July 1967

NEPHELOMETER STATIONS IN KURIL-JAPAN TRENCH

Station	Sounding (fm)	H (fm)	Nepheloid thickness	Intensity R	Date
C12-N129	3805	2400	1405	3.6	10 July 1968
C12-N128	3917	2550	1365	7.3	4 July 1968
C12-N111	3550	2350	1200	2.0	21 May 1968

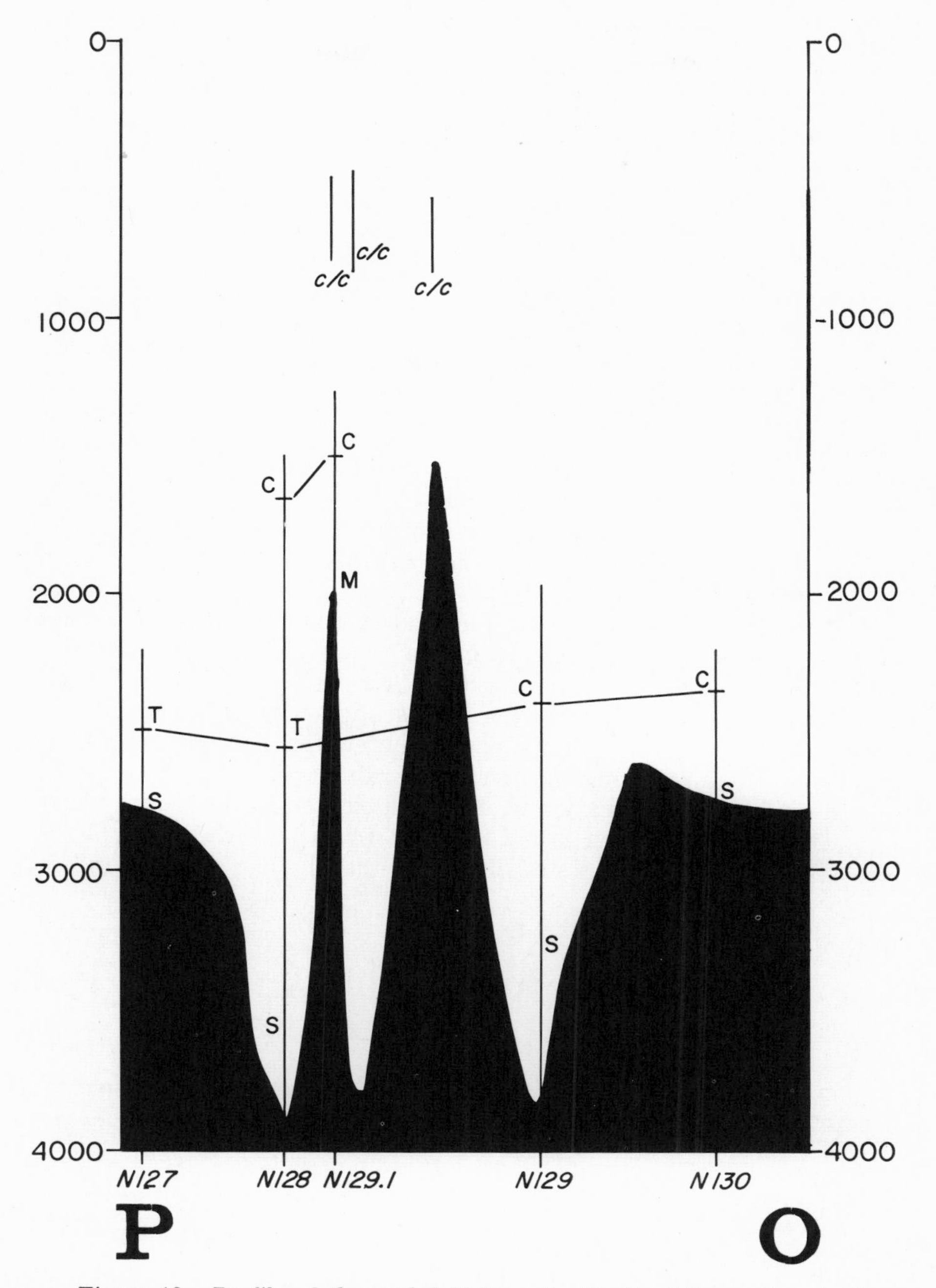

Figure 18. Profile of the nepheloid layer in the Kuril-Japan Trench along P-O (Fig. 1). Three crossings of the trench were made; C/C indicates course change. Symbols are as in Figure 5. S is strong intensity of scattering. Depths are in fm.

The station C11-N111 is on the west wall of the Japan Trench (Fig. 13). It appears that the nepheloid zone on the west wall may be higher than on the east wall of the trench. In addition, we have two stations (not published) on the north wall of the Aleutian Trench at which the clearest water is higher than on the south wall.

These observations are consistent with a westerly current over the Aleutian Trench and a southerly flow over the Kuril-Japan Trench. The current would appear higher on the north and west walls, respectively, in response to deflection to the right by the Coriolis force.

In the Aleutian Trench the dissolved oxygen concentration below a depth of 2200 to 2460 fm is 3.4 to 3.6 ml/l. No systematic gradient of oxygen concentration has been detected below sill depth of about 2500 fm. This value agrees with that at this depth in the adjacent part of the North Pacific Basin. Clearly, interchange of water with the trench and circulation within the trench are adequate for thorough mixing and replacement of the water before any appreciable consumption of dissolved oxygen has occurred.

The potential temperature and the salinity also appear to be constant below about 2200 fm (Fig. 2b). In order to maintain the adiabatic gradient, a lower limit is placed upon the rate of vertical mixing by the need to carry away the heat which is transferred to the water through the floor of the trench. Mixing in excess of this amount can be generated by the forces applied to the homogeneous trench water mass across its upper surface by currents above the sill depth. The resulting mean rate of mixing is adequate to maintain a uniform concentration of dissolved oxygen in the trench water.

The light scattering is not uniform in the trench water. From the sill depth it increases steadily to the trench floor, but the rate of

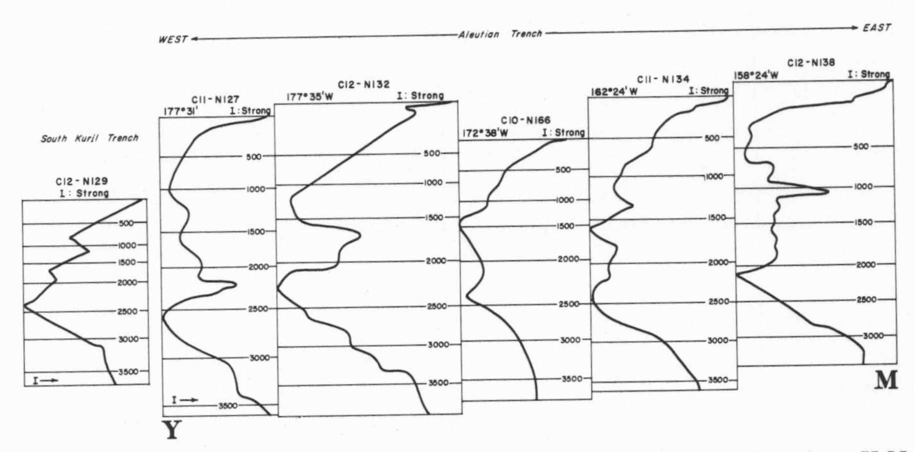

Figure 19. Scattering profiles in the Aleutian and Kuril Trenches along Y-M (Fig. 1). I is intensity of light scattering. Depths are in fm.

increase is small below about 3000 fm (Fig. 19). The vertical gradient of light scattering in the homogeneous trench water is interpreted as the result of a balance between the effect of the gravitational force, which causes the particles to settle to and concentrate at the bottom, and the effect of turbulent mixing, which tends to produce a uniform distribution of particles throughout the water mass. This gradient could be used to determine the rate of vertical mixing if the settling rate of the particles were known. Furthermore, this model implies that the nepheloid layer near the trench floor will be more intense than that in the abyssal waters in the adjacent basin from which the trench water is derived, even if the Aleutian Ridge is not a significant source of the type of particulate matter that forms the nepheloid layer.

An interesting aspect of the distribution of intensity of light scattering observed in all trenches is illustrated in Table 6. With one exception (C10-N166), the deepest stations show the largest value for the ratio R. This is an expected result considering the distribution of light scattering with depth at each station. The strongest scattering, and presumably the greatest concentration of suspended lutite, is immediately above the trench floor. At any higher level within the layer, scattering is less and a station taken at any level above the deepest sounding would be expected to measure a smaller value of light scattering at the bottom. More accurate data are needed to decide if there is station-to-station and trench-to-trench variation in nepheloid intensity.

An important question concerning the life span of a nepheloid layer arises in this study—is it a transient or steady-state phenomenon? After examining several hundred nephelometer records obtained on cruises of *Vema* and *Robert D. Conrad* from 1965 to 1968, we conclude that the nepheloid layer is both persistent in time and is to be found in many parts of the World Ocean. As illustrated by representative stations in Table 6, measurements made in the Aleutian Trench over a period of three years show a consistency strongly suggesting that the trench nepheloid layer is a steady-state feature. The general constancy of depth to the nepheloid layer with time implies that any disruption by such factors as turbidity currents, slumping, and tsunamis must be short-lived.

MARGINAL SEAS

A preliminary study was made of several basins bordering the western Pacific. We report here in a general manner our initial observations. There is an unexpected pattern of the nepheloid layer in the marginal seas. The Sulu and Celebes Seas, which are almost completely surrounded by islands and shallow shelves, have a thick turbidite sequence in their basins and no nepheloid layer at the bottom. This is in marked contrast to the Gulf of Alaska and the Bering Sea. The West Caroline Basin has a distinct nepheloid layer of generally moderate intensity beginning about 1700 fm.

The Philippine Sea has a nepheloid layer at the bottom which ranges regionally from weak to strong. The top of the nepheloid layer begins at about 1350 fm in the northern part of the basin south of Japan and at depths of about 1600 to 1800 fm in the deeper basin to the south.

Throughout all these basins a higher layer marked by a relative scattering maximum is often observed. Clearly some source is supplying sediment to the water at this level, and it is very likely that there is a single major source, for this upper layer is found at about the same level throughout the marginal seas. The important conclusion can be drawn that the circulation in all these basins must be connected at least to the depth of the upper scattering maximum.

The Aleutian Basin in the Bering Sea has a well-defined nepheloid layer 100 to 300 fm thick, with some stations showing a thicker layer. Intensity of light scattering is generally weak or moderate throughout this turbidite-filled basin. An interesting feature of this marginal sea is the occurrence above the nepheloid layer of one or two zones in which there is a relative maximum of light scattering.

It is apparent that islands contribute very little particulate matter which can remain in suspension sufficiently long to contribute substantially to the nepheloid layer. The principal source is continental, but the path of sediment transport to the seas is not immediately evident. For the western seas, we suspect that rivers emptying into the Yellow Sea, China Sea, and South China Sea, supply lutite. Such a source has been indicated by Griffin and others (1968). Suspended sediment may find its way into the Philippine Sea via the Luzon Strait, which probably has a sill depth between 1000 and 1500 fm. The particulate matter can serve as a valuable marker for the study of the transport of deep and intermediate waters.

CONCLUSION

A distinct steady-state nepheloid layer 100 to 1500 fm thick with regional variations in light scattering intensity is observed above the bottom of the deep basins and continental margin of the North Pacific. The nepheloid layer is characterized by relatively clear water at its top and the intensity of light scattering increasing to the bottom. This gradient is attributed to a balance between gravitational settling and turbulent diffusion of particulate matter in the homogeneous water. The level of clearest water generally coincides with the depth of the temperature minimum except in the central

North Pacific where the level of clearest water is conspicuously below the temperature minimum.

The generally weak scattering intensity of AABW flowing into the North Pacific suggests that the North American and Asian continents are the principal contributors of the light scattering particulate matter. This is further substantiated by the proximity of the most intense nepheloid layer observed to the continental margin of northwestern North America. The substantial contribution of atmospherically transported material to deep-sea sediments is acknowledged, but it is emphasized that the *in situ* size of the particulate matter is just as important as the number of suspended particles. Since light scattering is proportional to surface area, the more finely divided the particulate matter is, the more light it will scatter per unit mass or volume, and the more easily it can be held in suspension. We suggest that much wind transported and biogenic material settles directly to the ocean floor because of its particle size and therefore contributes only a constant factor to the light scattering because of its nearly uniform vertical distribution. As the larger particles would be essentially unaffected by near-bottom turbulence and hence have a relatively short residence time in the nepheloid layer, they would be expected in much smaller numbers than the finer suspended particles. Large particles that might be suspended in the nepheloid layer would therefore contribute much less to the gradient of light scattering than might be inferred from their rate of accumulation at the sea-floor. It is thus not unreasonable to find in certain areas, such as the Equatorial Pacific and the continental rise of the western United States, relatively high sedimentation rates associated with a moderate or weak nepheloid layer.

On the basis of the distribution of clay minerals in sediments and in water samples obtained from the nepheloid layer, from temperature, salinity and oxygen data, and from nephelometer data, we conclude that the most reasonable pattern of deep and bottom circulation is a system of counterclockwise gyres in the Northeast and Northwest Pacific Basins. Such gyres would transport particulate matter more or less uniformly over the basins; this would explain the generally uniform nepheloid layer observed over these large regions.

The Emperor Seamount Chain appears to be a permeable barrier to deep flow although differences in the mineralogy of the suspended particulate matter and in temperatures of the bottom water between the two basins suggest that it to some degree retards inter-basin exchange. If this is not the case and if we accept the generally claimed weak bottom currents, a rather long residence time for particles in the nepheloid layer is necessary to account for the regional uniformity of light scattering intensity. This residence time can be shortened appreciably by the existence of a closed gyre in each basin of the North Pacific. We offer this model only as a first approximation. It may be that a more complicated pattern exists as a result of one or more gyres in each basin and the possibly significant effect of tidal and transient currents. A complicated system of clock- and counter-clockwise gyres in the deep Pacific has been described by Burkov (1969).

The contribution of the nepheloid layer to deep-sea sedimentation in the North Pacific is not yet established because of lack of information about surface layer transport of such minerals as montmorillonite and the size distribution of the *in situ* particulate matter in deep-sea sediments and in the nepheloid layer.

ACKNOWLEDGMENTS

We gratefully acknowledge the financial support of the National Science Foundation (Grant GA-1615) and the Office of Naval Research (Contract N00014-67-A-0108-0004). We also thank the scientists and technicians who have gathered and processed the data, especially T. Aitken and L. Sullivan, and E. M. Thorndike for contributions to all aspects of the nephelometer program. J. D. Hays, A. L. Gordon, and G. Ross Heath have read the manuscript and offered valuable comments.

REFERENCES CITED

Burkov, V. A., 1969, The bottom circulation in the Pacific: Oceanology, v. 9, no. 2, p. 223-234.

Dudal, R., 1964, Correlation of soils derived from volcanic ash *in* Meeting on the classification and correlation of soils from volcanic ash: World Soil Resources Rept. F.A.O. 14, p. 134-138.

Eittreim, S., Ewing, M., and Thorndike, E. M., 1969, Suspended matter along the continental margin of the North American Basin: Deep-Sea Research, v. 16, p. 613-624.

Ewing, J., Ewing, M., Aitken, T., and Ludwig, W. J., 1968, North Pacific sediment layers measured by seismic profiling, *in* Knopoff, L., Drake, C. L., and Hart, P. J., *Editors,* The Crust and Upper Mantle of the Pacific Area: Am. Geophys. Union Mono. 12, p. 522.

Ewing, M., Eittreim, S. L., Ewing, J. I., and Le Pichon, X., 1970, Sediment transport and distribution in the Argentine Basin: Part 3, Nepheloid layer and processes of sedimentation: Physics and Chemistry of the Earth, v. 8, London, Pergamon Press (in press).

Ewing, M., and Thorndike, E. M., 1965, Suspended matter in deep ocean water: Science, v. 147, p. 1291-1294.

Gordon, A. L., and Gerard, R., 1970, North Pacific bottom potential temperature: Geol. Soc. America Mem. 126, p. 23-39.

Griffin, J. J., Windom, H., and Goldberg, E. D., 1968, The distribution of clay minerals in the World Ocean: Deep-Sea Research, v. 15, no. 4, p. 433-459.

Horn, D. R., Horn, B. M., and Delach, M. N., 1970, Sedimentary provinces of the North Pacific: Geol. Soc. America Mem. 126, p. 1-22.

Jerlov, N. G., 1968, Optical Oceanography: New York, Elsevier Publishing Co., 194 p.

Knauss, J. A., 1962, On some aspects of the deep circulation of the Pacific: Jour. Geophys. Research, v. 67, no. 10, p. 3943-3954.

Lynn, R. J., and Reid, J. L., 1968, Characteristics and circulation of the deep and abyssal water: Deep-Sea Research, v. 15, no. 5, p. 577-598.

Neumann, G., and Pierson, W. J., Jr., 1966, Principles of physical oceanography: Englewood Cliffs, New Jersey, Prentice-Hall, Inc., 545 p.

Norpac Committee, 1965, Oceanic observations of the Pacific, 1953: California Univ. Press, Berkeley and Los Angeles, 576 p.

—— 1965, Oceanic observations of the Pacific, 1954, The Norpac Data: California Univ. Press, Berkeley, 426 p.

—— 1960, Oceanic observations of the Pacific, The Norpac Data: Berkeley and Los Angeles, Tokyo, California Univ. Press and Tokyo Univ. Press, 532 p.

Nowroozi, A. A., Ewing, M., Nafe, J. E., and Fliegel, M., 1968, Deep ocean current and its correlation with the ocean tide off the coast of northern California: Jour. Geophys. Research, v. 73, no. 6, p. 1921-1932.

Olson, B. E., 1968, On the abyssal temperatures of the world oceans: Ph.D. dissert., Oregon State Univ., 144 p.

Opdyke, N. D., and Foster, J. H., 1970, The paleomagnetism of cores from the North Pacific: Geol. Soc. America Mem. 126, p. 83-119.

Reid, J. L., Jr., 1965, Intermediate waters of the Pacific Ocean: Johns Hopkins Univ. Oceanographic Studies, v. 2, 85 p.

Reid, J. L., Jr., Stommel, H., Stroup, E. D., and Warren, B. A., 1968, Detection of a deep boundary current in the western South Pacific: Nature, v. 217, p. 937.

Schwartzlose, R. A., and Isaacs, J. P., 1969. Transient circulation event near the deep ocean floor: Science, v. 165, p. 889-891.

Sverdrup, H. U., Johnson, M. W., and Fleming, R. H., 1942, The oceans: Their physics, chemistry and general biology: New York, Prentice-Hall, 1087 p.

Thorndike, E. M., and Ewing, M., 1967, Photographic nephelometers for the deep sea, *in* Deep-Sea Photography: Johns Hopkins Univ. Press, Baltimore, Maryland, p. 113-116.

Warren, B. A., 1968, General comments on the circulation of the South Pacific: Symposium on the scientific exploration of the South Pacific (in press).

Windom, H. L., 1969, Atmospheric dust records in permanent snowfields: Implications to marine sedimentation: Geol. Soc. America Bull., v. 80, no. 5, p. 761-782.

Wooster, W. S., and Volkmann, G. H., 1960, Indications of deep Pacific circulation from the distribution of properties at five kilometers: Jour. Geophys. Research, v. 65, no. 4, p. 1239-1249.

LAMONT-DOHERTY GEOLOGICAL OBSERVATORY CONTRIBUTION NO. 1485
MANUSCRIPT RECEIVED APRIL 10, 1969
REVISED MANUSCRIPT RECEIVED JULY 18, 1969

THE GEOLOGICAL SOCIETY OF AMERICA, INC.
MEMOIR 126, 1970

Paleomagnetism of Cores from the North Pacific

NEIL D. OPDYKE
AND
JOHN H. FOSTER

Lamont-Doherty Geological Observatory of Columbia University, Palisades, New York

ABSTRACT

A paleomagnetic study has been carried out on 114 piston cores from the Pacific Ocean north of 20° N. lat. The cores were sampled at 10 cm intervals, and the direction of magnetization was determined after partial demagnetization in fields of from 50 to 150 oersted peak field. The coercivity of the red clays from the central part of the North Pacific was found to decrease with depth in the cores so that in red clay cores magnetic stratigraphy is often undecipherable even after partial a.f. demagnetization. In other types of lithologies, a.f. demagnetization was successful in removing secondary components. The magnetic stratigraphy from the cores was interpreted in terms of the radiometrically derived time scale for the last 4.5 m.y. of earth history. Confusion has recently arisen as to whether the large event, which is often identified in the bottom part of the Matuyama series, is best correlated with the Olduvai or Gilsa event. The age of this event in 13 North Pacific cores was determined by interpolation between the Brunhes/Matuyama and Gauss/Matuyama boundary. The average age of the upper and lower boundaries of the event are 1.71 and 1.86 m.y. with a standard deviation exceeding $\pm$ 0.1 m.y. The original date on the Olduvai event is 1.91 $\pm$ 0.06 m.y. There is no significant difference between the range of the event as seen in marine sediments and the age of the Olduvai event at its type locality. Therefore, on the basis of priority, it seems proper to correlate this event with the Olduvai event.

Nine stratigraphic sections were constructed on the basis of magnetic stratigraphy, and from the cores used in these sections, rates of sedimentation were calculated using a date of 0.69 m.y. for the Brunhes/Matuyama boundary. These rates were then contoured and compared with the sediment distribution in the Pacific. The rates of sedimentation in the red clay area of the central Pacific are 3 mm/1000 yrs. The observed rates increased toward the margins of the Pacific in all cases, due to an increase in volcanic, biogenic and glacial detritus present in the cores. The effect of topography on rates of sedimentation is considered. Rates of sedimentation indicate a Miocene age for the time of cessation of turbidite deposition on the Aleutian abyssal plain. A comparison of the known rates of sedimentation with the sediment distribution favors the hypothesis of the mobility of the sea floor.

CONTENTS

INTRODUCTION

Previous paleomagnetic studies by Harrison (1966), Ninkovitch and others (1966), and Dickson and Foster (1966) on cores from the central and North Pacific region gave information of great value in regional stratigraphic studies and in evaluating the past behavior of the earth's magnetic field. This paper is an extension of these studies to the whole Pacific region north of 20° N. lat.

In recent years, the research vessels of Lamont-Doherty Geological Observatory have been directed on tracks which provide good coverage of large areas of the North Pacific. The most important of these are from cruises 20 and 21 of R/V *Vema* and cruises 10 and 11 of R/V *Robert D. Conrad.* Many of the cores from these cruises have been studied using standard paleomagnetic techniques and will be reported on here. Figure 1 shows the location of the cores used in this study. The location of the cores is given in Table 1.

LABORATORY PROCEDURE

Samples were taken from the desiccated cores at intervals ranging from 10 to 100 cm, using the methods and techniques described by Opdyke and others (1966). The direction toward the bottom of the core is marked; the specimen is then removed by cutting on a band saw. The specimens are approximately 2.54 cm in diameter and length. The direction of magnetization relative to the split face of the core is measured on a 5 cps magnetometer, described by Foster (1966). Since the eleventh cruise of *Conrad* and the twenty-fourth cruise of *Vema,* an attempt has been made to retain the internal orientation of the cores so that 180° changes in declination can be used to determine the magnetic stratigraphy as well as changes in the inclination. This allows the determination of magnetic stratigraphy in cores taken at low latitudes (Hays and others, 1969). In the present study, most interpretation is made from

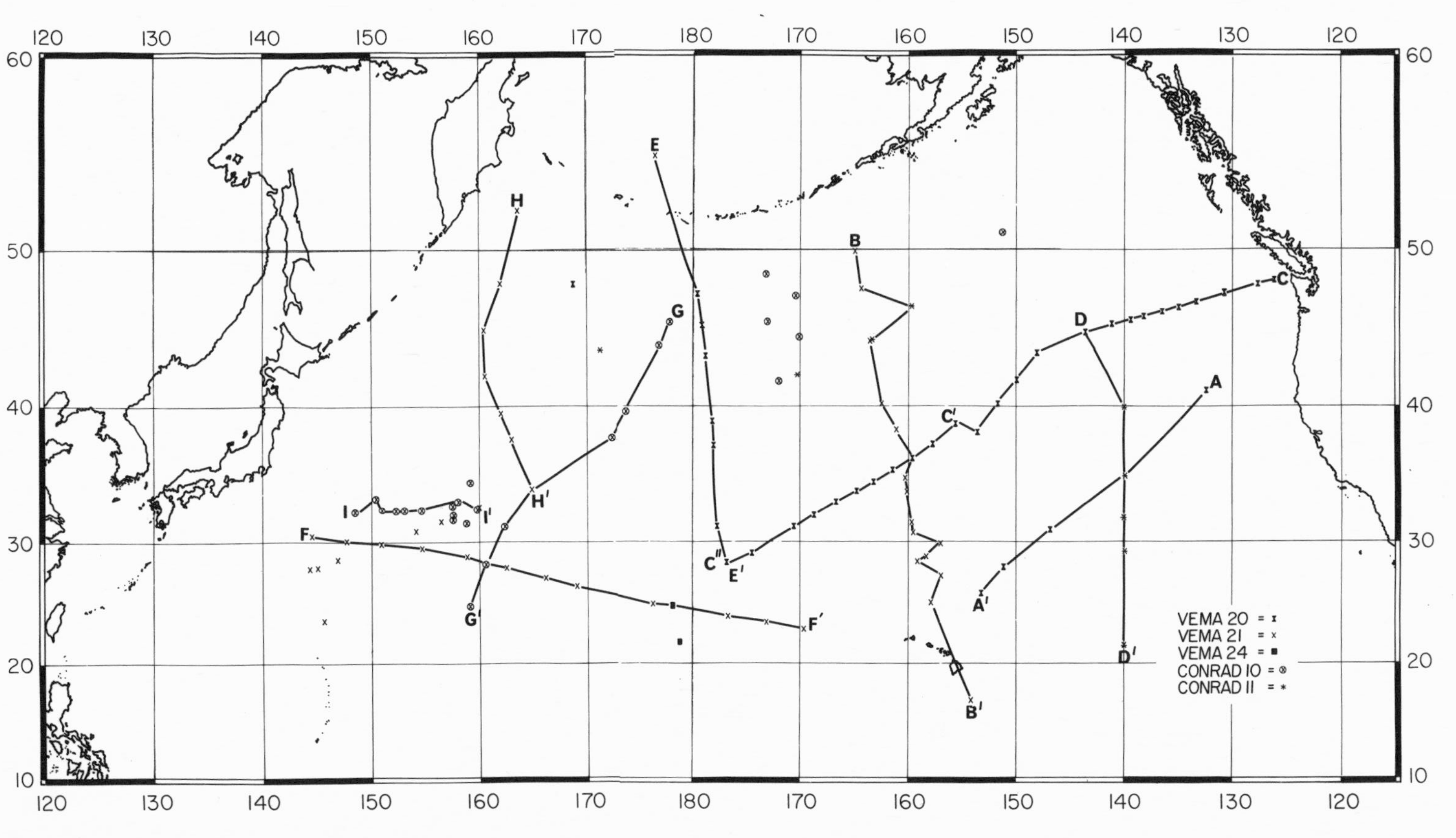

Figure 1. Map of core locations in the North Pacific. The lines of the stratigraphic sections shown in Figures 10-19 are indicated.

changes in inclination. However, even in cores which are not internally oriented, changes in declination often aid in interpretation.

STABILITY OF MAGNETIZATION

It has been shown previously that oceanic sediments, like continental rocks, contain unstable components of magnetization (Opdyke and others, 1966). It is therefore necessary to use alternating field partial demagnetization techniques to remove these secondary components. Figure 2 shows the plot of the inclination versus depth in core V21-173 for both the Natural Remanent Magnetization (NRM), and after cleaning in a peak field of 150 oersteds. A comparison of these plots demonstrates the remarkable decrease in dispersion often observed after a.f. demagnetization.

Figure 3A is a plot of a.f. partial demagnetization curves from V21-156 from the Bering Sea, which shows variations typical of fairly stable magnetization. Figure 3B shows curves from V21-173, which fall off steadily on a.f. demagnetization in successively higher fields. A small, unstable component seems to be present, which is eliminated by low fields, but the magnetization is of a partially stable type. Figure 3C shows three demagnetization curves from V21-141, two samples from the normal section of the core (depth 490 and 430 cm), and one from the reversed part of the core (377 cm). It can be seen that the a.f. demagnetization curve from the normally magnetized part of the core is similar to that for the normal specimen

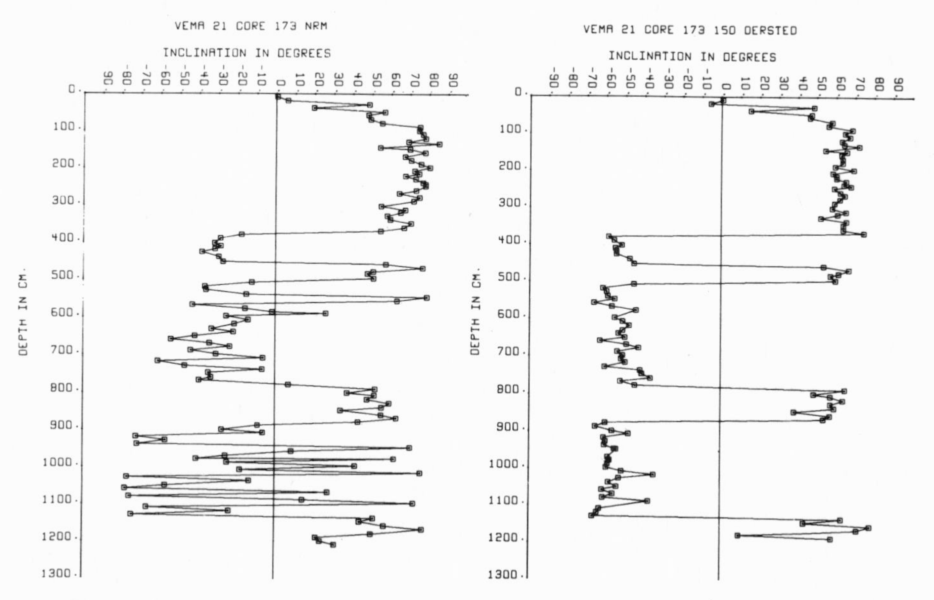

Figure 2. Magnetic inclination plots of V21-173 before and after a. f. demagnetization in 150 oersteds peak field.

TABLE 1. LOCATIONS OF CORES STUDIED PALEOMAGNETICALLY

Core	North Lat.	Long.	Core	North Lat.	Long.
RC10-157	24°47′	159°03′E	V20-95	33°53′	164°47′W
RC10-158	28°07′	160°36′E	V20-96	33°02′	166°42′W
RC10-159	31°13′	162°19′E	V20-97	32°04′	168°44′W
RC10-160	32°29′	159°50′E	V20-98	31°10′	170°35′W
RC10-161	33°05′	158°00′E	V20-100	29°05′	174°35′W
RC10-162	31°25′	158°48′E	V20-101	28°18′	176°57′W
RC10-163	32°43′	157°30′E	V20-102	31°11′	177°49′W
RC10-164	31°44′	157°30′E	V20-104	37°18′	178°10′W
RC10-167	33°24′	150°23′E	V20-105	39°00′	178°17′W
RC10-168	32°23′	148°26′E	V20-107	43°24′	178°52′W
RC10-169	32°31′	151°04′E	V20-108	45°27′	179°15′W
RC10-170	32°29′	152°14′E	V20-109	47°19′	179°39′W
RC10-171	32°29′	153°02′E	V20-119	47°57′	168°47′E
RC10-172	32°29′	154°38′E	V21-63	22°51′	169°41′W
RC10-174	32°04′	157°35′E	V21-64	23°27′	173°13′W
RC10-175	34°35′	159°10′E	V21-65	23°58′	176°51′W
RC10-178	37°48′	172°20′E	V21-67	24°58′	176°16′E
RC10-179	39°38′	173°43′E	V21-69	26°26′	169°02′E
RC10-181	44°05′	176°50′E	V21-70	27°05′	166°04′E
RC10-182	45°37′	177°52′E	V21-71	27°54′	162°31′E
RC10-201	48°32′	173°13′W	V21-72	28°47′	158°50′E
RC10-202	45°37′	173°00′W	V21-73	29°28′	154°36′E
RC10-203	41°42′	171°57′W	V21-74	29°51′	150°50′E
RC10-205	44°37′	170°03′W	V21-75	30°04′	147°41′E
RC10-206	47°13′	170°26′W	V21-76	30°25′	144°30′E
RC10-216	50°58′	151°10′W	V21-87	27°53′	145°03′E
RC11-166	43°46′	171°14′E	V21-89	23°35′	145°39′E
RC11-169	42°10′	170°14′W	V21-139	27°47′	144°18′E
RC11-170	44°29′	163°21′W	V21-140	28°33′	146°53′E

from V21-173. The specimen from the reversed part of the core shows an initial increase in intensity, probably caused by the removal of a secondary component in the direction of the present earth's field from the more stable reversed component. Figure 3D shows demagnetization curves of specimens from core V21-177, a non-biogenic red clay that shows a large initial drop in intensity of magnetization. The coercivity of this core apparently decreases markedly with depth.

In general, the cores which show the greatest amount of instability are those which have the red clay type of lithology. The radiolarian and diatomaceous muds are more stable, as is shown by the a.f. demagnetization curves shown in Figures 3A, B, C.

In most red clay type cores, a distinct decrease in the quality of the resolution of the magnetic stratigraphy occurs at depth. This is related directly to a decrease in the coercivity. This drop-off in

TABLE 1. (Continued)

Core	North Lat.	Long.	Core	North Lat.	Long.
RC11-171	46°36′	159°40′W	V21-141	30°48′	154°04′E
RC11-193	39°57′	140°03′W	V21-142	31°35′	156°25′E
RC11-194	35°00′	139°57′W	V21-144	32°41′	160°01′E
RC11-195	31°51′	139°59′W	V21-145	34°03′	164°50′E
RC11-196	29°11′	139°56′W	V21-146	37°41′	163°02′E
RC11-198	21°31′	140°00′W	V21-147	39°33′	162°05′E
V20-65	25°51′	153°12′W	V21-148	42°05′	160°36′E
V20-66	28°00′	151°10′W	V21-149	45°08′	160°28′E
V20-68	30°58′	146°48′W	V21-150	48°00′	162°01′E
V20-74	41°04′	132°22′W	V21-151	52°16′	163°38′E
V20-75	48°12′	126°10′W	V21-156	55°05′	176°20′E
V20-76	47°54′	127°39′W	V21-171	49°53′	164°57′W
V20-78	47°15′	131°02′W	V21-172	47°40′	164°21′W
V20-79	46°50′	133°18′W	V21-173	44°22′	163°33′W
V20-80	46°30′	135°00′W	V21-174	40°08′	162°30′W
V20-81	46°14′	136°30′W	V21-175	38°22′	161°06′W
V20-82	45°56′	138°14′W	V21-176	34°54′	160°19′W
V20-83	45°45′	139°24′W	V21-177	33°52′	160°08′W
V20-84	45°27′	141°11′W	V21-178	31°31′	159°42′W
V20-85	44°54′	143°37′W	V21-179	30°43′	159°34′W
V20-86	43°37′	148°06′W	V21-180	28°24′	159°11′W
V20-87	41°48′	149°55′W	V21-181	28°51′	158°21′W
V20-88	40°11′	151°39′W	V21-182	29°51′	157°02′W
V20-89	38°12′	153°35′W	V21-183	27°15′	157°00′W
V20-90	38°48′	155°37′W	V21-184	25°03′	157°54′W
V20-91	37°18′	157°42′W	V21-189	16°49′	154°11′W
V20-92	36°18′	159°38′W	V24-97	24°48′	178°04′E
V20-93	35°27′	161°28′W	V24-98	21°47′	178°47′E
V20-94	34°36′	163°14′W			

coercivity and simultaneous deterioration of the magnetic stratigraphy is shown graphically in Figure 4 for two cores, V21-177 and V20-98. Figure 4A gives the plot of intensity in emu for both the NRM and after partial a.f. demagnetization in a peak alternating field of 50 oersteds of V20-98. It can be seen that no large unstable component is present in the upper 475 cm of the core. However, at 475 cm the coercivity decreases abruptly and the curves diverge because of the large unstable component which is being removed in the relatively low field of 50 oersteds (peak). The inclination plot has been placed directly below the intensity data and on the same depth scale so that the two can be easily compared. It can be seen that the point where the internal consistency of the inclination data breaks down, 475 cm is the same point where the coercivity abruptly declines.

The magnetic behavior of V20-98 can be compared directly with

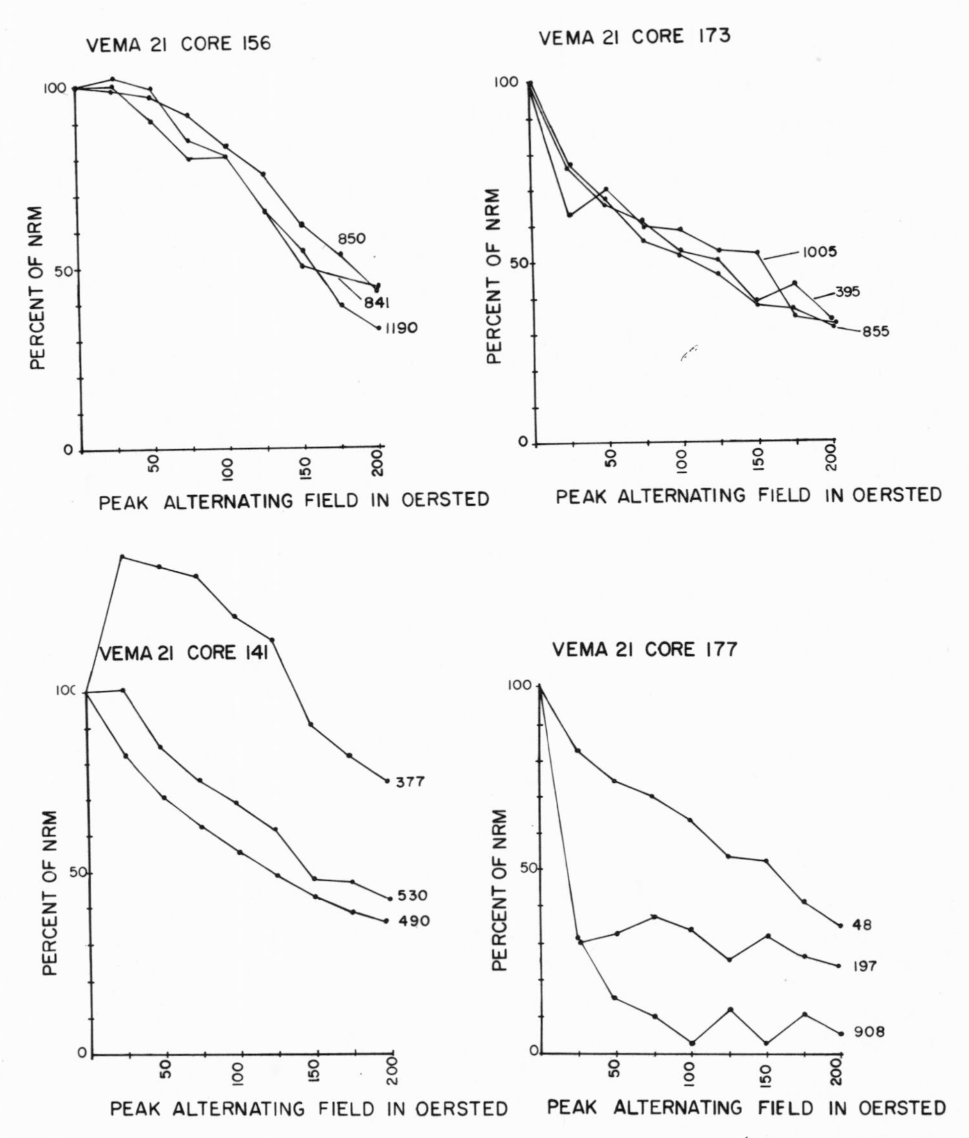

Figure 3. a.f. demagnetization curves for four cores of different lithologies showing different degrees of magnetic stability. Sample depths within each core are indicated beside the curves.

that of V21-177. In the latter, the divergence of the intensity plots for the NRM and after partial demagnetization occurs higher in the core, at a depth of 250 cm. At this point the inclination becomes positive and remains positive to the bottom of the core. This long period of normal polarity may be due to either the inability of our present cleaning technique to remove the large secondary component or to the original magnetization being completely destroyed.

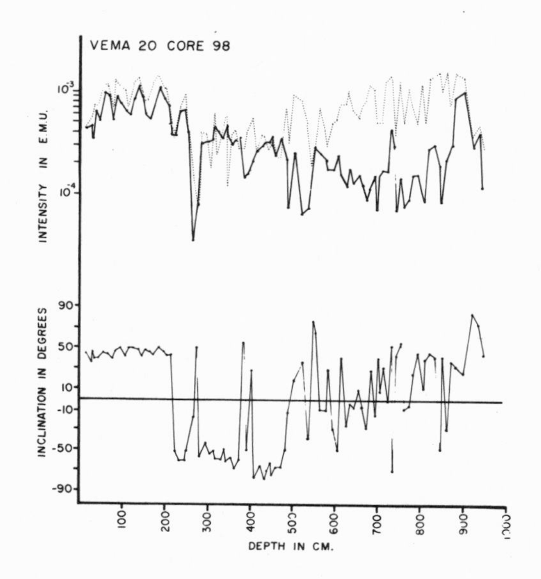

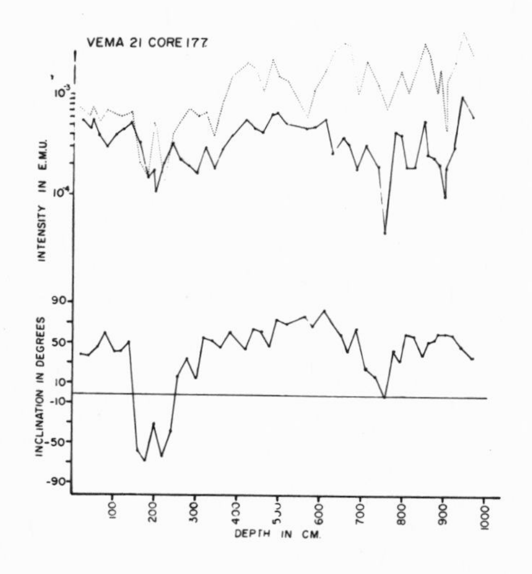

Figure 4. Composite plots of magnetic inclination (bottom plot) and intensity of magnetization (top plot) for V21-177 (a) and V20-98 (b). The dotted line indicates the samples NRM intensity in emu while the solid line shows the intensity after a.f. demagnetization in 50 oersteds.

A core with a similar type of stratigraphy, C10-158, is shown in Figure 7B. The magnetic stratigraphy of this core is similar to V21-177 and V20-98. In C10-158 the Brunhes and Matuyama series are present, along with what appears to be the Jaramillo and Olduvai events at 150 and 240 cm, respectively. At 300 cm the core passes into what is apparently the Gauss series; however, the core is normally magnetized from there to the bottom of the core at 1000 cm, which, if the rate of clay deposition is as slow at depth in the core as it is at the top, this normally magnetized section of core is much too long to represent only the Gauss series.

In most of the clay cores from the Central North Pacific Province of Horn and others (1970, this volume), a pronounced color change occurs at depth in these cores from red, or brown-red above, to very dark red (almost black) below. The decrease in coercivity observed in cores from the Central North Pacific Province roughly coincides with this color change. This change in color has been ascribed to an increase in manganese micronodules by Bramlette (1961), Horn and others (1970, this volume), and by Jacobs (1968, oral commun.). The decrease in magnetic stability seems to be related in some way to this increase in MnO_2. A more intensive study of the magnetic properties of these cores is under way.

Bramlette (1961) ascribed this color change to a change in bottom water temperature due to the onset of the Pleistocene glaciation. The apparent age of this color change varies from core to core in the Central Pacific Basin but certainly occurs in most cores within the upper Pliocene between 2 and 4 m.y. ago, a date which is consistent with Bramlette's hypothesis and with the data of apparent onset of ice rafting in the North Pacific (Conolly and Ewing, 1970, this volume).

It has been shown by Dickson and Foster (1966) from thermal demagnetization studies on red clay cores from the central Pacific,

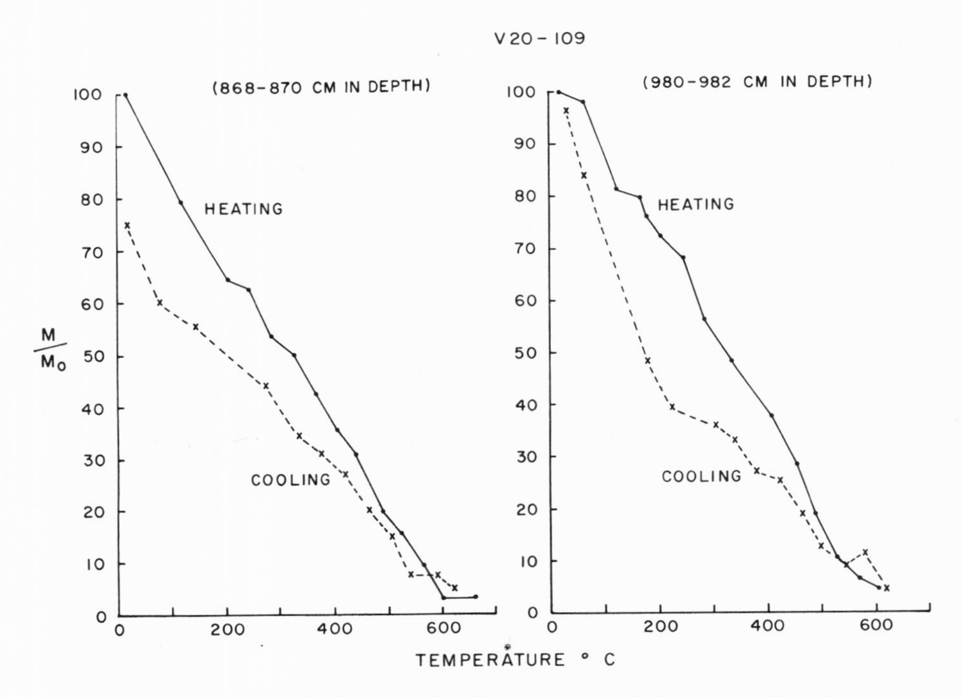

Figure 5. Thermal demagnetization curves of two samples from core V20-109.

that the remanent magnetization is distributed through several phases, including both magnetites and hematites. Curie point analysis of core V20-109 from the North Pacific, which is a radiolarian-diatomaceous mud, shows that the Curie points of the main magnetic constituents range only up to 600° C, indicating that most of the magnetization resides in the magnetites and titanomagnetites (Fig. 5). It seems reasonable to conclude that much of the remanent magnetization which is observed in the red clay cores is hematite of possible diagenetic origin, while in the more highly biogenic core of the Aleutian-Alaska province the origin of the magnetization is most probably due to a form of detrital remanent magnetization of magnetites and titanomagnetites. It is probable that in some cores these magnetizations exist together, both contributing to the remanent vector.

INTERPRETATION OF MAGNETIC STRATIGRAPHY

The interpretation of the ages of the magnetic stratigraphy follows the time scale based on K^{40}-Ar^{40} dating which has been built up over the years (Cox and others, 1963; McDougall and Tarling, 1963; McDougall and Chamalaun, 1966; Cox and others, 1968). The stratigraphic nomenclature used for the last 4.5 m.y. of earth history is based on that of Cox and others (1963) and recent modifications proposed by others. For magnetic stratigraphy older than the Gilbert reversed polarity epoch the number letter system presented by Hays and Opdyke (1967) is employed.

The standard magnetic stratigraphy is shown in Figure 6. The existence of very short magnetic events within the Brunhes Normal Epoch have recently been proposed by Bon Hommet and Babkine, (the LaChamp event, 1967) in France, and by Smith and Foster (the Blake event, 1969). The Blake event is short and appears to be of the order of 6000 years in length, occurring 100,000 years ago. Events of this short duration are difficult to observe in cores with rates of sedimentation of 1 cm/1000 yrs or less; therefore, in cores with low rates of sedimentation the most recent easily observable datum is the Brunhes-Matuyama boundary. Cox and Dalrymple (1967) consider that the most probable age for this boundary is .69 m.y. and this is used as the age of this boundary throughout this paper. The best estimate of the age and duration of the Jaramillo event is probably that of Opdyke (1969), which is obtained by extrapolation from the Brunhes-Matuyama boundary in cores where this event is recognized. The upper boundary is placed at .86 m.y. and the lower boundary at .92 m.y. with a duration of approximately 60,000 years. Two additional normal events have been named in the Matuyama epoch on the basis of K^{40}-Ar^{40} dates. These are the Gilsa event (McDougall and Wensink, 1966) dated at 1.6 m.y. on lavas from Iceland, and the second, the Olduvai event dated at 1.91 m.y. (Gromme and Hay, 1963). A more detailed examination of this event will be presented later; at present it is sufficient to note that a major event is often observed in marine sediments in the lower part of the Matuyama series. The age of the Matuyama-Gauss boundary is not well defined and estimates of its age range between 2.43 and 2.5 m.y. Cox (1968) prefers an age of 2.43 m.y. for this boundary and this age is used in this paper. Two reversed events are present in the Gauss normal epoch: one is dated at 2.8 m.y. and called the Kaena event by McDougall and Chamalaun (1966), and a second, called the Mammoth event and dated at 3.06 m.y. by Cox and others (1963). Both of these events are observed in marine cores (Foster and Opdyke, 1968; Hays and others, 1969). The age of the Gauss-Gilbert boundary is well established and is given as 3.32 m.y. by Cox and Dalrymple (1967). At least three events are known from the Gilbert reversed epoch. These were first observed in marine sediments by Hays and Opdyke (1967) and called Gilbert events "a," "b" and "c" in order of increasing age. Gilbert events "a" and "b" have subsequently been named the Cochiti and Nunivak events by Cox and Dalrymple (1967) with ages of 3.6 and 4.2 m.y. Gilbert event "c" has not yet been identified from work on land. Hays and Opdyke estimate the age of the beginning of the Gilbert reversed epoch at 5 m.y. by extrapolation from the Gauss-Gilbert boundary (Hays and Opdyke, 1967). Figure 6 shows the dates on which the paleomagnetic time scale is based.

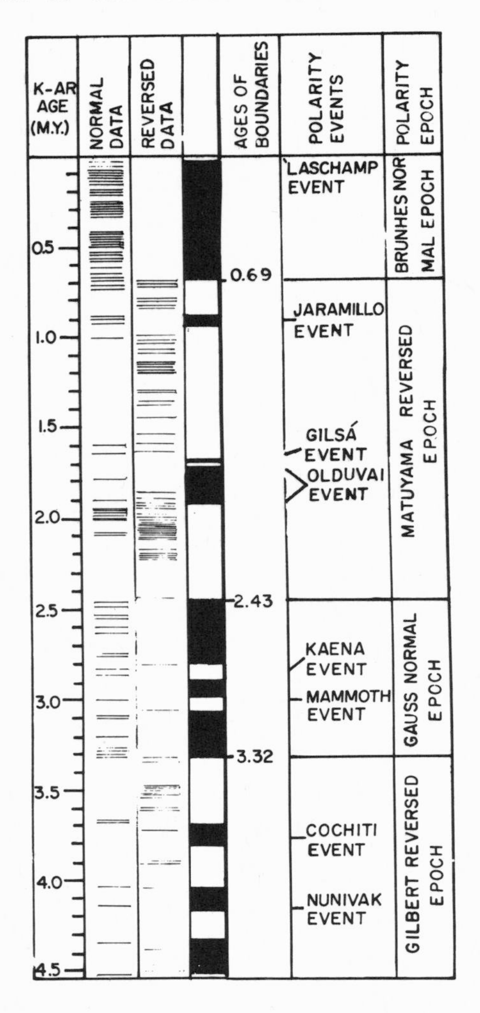

Figure 6. Magnetic time scale from Cox and others (1968) showing magnetic reversal sequence beside the dated polarity measurements from lava flows on land.

MAGNETIC RECORD AS DETERMINED FROM NORTH PACIFIC CORES

Figure 7 shows the plots of 14 cores from the North Pacific. The stratigraphies of these cores have several features in common. The short magnetic events which have been dated in the Brunhes series are not seen with certainty in these cores, presumably because the rates of sedimentation are sufficiently slow and integrate out events as short as 5000 years. There are two easily identifiable events in the Matuyama series; the first directly below the Brunhes-Matuyama boundary and the second in the lower part of the Matuyama series. The correlation of these events from core to core has been verified by paleontological zonation on which these age designations have been made. These are given in detail by Hays (1970) and Donahue (1970) in companion papers in this volume. The cores from the red clay areas contain no fossils so that correlation can only be effected by the sequence of magnetic reversals already confirmed. Some of the cores contain short spikes which may or may not represent short events; however, the longer events are often seen and are most important for stratigraphy.

Figures 7A, B, C and D show a series of inclination plots from cores of sites spread throughout the area of investigation. Core RC10-167 (Fig. 7A) has the longest complete Brunhes series section of any core in this study. The Brunhes-Matuyama boundary occurs at 1535 cm. In an attempt to identify any short events which might be present in the Brunhes series the core was sampled at close intervals. Two specimens were found to be reversely magnetized; however, both correlated with either a core pipe break or other physical break and so are unlikely to be real.

Core RC11-171 penetrates to the lower Matuyama series. Two large events are present at 575 and 900 cm. The younger one is interpreted to correlate with the Jaramillo event, and the lower and the larger one with the Olduvai event. A single normally magnetized specimen occurs between the Jaramillo event and the Brunhes-Matuyama boundary. Whether this represents a real short event is not known; however, similar small normally magnetized intervals have been found in other cores, lending support to the contention of Watkins (1968) that a short event occurs between the Jaramillo event and the Brunhes-Matuyama boundary.

Core V20-105 has a complete magnetic stratigraphy to the upper Gilbert series. The two major events (Jaramillo and Olduvai) are present in the Matuyama series. The sedimentation rate in this core increases with depth so that a reasonably detailed analysis of the Gauss series sediments is possible. Two reversed events are present;

the youngest one is correlated with the Kaena event and the older with the Mammoth event. The record of these two events in this core is the best that we have so far obtained. Interpolating between the boundaries of the Gauss series, the ages of the events in the cores would fall within the errors of the dates on land for these events discussed previously and shown in Figure 6. The length of the Kaena and Mammoth events in this core, using the rate of deposition during the Gauss series of 4 mm/1000 yr, would be 82,000 and 95,000 years, respectively.

Figure 7B shows the magnetic stratigraphy of three cores, all of which penetrate to the Gauss series or older sediments but have varying rates of sedimentation, decreasing from 4.9 mm/1000 yr for RC10-161, to 1.5 mm/1000 yr for RC10-158. In all these cores two prominent events are present in the Matuyama series, the youngest of which is correlated to the Jaramillo event and the older to the Olduvai event. Core RC10-161 terminates in the upper Gauss series. Conrad 10-159 apparently penetrates to below the Kaena event; however, below about 1050 cm the quality of the data decreases so that below this point the magnetic stratigraphy is untrustworthy. RC10-158 is discussed elsewhere, but it is worthwhile noting that the core penetrates to the Gauss series at 325 cm and below that point the magnetic inclination is positive and does not reflect any changes in field direction.

Figure 7C shows inclination plots of three cores from the northeast Pacific Ocean. V20-85 is very similar to plots described previously and ends in sediments of the upper Gauss series. However, Conrad 11-193 is more complicated and more interesting. Within the Brunhes series a single point at 70 cm is reversely magnetized; it is not possible to tell whether this represents a short magnetic event or disturbance within the core. In this core, instead of two events within the Matuyama series, there are three. The large event at 475 cm is most probably the Olduvai event, and that at 250 cm is clearly the Jaramillo event; therefore, an apparent event occurs in this core between the Olduvai and the Gauss series. A short event or events has been predicted at about this time by Cox (1969) and Vine (1968), largely on the basis of the work on magnetic profiles at sea (Pitman and Hirtzler, 1966). As can be seen from an inspection of the other profiles a short event in this interval still awaits confirmation. RC11-193 apparently ends in sediments of the Gilbert series after passing through the Gauss series in which the Kaena and Mammoth events are present. An inclination plot of RC11-170 is shown just below RC11-193. The magnetic stratigraphy is unambiguous to the Olduvai event at 700 cm. The interval between the Olduvai event and Gauss Normal series contains several short spikes which one might be tempted to correlate with the event shown at

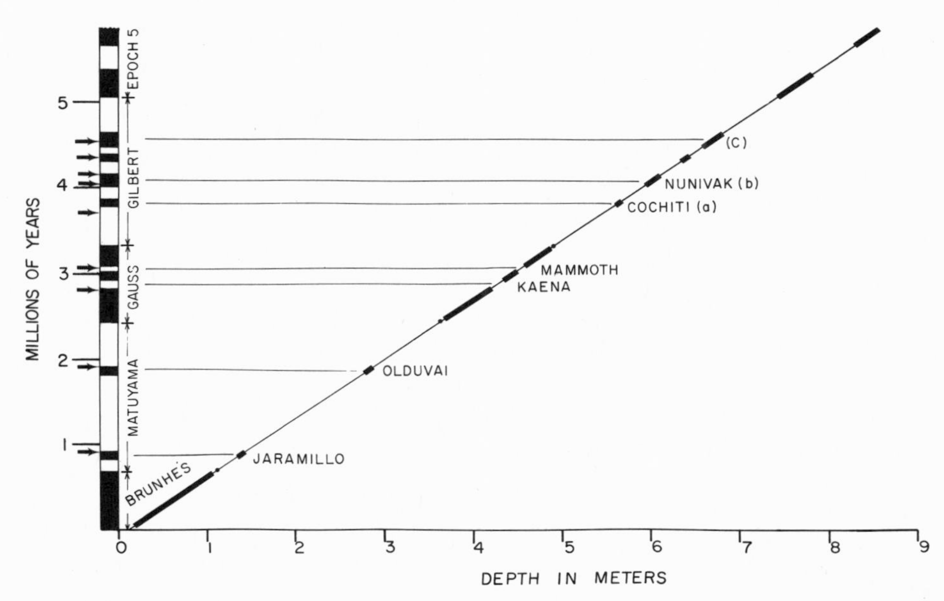

Figure 8. Time versus depth plot for core V20-88 showing an almost constant sedimentation rate. The arrows to the left of the magnetic time scale indicate the original dates from lavas used in defining each magnetic event.

the X axis at 12 cm, very close to the origin. The dates of the Jaramillo, Olduvai, Kaena and Mammoth events, as determined by this slope, coincide (within the errors) with the dates which originally defined them. This is also true of the "a" (Cochiti) and "b" (Nunivak) event. "C" event of the Gilbert series is centered about 4.5 m.y. and the dates of two normally magnetized basalts from Nunivak Island fall within this interval. Extrapolated age of the Gilbert epoch 5 boundary is 5.05 m.y., which is extraordinarily close to the age of 5 m.y. estimated for this boundary from Antarctic cores (Hays and Opdyke, 1967). The estimated age of the upper boundary of event "a," epoch 5, is 5.45 m.y., and the lower boundary would be placed at 5.75 m.y., making the length of this event 300,000 yrs long. The agreement of the age of the Gilbert epoch 5 boundary between this core and the Antarctic cores makes the 5 m.y. date for this boundary very plausible.

THE OLDUVAI EVENT

The most important magnetic event from a stratigraphic viewpoint is the large event in the lower part of the Matuyama series, which in previous studies on deep-sea sediments has been correlated with the Olduvai event (Opdyke and others, 1966; Dickson and Foster, 1966; Hays and Opdyke, 1967; Berggren and others, 1967; Hays and others, 1969). It is the event in which the Pliocene/Pleisto-

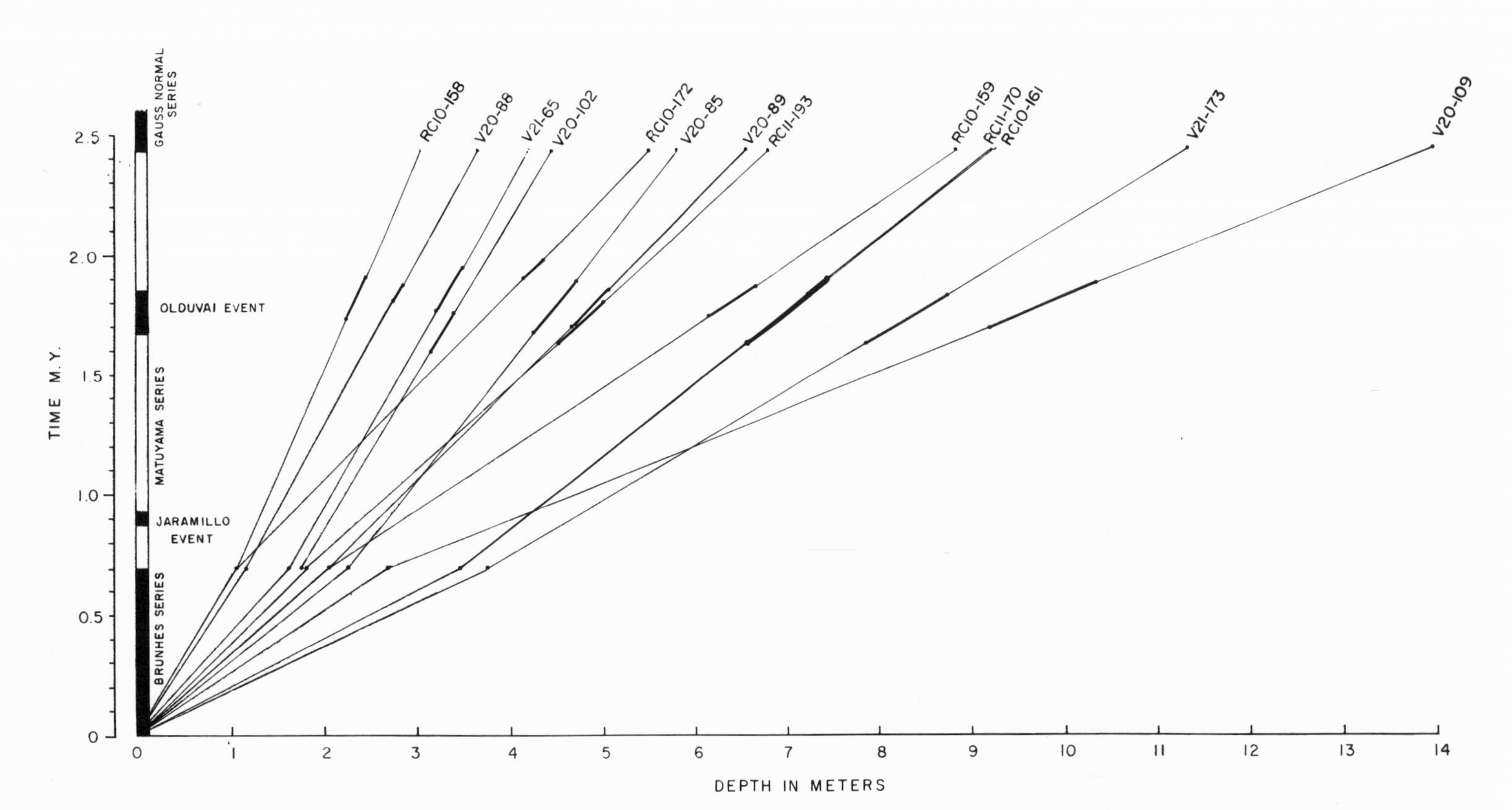

Figure 9. Time versus depth plots for 13 cores showing the duration of the Olduvai event assuming an average sedimentation rate during the Matuyama Epoch based on the Brunhes-Matuyama boundary at .69 m.y. and Matuyama-Gauss boundary at 2.43 m.y. The average duration of the Olduvai event figured in this manner is 150,000 years from 1.71-1.86 m.y.B.P.

cene boundary occurs (Berggren and others, 1967). Confusion has arisen due to the fact that some workers believe the large event is not the Olduvai but is instead the younger Gilsa event (Cox, 1969; Vine, 1968). In an effort to more closely delineate the age of the event as seen in deep-sea cores, a time versus depth plot was constructed from those cores which contain this large event and also pass through to the Upper Gauss normal series. In this study 13 cores are available that have these qualifications. An age of .69 m.y. has been used for the age of the Brunhes-Matuyama boundary, and an age of 2.43 m.y. is chosen for the Gauss-Matuyama boundary. Assuming that the rate of sedimentation is constant between the two points, a mean age for the event can be obtained, as shown in Figure 9. The age of the upper boundary of the event ranges between 1.59 and 1.90 m.y., with the average age for the upper boundary at 1.71 ± .26 m.y. The lower boundary ranges in age between 1.75 and 1.98 m.y., with an average of 1.86 ± .19 m.y. The average length of the event would be 150,000 ± 47,000 years.

It is now probably worthwhile to return and examine the data on which the Olduvai (Gromme and Hay, 1963, 1967) and Gilsa (McDougall and Wensink, 1966) events were first established. The age originally quoted as the age of the Olduvai event is 1.92 ± 0.6 m.y., obtained on a sample from the pahoehoe flow at Olduvai Gorge. Other samples of the same flow gave ages of 1.60 and 4.4 m.y. Evernden and Curtis (1965) regard the age of 1.92 as being reliable. Samples from the underlying aa flow give an age of 1.76 m.y., which Evernden and Curtis believe to be too young. Feldspar separated from tuffs overlying the normally magnetized lava flows gave ages of 1.78, 1.85 and 1.86 m.y. The most concordant results from the Olduvai Gorge are from a welded tuff 3 to 6 feet above the pahoehoe basalt, which give six ages ranging from 1.70 to 1.76 m.y. Anorthoclase and plagioclase were both separated from a tuff underlying the normally magnetized basalts, and ages of 1.85 and 1.91 were obtained; both dates are regarded as reliable by Evernden and Curtis. The age of the normally magnetized lavas would be bracketed between 1.75 and 1.92 m.y. It should be noted that neither the upper nor the lower boundaries of the Olduvai event can be determined from its type locality. It can be seen that the age limits of the basalt overlap the age of the large event, as determined from marine cores, whether or not an age of 2.43 or 2.5 m.y. is placed on the Matuyama-Gauss boundary.

The Gilsa event was named by McDougall and Wensink (1966) from Icelandic rocks. In the middle of a reversed sequence of lava flows of the Matuyama epoch, three lava flows occur; the upper flow and the lower flow are normally magnetized and the middle flow is reversely magnetized. The K^{40}-Ar^{40} date on the upper flow is 1.6 ±

.05 m.y. This age is apparently too young to represent the large event seen in marine cores, since the average age of the upper boundary of the event has a date of 1.71 m.y. The question of which of these events is seen in the cores remains. Taking into account the overlap of ages between the type section of the Olduvai event in Olduvai Gorge, and the ages obtained from marine cores by interpolation, it seems reasonable on this basis alone to equate the Olduvai event with the large event seen in cores. It appears this would also be proper on the basis of priority. We think it is reasonable to equate the Gilsa with the split seen at the top of the Olduvai in V20-109, which was previously suggested by Ninkovitch and others (1966). We contend that this best reconciles the different data, keeping in mind the errors involved in the various techniques.

Dymond (1969) has published K^{40}-Ar^{40} dates on volcanic ash from three cores included in the present study. Two of the cores, V21-145 and V21-173, in which the dated ash occurs also contain the Olduvai event. The ash occurs in these cores between the Jaramillo and Olduvai events at 817 and 720 cm, respectively (Hays and Ninkovitch, 1970, this volume). The date on the ash in V21-173 is 1.62 $\pm$ 0.08 m.y. Using a date of .69 m.y. for the Brunhes-Matuyama boundary and extrapolating the rates of sedimentation, it is possible to estimate the upper boundary of the Olduvai event. In the case of V21-145 this age is 1.86 $\pm$ .13 m.y., and for V21-173 it is 1.78 $\pm$ .09 m.y. Therefore, in both of these cases the K-Ar dates are in accord with the correlation of the big events in the core, with the date from the Olduvai Gorge rather than that of the younger Gilsa event. The K-Ar date from the third core was obtained from V20-108 in which the Olduvai event does not occur. Ninkovitch and others (1966) have, however, correlated this ash with an ash which occurs in V20-107 and V20-109. Assuming this correlation is correct and extrapolating to the top of the large event, the age would be 2.06 $\pm$ .39 m.y., and again the correlation of the large event with the Olduvai event would be favored.

Several K^{40}-Ar^{40} dates have been obtained on normally magnetized lava flows which fall in the region of from 2.1 to 2.3 m.y. These dates appear to be much too old to be equated with the Olduvai event, as seen in marine cores, or with the date from the type locality. However, in several cores short normally magnetized sections occur which precede the Olduvai event. A short event may be present at this time. A suggestion of short magnetic events of this age appears on magnetic profiles from the *Eltanin* in the South Pacific (Pitman and Heirtzler, 1966). It should be noted that in marine sediments with rates of sedimentation below 1cm/1000 yrs, it is very difficult to consistently identify events with durations as short as 2 X 10^4 years. In cores V21-173 and RC11-193, there appears to be a short

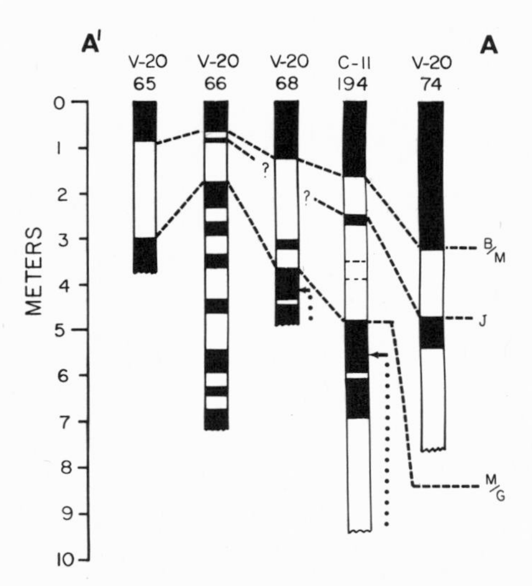

Figure 10. Stratigraphic section A-A′ located as in Figure 1 NE. of the Hawaiian Islands. Correlations between cores are shown in dashed lines. Magnetic epochs and events abbreviations are: (B/M) Brunhes-Matuyama boundary; (J) Jaramillo Event; (O) Olduvai Event; (M/G) Matuyama-Gauss boundary. The arrows and dotted lines beside the lower portion of several of the cores indicate that section where the magnetic stratigraphy becomes unreliable.

magnetic event preceding the Olduvai and may represent this early Matuyama series event.

MAGNETIC STRATIGRAPHY

The correlation of the piston cores along seven sections based on the magnetic stratigraphy is shown in Figures 10 to 19. The magnetic interpretation is based primarily on the inclination data, as previously described. The drop-off in coercivity with depth in red lutite cores, which was detailed earlier, gives unreliable magnetic stratigraphy. The region where this takes place in the cores is indicated on each core of the profile although the complete data is pre-

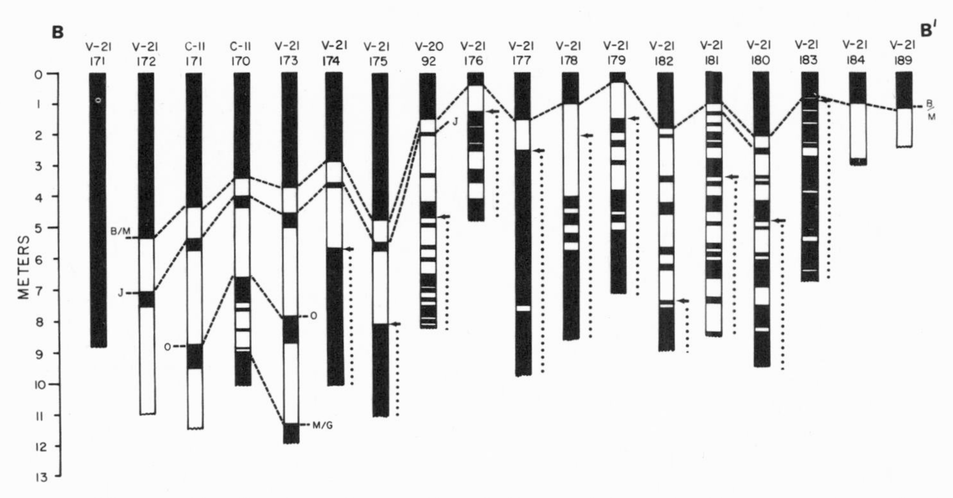

Figure 11. Stratigraphic section B-B′ located as in Figure 1, from just south of the Aleutian Arc to Hawaii. Abbreviations are the same as for Figure 10.

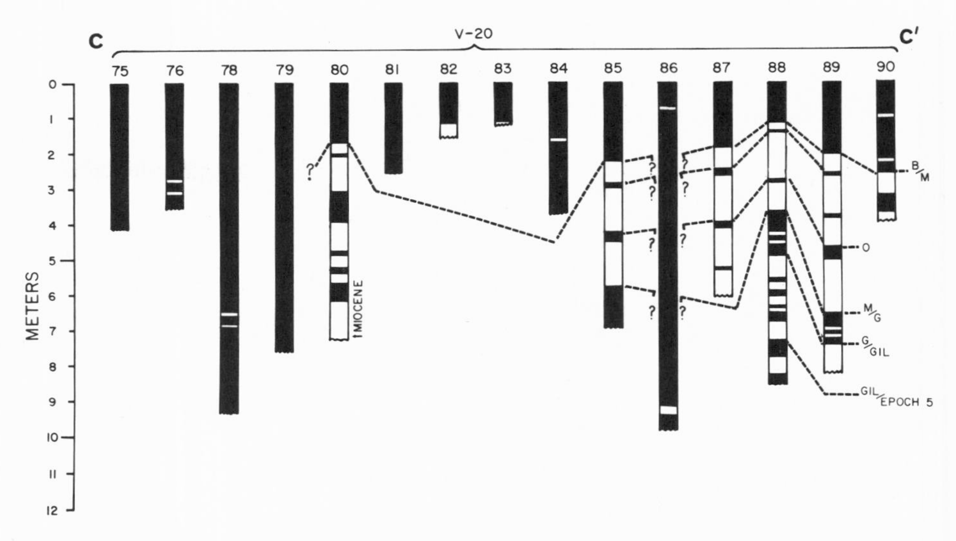

Figure 12. Stratigraphic section C-C′ located as in Figure 1. Abbreviations and correlations are the same as in Figure 10 with the addition of the following: (G/GIL) Gauss-Gilbert boundary; (GIL/Epoch 5) Gilbert-Epoch 5 boundary. The base of V20-80 is Miocene as indicated.

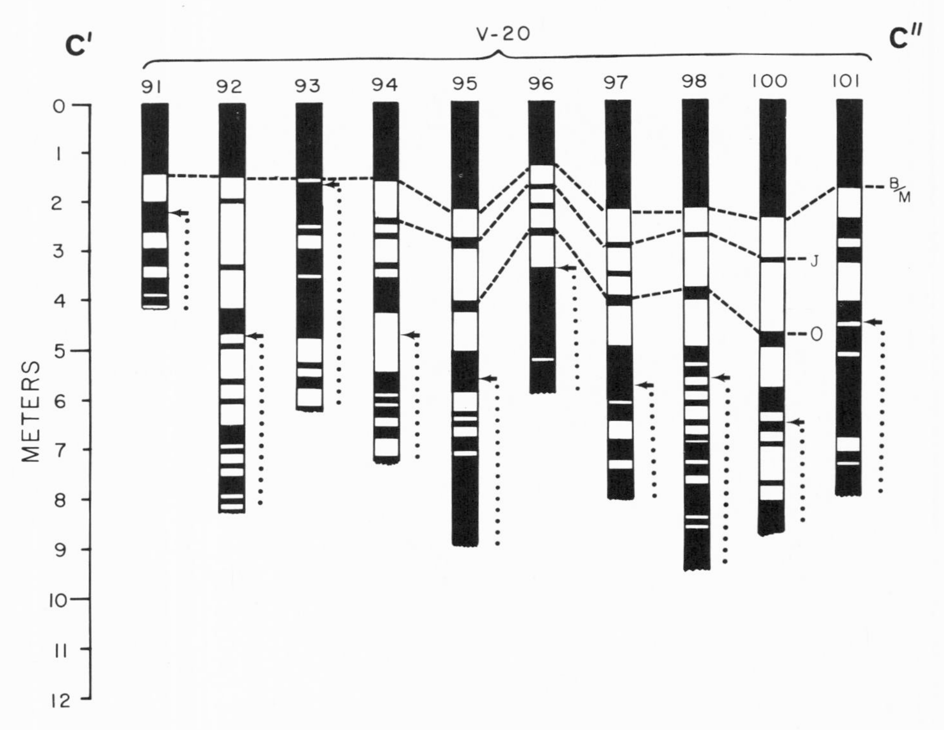

Figure 13. Stratigraphic sections C′-C″ from the *Vema* 20 track located as indicated in Figure 1. Abbreviations and correlations are the same as for Figure 10.

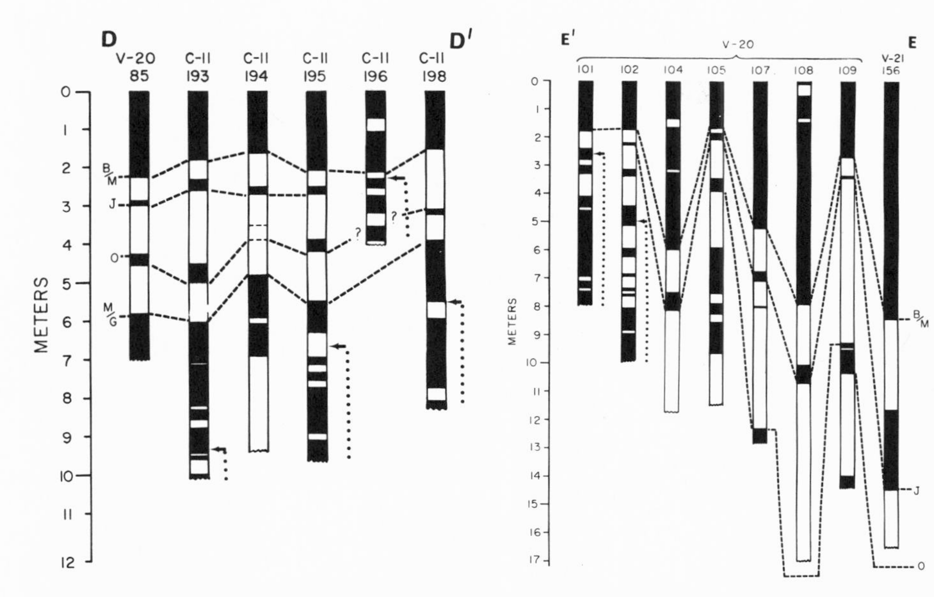

Figure 14. Stratigraphic section D-D′ located as indicated in Figure 1. Abbreviations and correlations as in Figure 10.

Figure 15. Stratigraphic section E-E′ located as indicated in Figure 1. Abbreviations and correlations as in Figure 10.

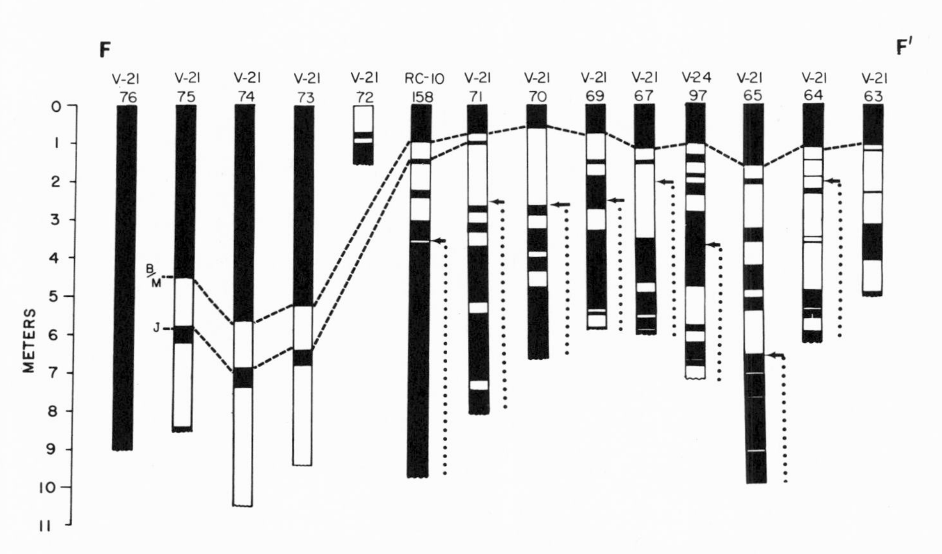

Figure 16. Stratigraphic section F-F′ located as indicated in Figure 1. Abbreviations and correlations as in Figure 10.

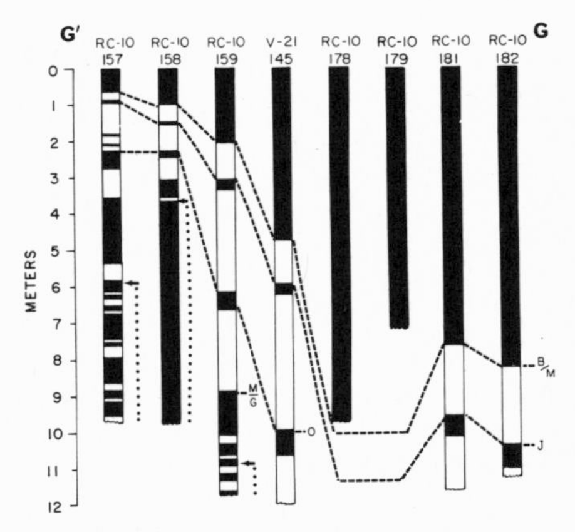

Figure 17. Stratigraphic section G-G′ located as indicated in Figure 1. Abbreviations and correlations as in Figure 10.

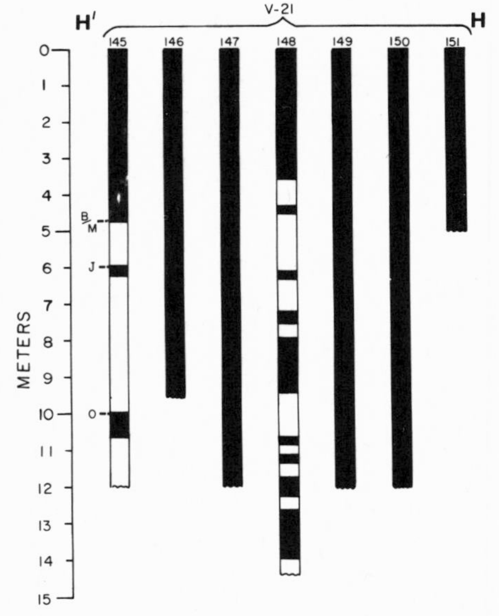

Figure 18. Stratigraphic section H-H′ from the V-21 cruise track located as indicated in Figure 1. Abbreviations and correlations as in Figure 10.

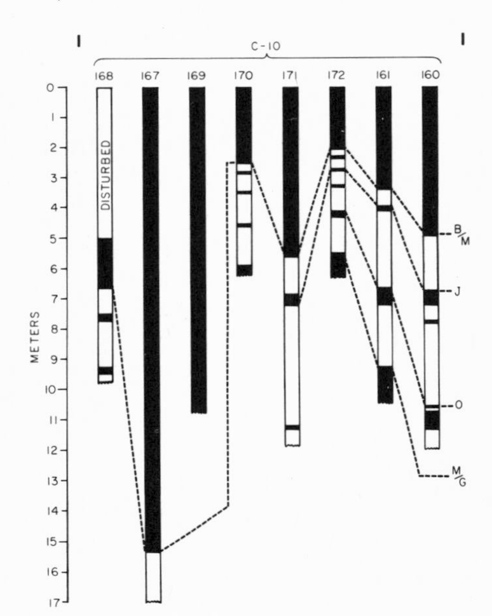

Figure 19. Stratigraphic section I-I′ from RC-10 track located as indicated in Figure 1. Abbreviations and correlations as in Figure 10. The uppermost barrel of RC-10-168 was disturbed in shipboard handling as indicated; the data quality for the rest of the core is good.

sented. We strongly suggest that no reliance be placed on data from these parts of the cores. In most cases the top of the core is normally magnetized and this is presumed to represent the Brunhes series. Where fauna are present in the core this has been used to aid in the interpretation presented. In only one case, V21-72, does the core begin reversely magnetized, and in this case, a manganese crust is present at the top of the core, indicating a long period of non-deposition. The reversed section in this core is correlated with the Matuyama series, although it may be older since fossils are not present.

It is apparent from the profiles that the rates of sedimentation increase toward the margin of the Pacific. This increase in rates of sedimentation is probably due to a combination of factors. It is reasonable to expect that a greater amount of detritus is being provided from erosion of the surrounding continental areas. This is especially true off the northwest coast of North America where deposition through the action of turbidites is common. Around the great arc of the Pacific from Alaska through Japan, significant amounts of volcanic detritus are present in the sediments. Glacially derived debris is common in many regions in the northern part of the Pacific north of 45°. In regions of increased biogenic productivity, the skeletal portions of diatoms and radiolarians become an important part of the bulk sediments.

Figure 20 shows the rates of sedimentation for cores throughout the North Pacific plotted over the sediment thickness map of Ewing and others (1968). An attempt has been made to contour the rates of sedimentation based on the depth of the Brunhes-Matuyama boundary. The outstanding feature revealed by these contours is that rates of sedimentation throughout the central basin of the Pacific Ocean are below 3mm/1000 yrs, a rate which is in harmony with the very thin layer of transparent sediment observed in the region by Ewing and others (1968). The rates of sedimentation increase from this central area toward the northeast, north and west.

The increasing rate of sedimentation toward the northeast can be seen clearly in profiles A′ and C′. Profile C-C′ begins just outside Vancouver and goes to Midway Island. The cores between Vancouver and the Juan de Fuca Rise do not penetrate to the Brunhes-Matuyama boundary and they cannot be expected to do so since the rate of sedimentation in this region is above 10 cm per/1000 yrs with turbidites commonly occurring. Cores 78 and 79, near the crest of the rise, have rates of sedimentation exceeding 10 mm/1000 yrs. V20-80 does have changes of polarity down the core but is Miocene at the bottom and is probably from a fault scarp, so that rates of sedimentation calculated from the core would have little meaning.

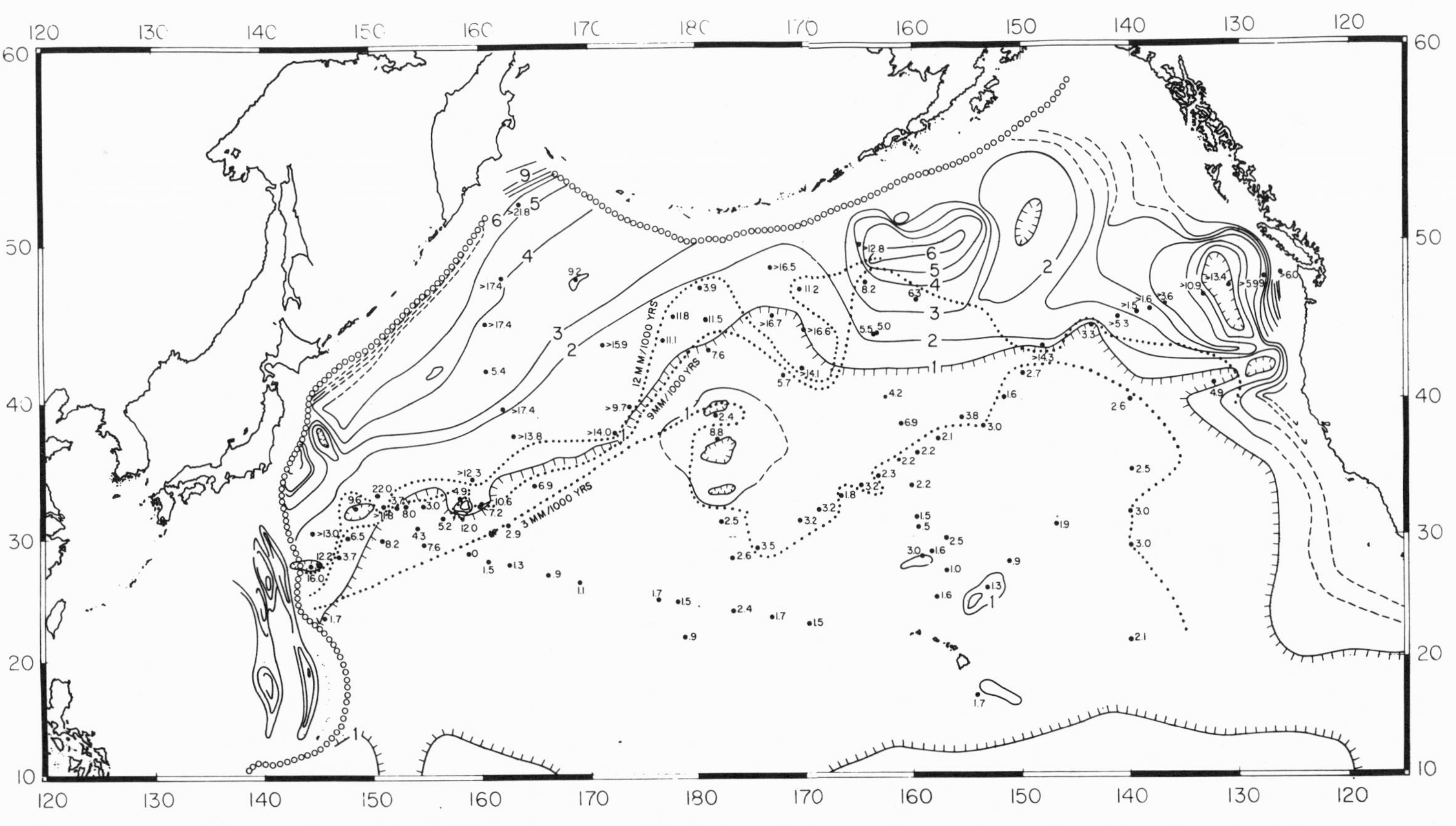

Figure 20. Map on North Pacific sedimentation rates during the Brunhes epoch superimposed on the sediment thickness map of Ewing and others (1968). Sedimentation rate contours (9 and 12mm/1000 years) are shown in dotted lines.

The next three cores, V20-81, 82, and 83 in this profile, contain turbidite horizons so that the rates of sedimentation are probably high. The next core in the profile, V20-85, is instructive in that turbidites are not present and a good magnetic stratigraphy is obtained from the core. The next core, however, is again a turbidite core and the Brunhes-Matuyama boundary is not penetrated. The rest of the cores along this profile are red clays and have low rates of sedimentation.

Profile B-B′ illustrates the general increase in sedimentation which occurs toward the Aleutian arc. This increase is not due to turbidite deposition toward the Aleutian arc since turbidity current deposits are not recorded from the region. The increased rate of sedimentation in this area is probably due to increased amount of siliceous organisms (radiolarians and diatoms) which are present in the sediments north of the 40th parallel and also to an increasing amount of volcanic ash and glacial material.

In the region of the north-central Pacific south of the Aleutian arc, there are some interesting variations of sedimentation rates with longitude as well as latitude. The contours of high rates of sedimentation above 12mm/1000 yrs swing to the south at 170° W. long., then swing back to the north at 180° long. West of 180° the contours swing sharply to the south to the region of the Shatsky Rise. Northwest and west of the Shatsky Rise the rates of sedimentation exceed 12mm/1000 yrs. The one exception to this is V21-148, which contains a very long magnetic stratigraphy. An examination of the profiler records shows that the layers of sediment thin in this region, and although the reason for this thinning is not clear, it is apparently a result of a decrease in the rate of sedimentation.

It is of some interest to compare the rates of sedimentation obtained through paleomagnetic studies with those previously obtained by Goldburg and Koide (1962), using th^{232}-th^{230} methods. Harrison and Somayajulu (1966) and Ku and others (1968) have shown that rates of sedimentation derived from geochemical methods and those obtained by paleomagnetic methods are essentially in agreement. Rates given by Goldburg and Koide (1962) have been corrected for $CaCO_3$ and opal content. In the central Pacific west of Hawaii they give rates of 1mm/1000 yrs, which is in essential agreement with what we obtain in this region; however, in the central North Pacific they show rates of .6 to 1.5 mm/1000 yrs, which seems much too low, even on an opal-free basis.

EFFECT OF TOPOGRAPHY AND RELIEF

In regions where sediment cover is uniform and the relief is low, cores usually show a reasonably continuous record of magnetic stratig-

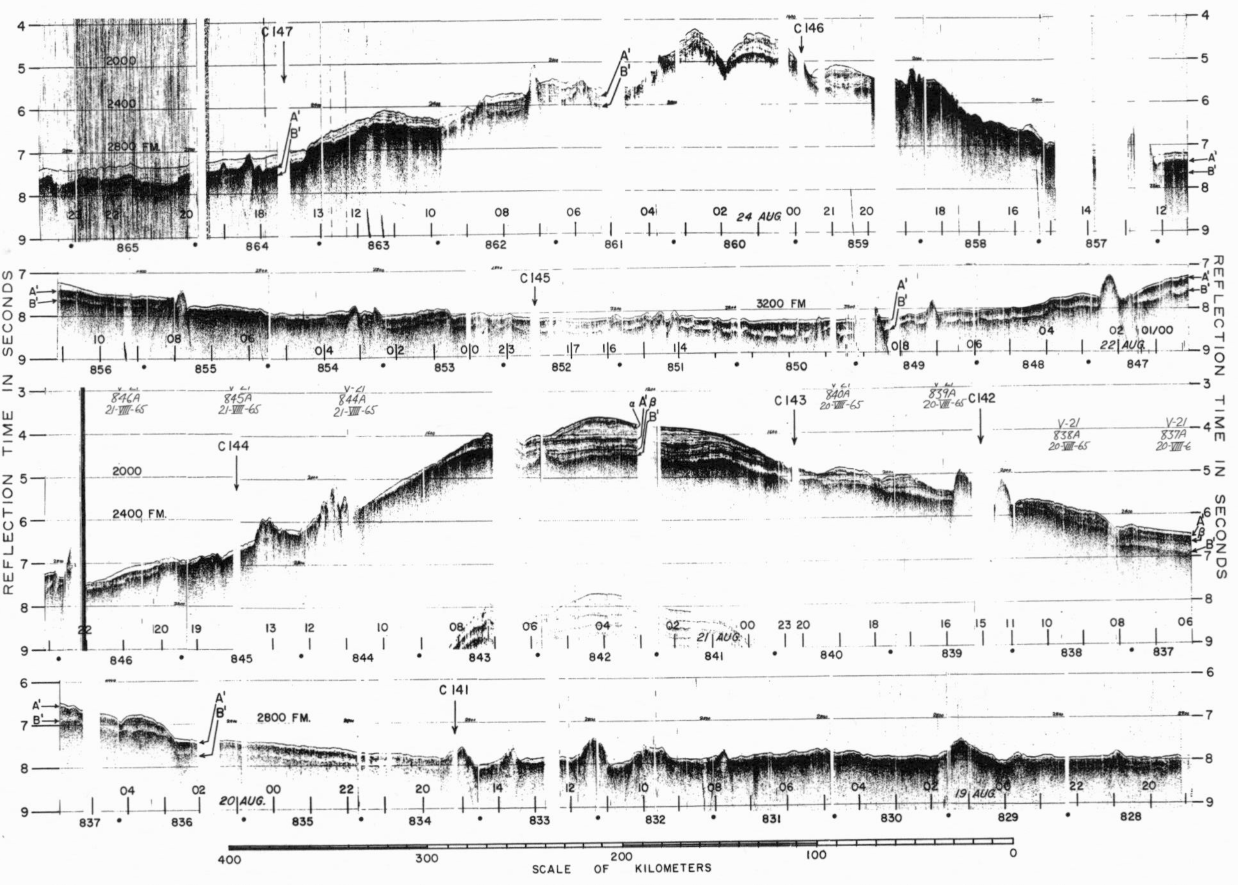

Figure 21. Profile of V-21 crossing of the Shatsky Rise from Ewing and others (1966), showing sediment layers and core locations for those cores pictured in Figure 22.

raphy, while in areas of rough topography, unconformities are a common occurrence. Figure 21 shows the profiler record of the *Vema* 21 track over the Shatsky Rise. These profiler records were previously published by Ewing and others (1966). The magnetic stratigraphy of four of the cores taken along this track is shown in Figure 22. It can be seen that cores such as V21-144 and V21-145, which were taken in regions where the transparent zone is thin, have complete magnetic stratigraphies. V21-142, however, which was taken in an area of rough topography between two peaks, has a magnetic stratigraphy abbreviated with normal and reversed sections but apparently no Jaramillo or Olduvai event. V21-141, taken on a small hill, seems to have a more or less normal stratigraphy. Cores V21-146 and 147 taken from areas of a thick zone of transparent sediment do not show reversals, and plots of inclination are not shown.

COMPARISON OF SEDIMENTATION RATES WITH SEDIMENT DISTRIBUTION

Ewing and others (1968) presented an isopach map of the sediments of the North Pacific shown in Figure 20. The isopachs were constructed on the transparent sediments above the basement and above the layer "A" where it is present. Layer "A" is found to be Upper Cretaceous in age where it appears as outcrops on the Shatsky Rise. The thickness of sediments is probably less than 100 m in the central Pacific region, and increases toward the equator and the margins of the Pacific. In general, the isopachs based on rates of sedimentation parallel those based on sediment thickness. However, there do seem to be some rather important differences in detail, particularly in the region south of the Aleutian arc. At latitude 170° W. the seismic isopachs show sediments apparently not much in excess of 200 m approaching the Aleutian Trench. Just to the east of 160° W., however, a thick wedge of sediment in excess of 600 m is present south of the trench. These sediments are discussed in detail later in the paper. The contours on rates of sedimentation appear not to follow this pattern but dip to lower latitudes at 170° W. and toward the trench at 180°. It must be remembered that the rates are only averaged over the last 690,000 years and may have been recently established because of the increased intensity of glaciation during the last 690,000 years, which probably caused increased productivity south of the Aleutian arc. In some cores in profile B-B′ north of 40°, there are siliceous oozes overlying barren red clays. These indicate a recent movement of a higher productivity zone toward the south in this region. This zone of higher productivity may be caused by the convergence of Antarctic Bottom Water in this

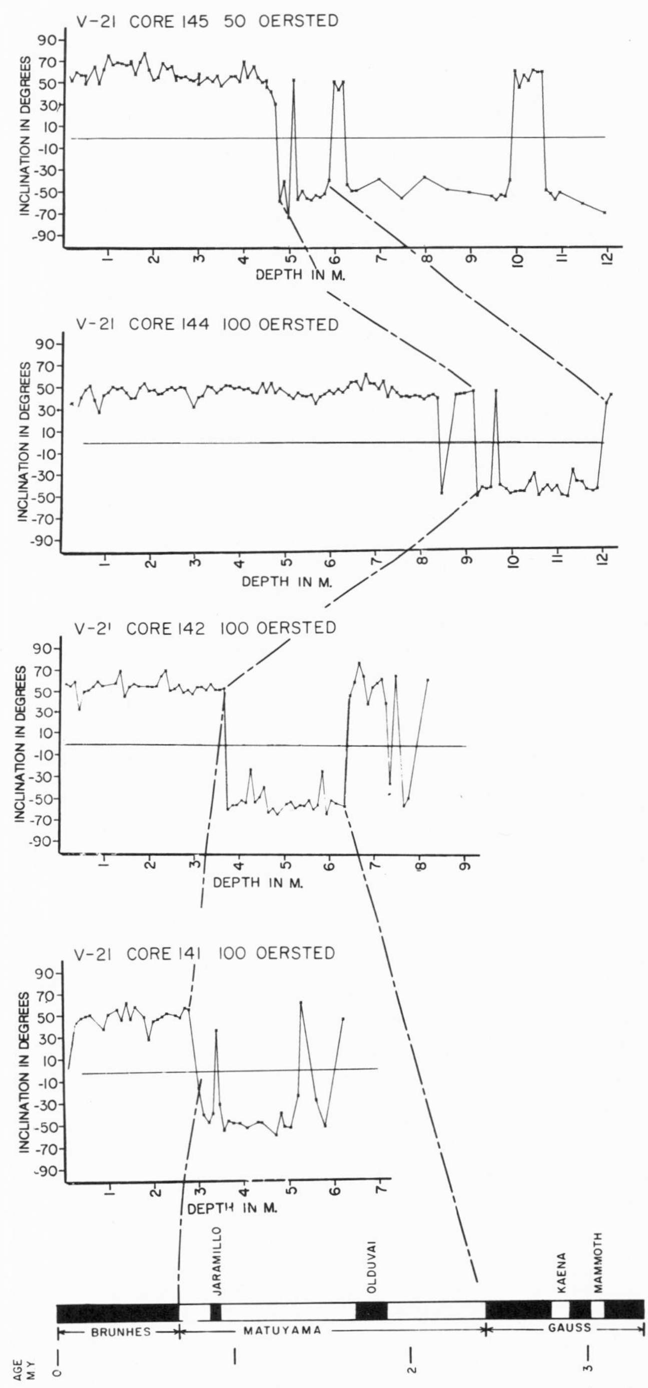

Figure 22. Plots of magnetic inclination versus depth of Shatsky Rise cores located as in Figure 21.

region, as shown by Gordon and Gerard (1970, this volume). This serves as a graphic warning that the bulk rates of sedimentation can change radically through time. Nevertheless, one can say that the isopachs on sediment thickness and the isopaths on rates show overall agreement; that is to say, where the sediment is thick the rates are high, and vice versa.

Part of the region which has been included in this study is underlain by layer "A," which is a prominent reflecting horizon in the western Pacific Ocean. As mentioned previously, the age of this horizon on the Shatsky Rise has been shown by Ewing and others (1966) to be Upper Cretaceous in age. In principle, one should be able to date this horizon by extrapolation of the rates of sedimentation, if these have remained constant. It seems certain that the rates obtaining during the Brunhes epoch have not persisted throughout the Tertiary; higher rates due to the increased erosion and increased productivity due to changing climate of the glacial Pleistocene are probable. Most cores which penetrate to the Gauss normal series show a decrease in sedimentation rate relative to the Brunhes normal series during the Matuyama reversed series. However, it is difficult to know how widespread this rate change is, since in areas of constant sedimentation the rates are high enough so that the Gauss series cannot be reached in cores of normal length. The samples are therefore biased toward cores which show low rates of sedimentation during the Matuyama. Several cores (V20-109 and V20-105) show increased rates at depth. It seems reasonable to believe that the rates of sedimentation should be most uniform in the region of red lutite in the central Pacific region where low rates are observed. In some cores from this region that apparently have the best magnetic stratigraphy, the rates appear to be fairly constant, such as in cores V21-65 and V20-88. Therefore, extrapolated ages to the basement might be most meaningful in the region of the central Pacific. The extrapolated age to basement in these areas give ages of about 70 m.y. for layer "A," which is in reasonable agreement with its presumed Upper Cretaceous age. This, of course, must be considered an approximation. In the region north and east of the Shatsky Rise, however, extrapolated ages to the top of layer "A" give very young ages of about 20 m.y. This age seems much too young and may only represent a greatly increased productivity during the upper Cenozoic.

AGE OF THE ALEUTIAN ABYSSAL PLAIN

South of the Aleutian arc, centered at about 160° W. long., is a large accumulation of sediment up to 400 m in thickness, which is composed of horizontally bedded turbidites overlain by a transparent layer of variable thickness ranging from 43 m on the eastern side, to 96 m thick about 165° W. long. (Hamilton, 1967; Ewing and others, 1968). Hamilton (1967) presents strong arguments for the derivation of the turbidites from the north; relic turbidite channels recently discovered support this conclusion (Grim and Naugler, 1968).

The formation of the present Aleutian Trench evidently cut off the source of supply from the north, and since that time only pelagic sediments have accumulated in the area. If the rate of accumulation of this pelagic clay is known, then it is possible to estimate the time at which the Aleutian Trench came into existence. Five of the cores of this study are in the pelagic sediments in question and the rates of sedimentation range from a high of greater than 12mm/1000 yrs in the north, to a low of 5mm/1000 yrs in the south for the Brunhes series. It may be, as has been previously stated, that these rates are too high because of an increase in biogenic activity during the Pleistocene. However, in cores to the south, which are barren, rates for this interval are regularly as high as 3mm/1000 yrs. Therefore, a rate as high as 5mm/1000 yrs is probably a good compromise. If this is the case, the thickest accumulation of pelagic sediment (96 m) shown by Hamilton would give an age of 19 m.y. for the base of the pelagic sediments, or a Miocene age. At the point where the cores were obtained, direct comparison with Hamilton's profile yields a date of 10 m.y. for the beginning of pelagic sedimentation and hence for the formation of the Aleutian Trench. Even if the rates were the minimum expected, 3mm/1000 yrs, the base would be 35 m.y., or Oligocene in age. It seems most likely that these pelagic sediments have accumulated since sometime in the Miocene. Since turbidites accumulate more rapidly, it is entirely possible that the base of the turbidite sequence is no older than Oligocene and indeed the whole thickness may have accumulated since the Miocene. An Eocene age for the base of the pelagic sediments was preferred by Hamilton on the basis of the rates of sedimentation given by Goldburg and Koide (1962). In the light of this study it is very unlikely that the base of the pelagic sediment is this old.

In order to show schematically the variation in rates of sedimentation with latitude and longitude, two profiles of sedimentation rate versus latitude and longitude were constructed. Figure 3 is a composite profile showing the variation of the rates of sedimentation with latitude. The equatorial portion of the profile is taken from the *Conrad* 11 track along 140° W. long. from data reported by Hays and others (1969). The remainder of the profile is taken from profile B-B′ of this study. The major features are the equatorial high with sedimentation rates of over 15mm/1000 yrs, which falls off rapidly with distance north from the equator, the low rates be-

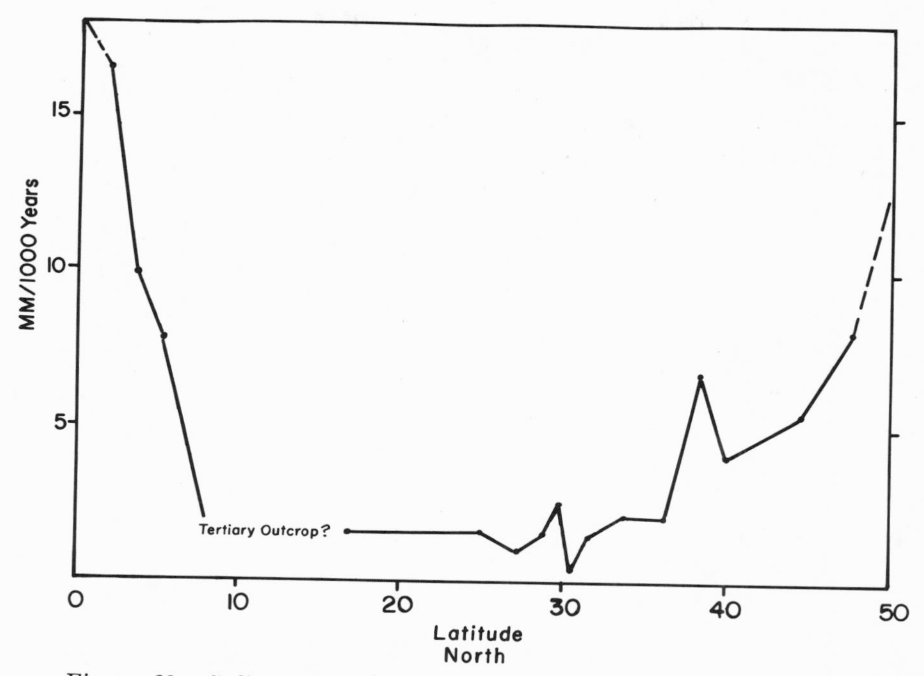

Figure 23. Sedimentation rate profile showing the variation of sedimentation rate with latitude. The profile runs along 140° W. long. to 45° N. lat. then along profile C′-C.

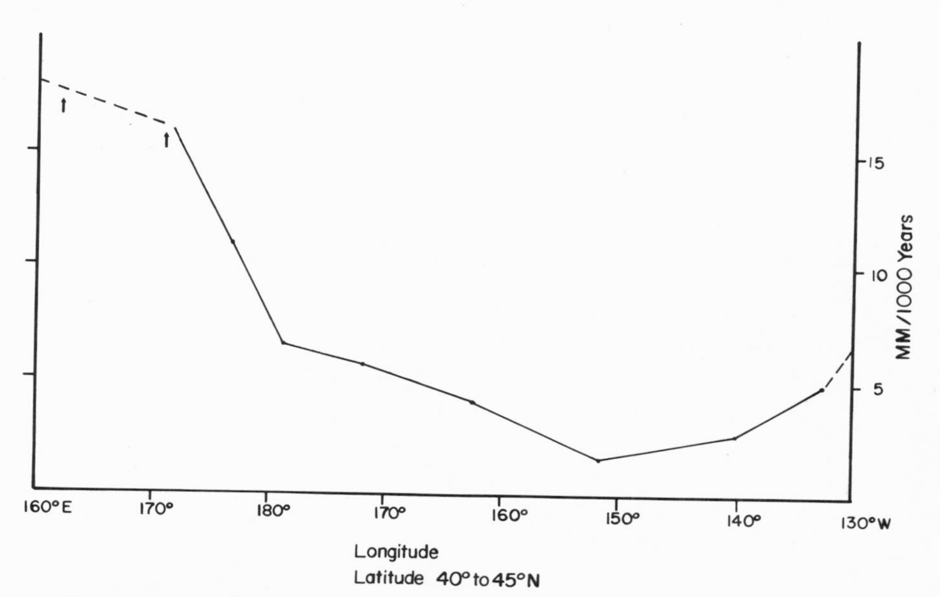

Figure 24. Sedimentation rate profile illustrating variation with longitude. The profile runs between 40° and 45° N. from 130° E. long. Where cores fail to penetrate the Brunhes-Matuyama boundary, a minimum rate is shown by an arrow.

tween 8° and 35° N. lat., and the steady rise in rates of sedimentation north of 35° N.

The east-west profile shown in Figure 24 is a composite profile constructed from cores between 40° and 45° N. lat. between 130° W. long. and 160° E. long. This profile shows a mid-Pacific minimum at about 150° W. long., rather than a steady increase in rates of sedimentation west of 180°. Two long cores west of 170° E. fail to penetrate to the Brunhes-Matuyama boundary. The arrows show the minimum rates; the true rates may be much in excess of the minimum shown.

RATES OF SEDIMENTATION AND AGE OF THE PACIFIC BASIN

The greatest thickness of sediment measured in the Pacific by Ewing and others (1968) is about 400 to 500 m near the Shatsky Rise. Thicker sections of sediment exist in other areas, such as the Aleutian abyssal plain discussed earlier, the region on the equator discussed by Hays and others (1969), and the area near the continental margin off the west coast of the United States and the east coast of Asia. However, the region near the Shatsky Rise contains what is probably the greatest thickness of pelagic sediments. Outcrops in the region have at least a Cretaceous age (Ewing and others, 1966). The lowest rates of sedimentation that one can reasonably ascribe to the North Pacific is about 1mm/1000 yrs, and in many areas is shown to be much higher. In all areas where layer 2 outcrops or has been drilled by the JOIDES operation, it has been shown to be basalt. Therefore, it is reasonable to believe that layer 2 in the Pacific is basalt and that the sediment above it represents the total accumulation. Using a minimum rate of sedimentation of 1mm/1000 yrs, one would not expect the oldest sediment to be older than 400 or 500 m.y. old, or lower Paleozoic. (We do not suggest that the sediments are this old but we doubt they can be older than this.) If the Pacific Ocean has been in its present position, unchanged since the Precambrian, one would expect an order of magnitude of more sediment to be present. We conclude, therefore, that the sediment distribution and known rates of sedimentation overwhelmingly favor the modern theories of sea-floor spreading and are contrary to a hypothesis of a static ocean floor.

ACKNOWLEDGMENTS

The writers wish to thank the Director, Dr. M. Ewing, and the Chief Scientists on the Lamont-Doherty ships, without whose help this study could not have been done. They also wish to thank Mrs. K. Henry for her assistance in the preparation of the manuscript, and Dr. G. Dickson for the thermo-magnetic determinations. Doctors J. Hays and D. Hayes read the manuscript and made many helpful suggestions. This work was carried out under National Science Foundation grants GA-824, GA-1193, GA-558, GA-861, and grant N-00014-67-A-0108-0004 of the Office of Naval Research.

REFERENCES CITED

Berggren, W. A., Phillips J. D., Bertels, A., and Wall, D., 1967, Late Pliocene-Pleistocene stratigraphy in deep-sea cores from the south-central North Atlantic: Nature, v. 216, p. 253-254.

Bon Hommet, N., and Babkine, J., 1967, Sur la presence d' aimantations inversees dans la chaine des Puys: Acad. Sci. Comptes Rendus, v. 264, Ser. B, p. 92.

Bramlette, M. N., 1961, Pelagic sediments, *in* Sears, Mary, *Editor,* Oceanography: Am. Assoc. Adv. Sci. Pub. 67, p. 345-366.

Conolly, J. R., and Ewing, M., 1970, Ice rafted detritus in northwest Pacific deep-sea sediments: Geol. Soc. America Mem. 126, p. 219-231.

Cox, Allan, 1968, Lengths of geomagnetic polarity intervals: Jour. Geophys. Research, v. 73, no. 10, p. 3247-3260.

——1969, Geomagnetic reversals: Science, v. 163, no. 3864, p. 237-245.

Cox, Allan, and Dalrymple, G. Brent, 1967, Geomagnetic polarity epochs: Nunivak Islands, Alaska: Earth and Planetary Sci. Letters, v. 3, p. 173-177.

Cox, A., Doell, R. R., and Dalrymple, G. B., 1963, Geomagnetic polarity epochs and Pleistocene geochronometry: Nature , v. 198,p. 1049-1051.

——1968, Radiometric time-scale for geomagnetic reversals: Geol. Soc. London Quart. Jour., v. 124, p. 53-66.

Creer, K. M., 1962, On the origin of the magnetization of red beds: Jour. Geomagnetism and Geoelectricity, v. XIII, nos. 3, 4, p. 86-100.

Dickson, G. O., and Foster, J. H., 1966, The magnetic stratigraphy of a deep sea core from the North Pacific Ocean: Earth and Planetary Sci. Letters, v. 1, p. 458-462.

Donahue, J., 1970, The stratigraphy and distribution of North Pacific diatoms: Geol. Soc. America Mem. 126, p.121-138.

Dymond, J., 1969, Age determination of deep-sea sediments: A comparison of three methods: Earth and Planetary Sci. Letters, v. 6, p. 9-14.

Evernden, J. F., and Curtis, G. H., 1965, The potassium-argon dating of late Cenozoic rocks in East Africa and Italy: Current Anthropology, v. 6, no. 4, p. 343-385.

Ewing, J., Ewing, M., Aitken, T., and Ludwig, W. J., 1968, North Pacific sediment layers measured by seismic profiling, *in* Knopoff, L., Drake, C. L., and Hart, P. J., *Editors,* the Crust and Upper Mantle of the Pacific Area: Am. Geophys. Union Mono. 12, p. 522.

Ewing, M., Saito, T., Ewing, J., and Burckle, L. H., 1966, Lower Cretaceous sediments from the northwest Pacific: Science, v. 152, p. 751-755.

Foster, J. H., 1966, A paleomagnetic spinner magnetometer using flux-gate gradiometer: Earth and Planetary Sci. Letters, v. 1, p. 463-466.

Foster, J. H., and Opdyke, N. D., 1968, Evidence for the existence of the Kaena and Mammoth events in deep-sea cores (abs.): Am. Geophys. Union, v. 49, no. 1, p. 156.

Goldburg, E., and Koide, M., 1962, Geochronological studies of deep-sea sediments by the ionium/thorium method: Geochim. et Cosmochim. Acta, v. 26, p. 417-450.

Gordon, A. C., and Gerard, R., 1970, North Pacific bottom potential temperature: Geol. Soc. America Mem. 126, p. 23-39.

Grim, P. J., and Naugler, F. P., 1968, A fossil deep-sea channel on the Aleutian abyssal plain (abs.): Geol. Soc. America Spec. Paper **121**, p. 118-119.

Gromme, C. S., and Hay, R. L., 1963, Magnetization of basalt of Bed I, Olduvai Gorge, Tanganyika: Nature, v. 200, p. 560-561.

——1967, Geomagnetic polarity epochs: new data from Olduvai Gorge, Tanganyika: Earth and Planetary Sci. Letters, v. 2, p. 111-115.

Hamilton, E. L., 1967, Marine geology of abyssal plains in the Gulf of Alaska: Jour. Geophys. Research, v. 72, p. 4189-4214.

Harrison, C. G. A., 1966, The paleomagnetism of deep-sea sediments: Jour. Geophys. Research, v. 71, no. 12, p. 3037-3043.

Harrison, C. G. A., and Somayajulu, B. L. K., 1966, Behaviour of the Earth's magnetic field during a reversal: Nature, v. 212, p. 1193-1195.

Hays, J., 1970, The stratigraphy and evolutionary trends of Radiolaria in North Pacific deep-sea sediments: Geol. Soc. America Mem. 126, p. 185-218.

Hays, J. D., and Ninkovich, D., 1970, North Pacific deep-sea ash chronology and age of present Aleutian underthrusting: Geol. Soc. America Mem. 126, p. 263-290.

Hays, J. D., and Opdyke, N. D., 1967, Antarctic Radiolaria, magnetic reversals and climatic change: Science, v. 158, p. 1001-1011.

Hays, J. D., Saito, T., Opdyke, N. D., and Burckle, L. H., 1969, Pliocene-Pleistocene sediments of the equatorial Pacific—their paleomagnetic, biostratigraphic and climatic record: Geol. Soc. America Bull., v. 80, p. 1481-1514.

Horn, D. R., Horn, B. M., and Delach, M. N., 1970, Sedimentary provinces of the North Pacific: Geol. Soc. America Mem. 126, p. 1-21.

Ku, T., Broecker, W., and Opdyke, N. D., 1968, Comparison of sedimentation rates measures by paleomagnetic and the ionium methods of age determination: Earth and Planetary Sci. Letters, v. 4, p. 1-16.

McDougall, I., and Chamalaun, F. H., 1966, Geomagnetic polarity scale of time: Nature, v. 212, p. 1415-1418.

McDougall, I., and Tarling, D. H., 1963, Dating of polarity zones in the Hawaiian Islands: Nature, v. 200, p. 54-56.

McDougall, I., and Wensink, H., 1966, Paleomagnetism and geochronology of the Pliocene-Pleistocene lavas in Iceland: Earth and Planetary Sci. Letters, v. 1, p. 232-236.

Ninkovich, D., Opdyke, N. D., Heezen, B. C., and Foster, J. H., 1966, Paleomagnetic stratigraphy, rates of deposition and tephrachronology in

North Pacific deep-sea sediments: Earth and Planetary Sci. Letters, p. 476-492.

Opdyke, N. D., 1969, The Jaramillo event as detected in oceanic cores, *in* Runcorn, S. K., *Editor,* The Application of Modern Physics to the Earth and Planetary Interiors: Wiley-Interscience, London, p. 549-554.

Opdyke, N. D., Glass, B., Hays, J. D., and Foster, J., 1966, Paleomagnetic study of Antarctic deep-sea cores: Science, v. 154, p. 349-357.

Pitman, W. C., and Heirtzler, J. R., 1966, Magnetic anomalies over the Pacific Antarctic ridge: Science, v. 154, p. 1164.

Smith, J. D., and Foster, J. H., 1969, A short geomagnetic reversal in the Brunhes Normal Polarity Epoch: Science, v. 163, p. 565-567.

Vine, F. J., 1968, Magnetic anomalies associated with mid-ocean ridges, *in* Phinney, R. A., *Editor,* The History of the Earth's Crust, A Symposium: Princeton Univ. Press, Princeton, N. J., p. 61-72.

Watkins, N. D., 1968, Short period geomagnetic polarity events in deep-sea sedimentary core: Earth and Planetary Sci. Letters, v. 4, no. 5, p. 341-349.

LAMONT-DOHERTY GEOLOGICAL OBSERVATORY CONTRIBUTION NO. 1491
PRESENT ADDRESS: (FOSTER) THE OHIO STATE UNIVERSITY, COLUMBUS, OHIO
MANUSCRIPT RECEIVED APRIL 10, 1969
REVISED MANUSCRIPT RECEIVED JULY 18, 1969

THE GEOLOGICAL SOCIETY OF AMERICA, INC.
MEMOIR 126, 1970

Pleistocene Diatoms as Climatic Indicators in North Pacific Sediments

JESSIE G. DONAHUE

Lamont-Doherty Geological Observatory
of Columbia University
Palisades, New York

ABSTRACT

Three biostratigraphic zones in the late Pliocene-Pleistocene sediments of the North Pacific are defined using diatoms with limited ranges. The extinction of one species is examined; in five out of six cores it occurs at the lower boundary of the Jaramillo event (.87 to .89 m.y.B.P.). Diatom temperature values (Td values) based on warm and cold water species are calculated so that low Td values indicate lower temperature. The Td values reveal two major divisions within the Pleistocene: 2 m.y. to 700,000 years B.P. is a time of relatively stable Td values; 700,000 years B.P. to the Recent has a lower average Td value than the 2 m.y. to 700,000-year interval and shows greater fluctuations in the Td curve. One core contains nine Td value minima in the last 700,000 years.

CONTENTS

Figure

Plate

Table

INTRODUCTION

Studies of diatoms from North Pacific sediment cores have recently been carried out by several investigators. Kanaya and Koizumi (1966) studied 86 core tops from all latitudes in the North Pacific and determined geographic distributions of groups of species. They studied the diatom stratigraphy of one core, V20-130, in the northwest Pacific and plotted a diatom temperature curve for this core.

Jousé (1962) published a monograph summarizing much of her earlier work and gave new data on diatoms from the Bering Sea, the Sea of Okhotsk, and the northwest part of the Pacific. She presented charts on geographic distributions of neritic and littoral diatoms in the Sea of Okhotsk and along the Kamchatka Peninsula, as well as summarizing her findings on diatom stratigraphy. She finds five horizons based largely on relative abundances of warm and cold water species of diatoms.

Kozlova and Mukhina (1967), in a broad regional study of the Pacific Ocean, compared species composition of diatoms in the plankton and the sediments. They sampled the water column from the surface to the bottom to determine the effects of solution. They concluded that the distribution of species in the bottom sediments closely reflects their distribution in the plankton, with the exception of weakly silicified forms which were dissolved before deposition.

The present study was conducted to add to the knowledge of diatom stratigraphy in deep-sea cores from the North Pacific. Long cores are available in the Lamont-Doherty Geological Observatory collection, and time control through magnetic stratigraphy greatly improves chances for accurate correlation from core to core. The role of diatoms in defining stratigraphic zones and as climate indicators is developed and strengthened.

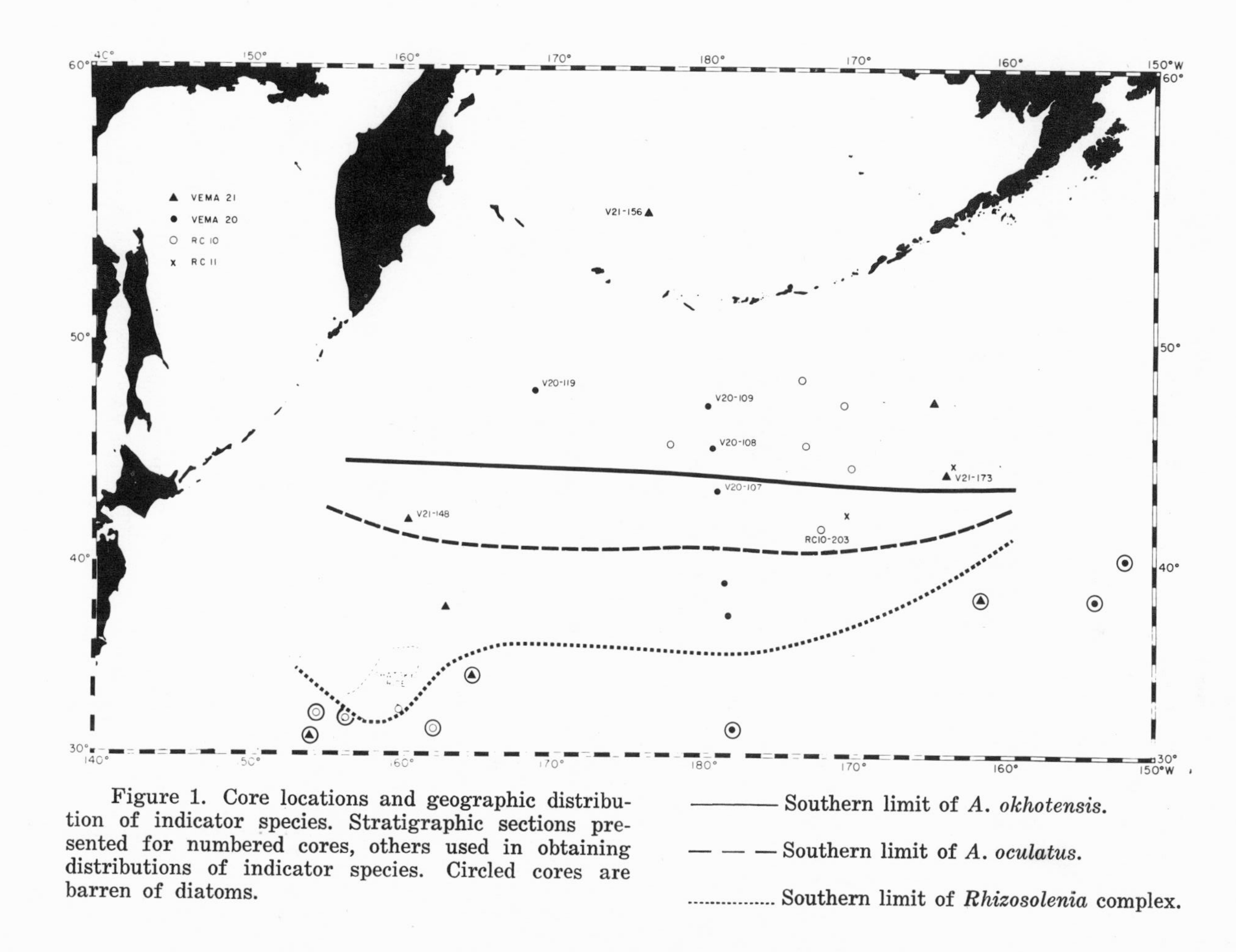

Figure 1. Core locations and geographic distribution of indicator species. Stratigraphic sections presented for numbered cores, others used in obtaining distributions of indicator species. Circled cores are barren of diatoms.

——— Southern limit of *A. okhotensis*.

— — — Southern limit of *A. oculatus*.

.... Southern limit of *Rhizosolenia* complex.

STRATIGRAPHY

The oldest sediments examined were not included in a zone. They are upper Pliocene in age and are characterized by *Coscinodiscus mukhinae* (Mukhina) Donahue (Pl. 3, a-f) and by abundant *Rhizosolenia curvirostris* var. *inermis* Jousé (Pl. 1. b-c). Three stratigraphic zones are defined in sequence from oldest to youngest.

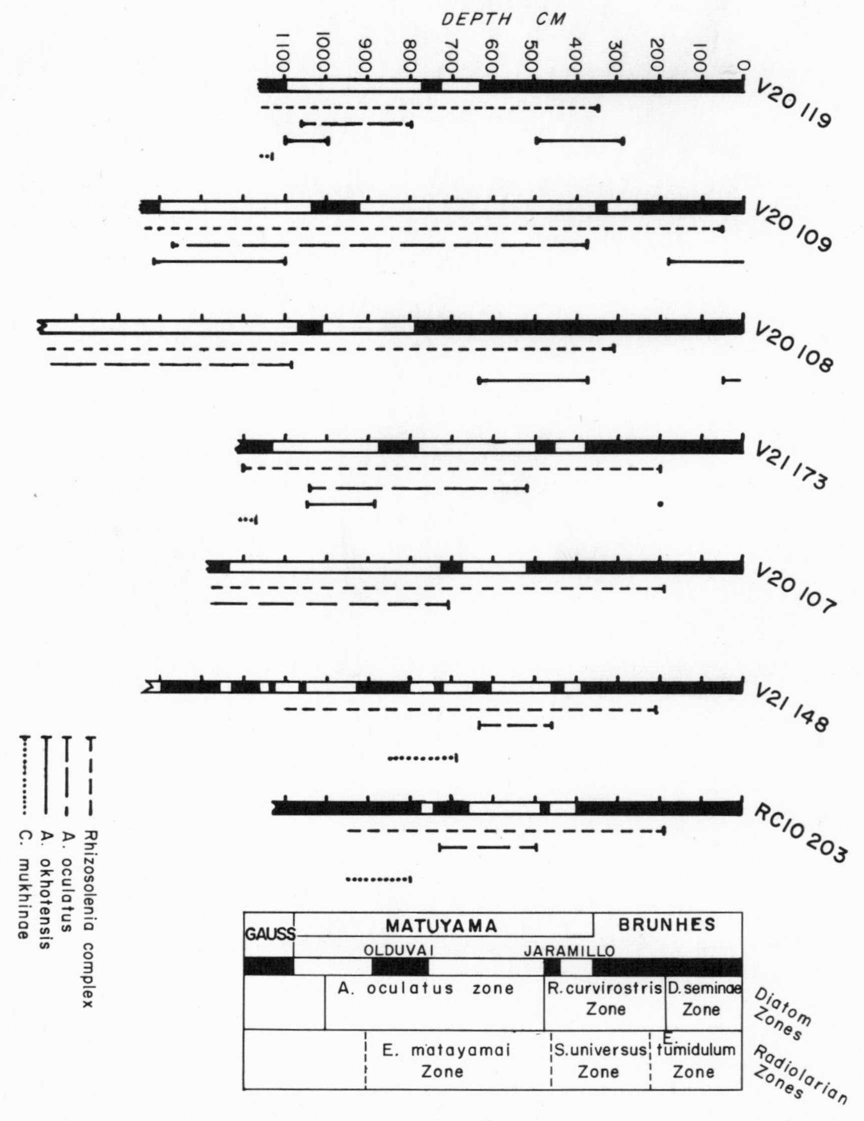

Figure 2. Stratigraphic ranges of indicator species and comparison of diatom and radiolarian zones.

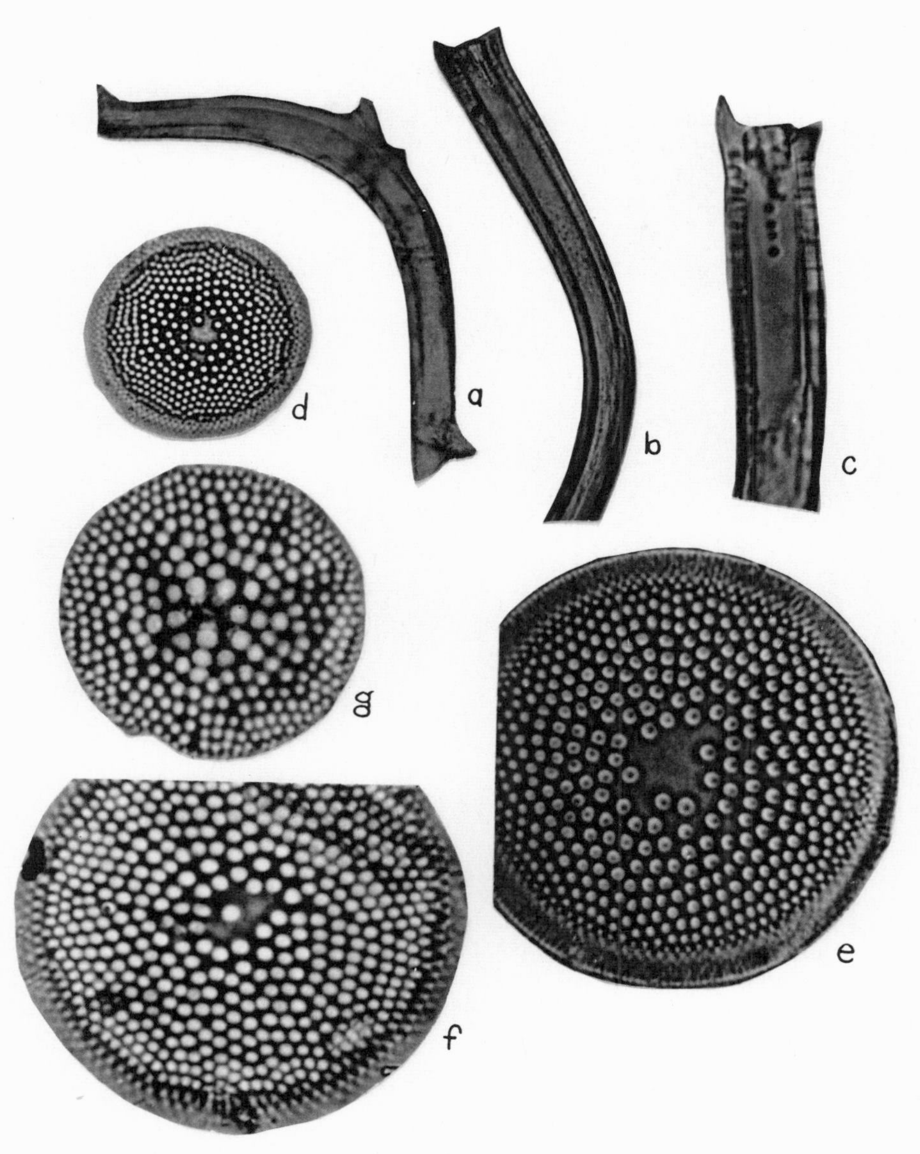

SPECIES USED TO DEFINE STRATIGRAPHIC ZONES

(a) *R. curvirostris,* V21-156 1098 cm, w=13μ, 1=71μ.
(b) *R. curvirostris* var. *inermis,* RC10-203 723 cm, w=14μ, 1=100μ.
(c) *R. curvirostris* var. *inermis,* V20-119 1137 cm, w=10μ.
(d) *A. oculatus,* V20-119 888 cm, diameter 60μ.
(e) *A. oculatus,* RC10-203 686 cm, diameter 50μ.
(f) *A. oculatus,* V20-109 440 cm, diameter 48μ.
(g) *A. oculatus,* V20-109 440 cm, diameter 33μ.

DONAHUE, PLATE 1

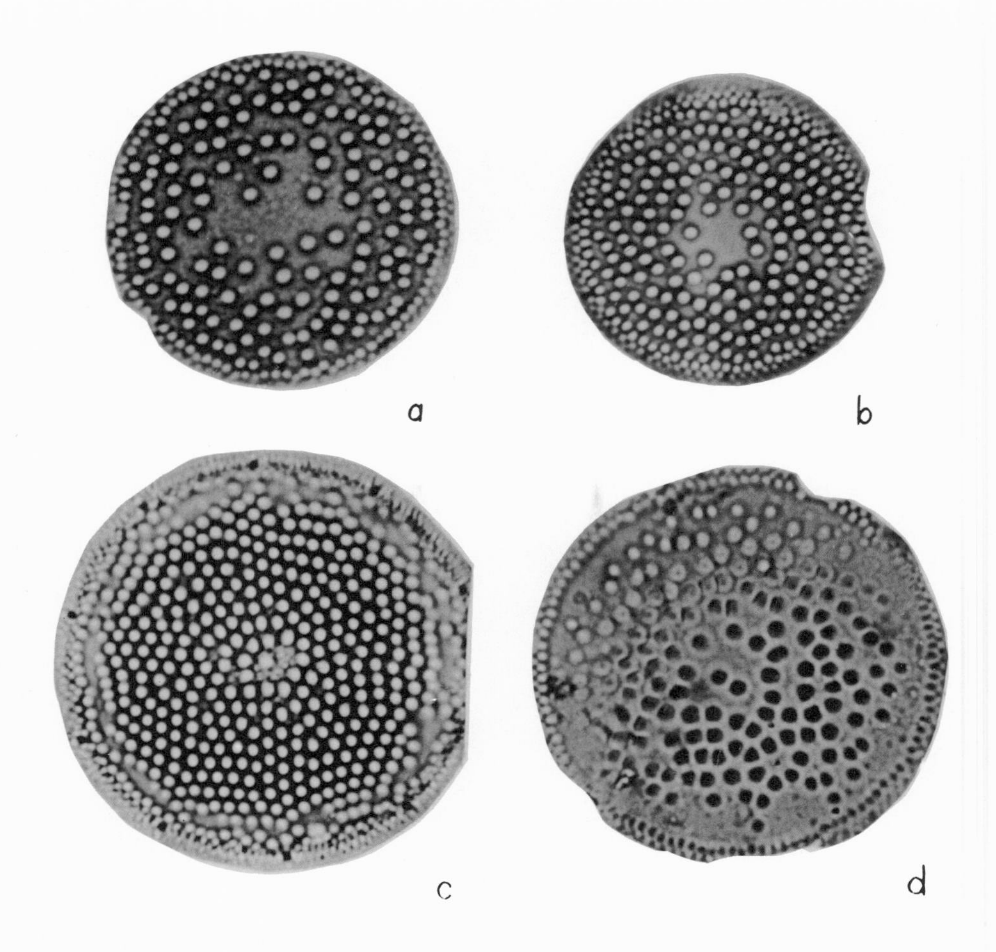

SPECIES USED AS CLIMATIC INDICATOR

(a) *A. okhotensis,* V20-119 1048 cm, diameter 36μ.
(b) *A. okhotensis,* V20-119 1048 cm, diameter 34μ.
(c) *A. okhotensis,* V20-119 151 cm, diameter 46μ.
(d) *A. okhotensis,* V20-109 1100 cm, diameter 44μ.

DONAHUE, PLATE 2
Geological Society of America Memoir 126

SPECIES USED TO IDENTIFY UPPER PLIOCENE SEDIMENTS→

(a) *C. mukhinae,* V20-119 1137 cm, diameter 37μ (focus on valve center).
(b) same as (a), (focus on valve face).
(c) same as (a), (focus on valve margin).
(d) *C. mukhinae,* V20-119 1137 cm, diameter 36μ, (focus on valve center).
(e) same as (d), (focus on valve face).
(f) same as (d), (focus on valve margin).

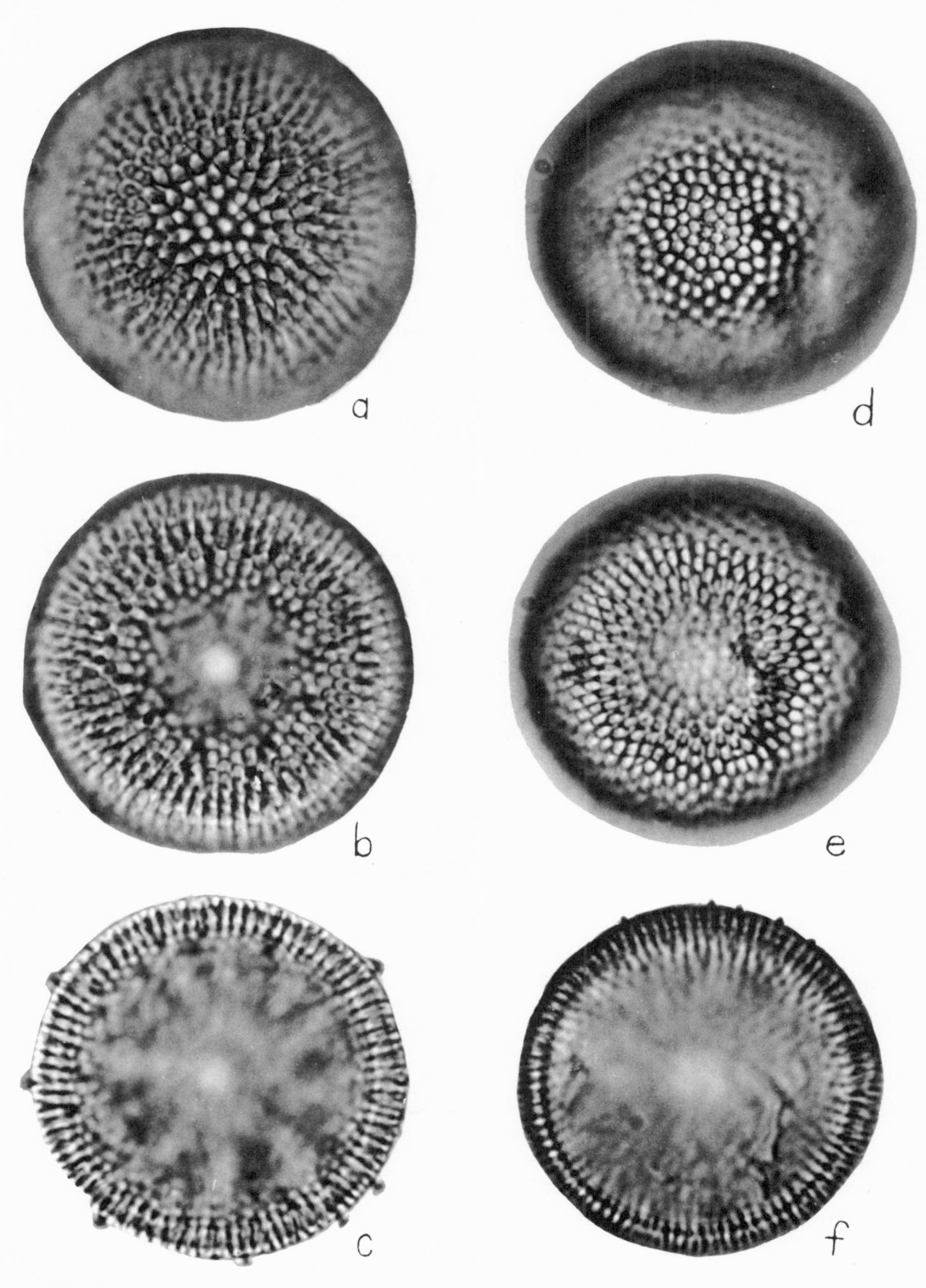

SPECIES USED TO IDENTIFY UPPER PLIOCENE SEDIMENTS

***Actinocyclus oculatus* Zone**

The oldest zone is defined by the geologic range of *Actinocyclus oculatus* Jousé (Pl. 1, d-g); uppermost Pliocene to middle Pleistocene in age. It corresponds to the *Eucyrtidium matuyamai* Zone and the upper portion of the *Lamprocyclas heteroporos* Zone of Hays (1970, this volume). Also present throughout the zone are *Rhizosolenia curvirostris* Jousé (Pl. 1, a), and *R. curvirostris* var. *inermis*. *Actinocyclus okhotensis* Jousé (Pl. 2, a-d) is found in the lower half of the zone.

***Rhizosolenia curvirostris* Zone**

The overlying zone is defined by the occurrence of *R. curvirostris* and *R. curvirostris* var. *inermis* (hereafter referred to as the *Rhizosolenia* complex), subsequent to the extinction of *A. oculatus;* middle to upper Pleistocene in age. It corresponds to the *Stylatractus universus* Zone and the lower one-third of the *Eucyrtidium tumidulum* Zone (both of Hays, 1970, this volume). The species *R. curvirostris* first appears in the lower Pleistocene and increases in abundance to about 33 percent of the *Rhizosolenia* complex near the top of the *R. curvirostris* Zone. *R. curvirostris* var. *inermis* occurs in the present material from the upper Pliocene to the top of the *R. curvirostris* Zone. *A. okhotensis* is found in the upper half of this zone.

***Denticula seminae* Zone**

The youngest zone is defined by the occurrence of *Denticula seminae* Simonsen & Kanaya subsequent to the extinction of the *Rhizosolenia* complex; late Pleistocene to Recent in age. It corresponds to the upper two-thirds of the *E. tumidulum* Zone of Hays (1970, this volume), and is characterized by abundant *D. seminae,* with *A. okhotensis* occurring near the top.

EXTINCTION OF *Actinocyclus oculatus*

The disappearance of *A. oculatus* was studied in detail by counting 100 diatoms at 10 cm intervals in six cores through the last occurrence of this species. The data are plotted against depth in the core (Fig. 3a) and against time (Fig. 3b), assuming .69 m.y. = the age of the lower boundary of the Brunhes epoch, and .92 m.y. = the age of the lower boundary of the Jaramillo event (Opdyke and Foster, 1970, this volume).

The exponential decline in abundance just prior to the last occurrence of this species is most reasonably explained by vertical mixing of burrowing organisms. Berger and Heath (1968) proposed a theoretical curve describing the upward decrease in abundance in a deep-sea core of a species that disappears suddenly and is a con-

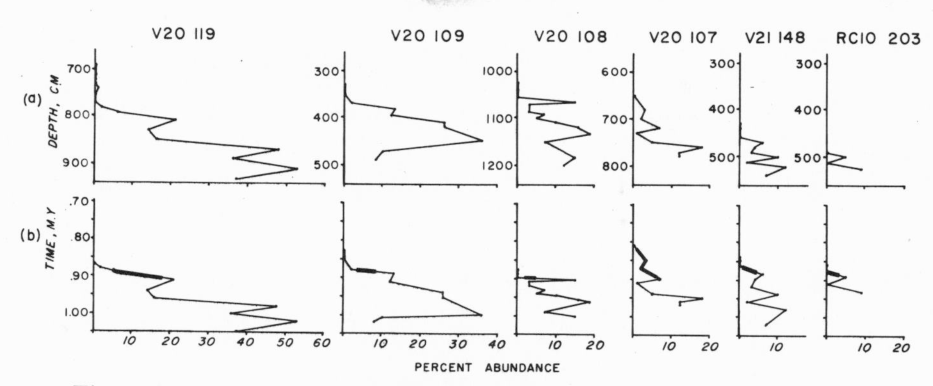

Figure 3. Curves of final decrease in abundance of *A. oculatus*: (a) plotted against depth, cm; (b) plotted against time, m.y. The heavy black bar represents the limits between which the extinction point falls.

stant percent of the sediment during its geologic range. They suggest that the point of extinction is represented in the sediment by the level at which the species is 37 percent of its maximum. One of the basic assumptions of Berger and Heath (1968), namely, that the abundance of the species in question remains constant, is not met. Fluctuations in abundance of *A. oculatus* through its geologic range are probably caused by climatic fluctuations. Therefore, instead of calculating a point at which the species became extinct, upper and lower limits were determined between which the species *probably* became extinct. The upper limit was defined as 37 percent of the maximum abundance of *A. oculatus* in a zone 100 cm or less beneath the final decrease in abundance. The lower limit was similarly defined as 37 percent of the minimum abundance within this same zone. These limits are plotted on Figure 3a. In five cores, the extinction point occurs between .87 and .91 m.y.B.P., which is just at or above the lower boundary of the Jaramillo event. In core V20-107, the age of extinction falls in the range .82 to .90 m.y.B.P., which is younger than the other values.

EXTINCTION OF THE *Rhizosolenia* COMPLEX

The age of the extinction point of the *Rhizosolenia* complex[1] is less consistent and varies between .45 m.y.B.P. and .25 m.y.B.P. Three factors may contribute to this erratic pattern: (1) Both *R. curvirostris* and *R. curvirostris* var. *inermis* were counted together. The present study suggests that *R. curvirostris* was more abundant than *R. curvirostris* var. *inermis* in higher latitudes. If the extinctions of *R. curvirostris* and *R. curvirostris* var. *inermis* were not synchronous, plotting the two together could result in a greater spread

[1]This age has since been more accurately determined at about .26 m.y.B.P.

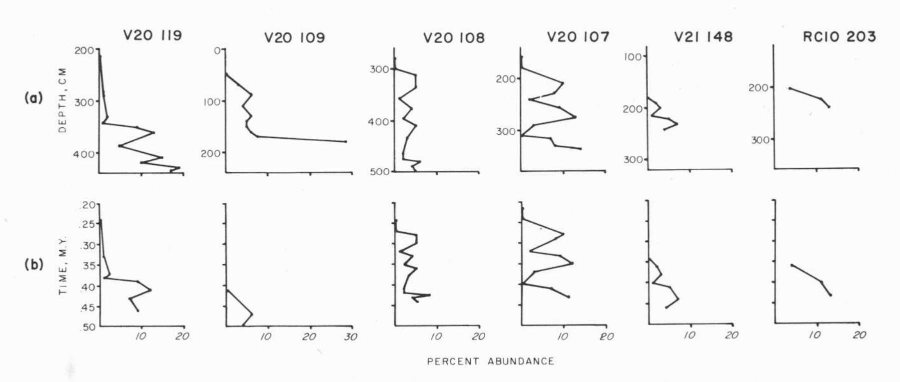

Figure 4. Curves of final decrease in abundance of *Rhizosolenia* complex: (a) plotted against depth, cm; (b) plotted against time, m.y.

in the age of extinction point from core to core. (2) The extinction of the *Rhizosolenia* complex in the middle of the Brunhes normal epoch results in less exact control by magnetic stratigraphy. Unconformities and changes in rates of sedimentation would be difficult to detect, and could occur either above or below the extinction point. (3) The extinction of the *Rhizosolenia* complex may have been relatively slow, with individuals persisting much longer in some areas than others.

DIATOM TEMPERATURE CURVES

A diatom temperature value (Td) can be calculated for any sediment sample containing diatoms by using the following formula:

$$Td = \frac{Xw \cdot 100}{Xw + Xc}$$

where

Xw = the total number of a selected group of warm water species,
Xc = the total number of a selected group of cold water species.

The resulting number can vary from zero (when Xw = 0) to 100 (when Xc = 0).

The thanatocoenoses of diatom species in core tops in the North Pacific have been studied by Kanaya and Koizumi (1966). The two groups of species, Xw and Xc, have been selected using Kanaya and Koizumi (1966) as a guide. The cold water taxa are those species common in the sediments underlying the subarctic water mass, while the warm water taxa are those from the sediments underlying the North Pacific central water mass. All species are extant.

The Xc group consists of:

**Denticula seminae* Simonsen & Kanaya
**Rhizosolenia hebetata forma hiemalis* Gran
†*Coscinodiscus marginatus* Ehrenberg

Actinocyclus okhotensis Jousé
Actinocyclus curvatulus Janisch
Actinocyclus divisus (Grunow) Hustedt
Thalassiosira gravida Cleve
Thalassiosira spp.

The Xw group consists of:

**Thalassiosira oestrupii* (Ostenfeld) Proskina-Lavrenko
**Nitzschia marina* Grunow
†*Coscinodiscus nodulifer* A. Schmidt
†*Hemidiscus cuneiformis* Wallich
Coscinodiscus radiatus Ehrenberg
Coscinodiscus lineatus Ehrenberg
Pseudoeunotia doliolus (Wallich) Grunow

*These species are consistently abundant in the Pleistocene samples but are much less abundant in the Pliocene samples.

†These species occur throughout the section but are consistently abundant in the Pliocene samples.

Counts of 250 individuals were made from strewn slides and were used to obtain the numbers of Xw and Xc. Xw + Xc varied between 35 and 100 but were generally greater than 50. Each species used in either group appeared at least eight times on at least one slide; 8 out of 250 is the smallest number significantly different from zero (Dixon and Massey, 1957). In this way, groups are made as comprehensive as possible.

Three cores were chosen on which to calculate Td values. V21-148 is characterized by a mixture of both southern and northern floras and has a long magnetic stratigraphy; the details of climatic fluctuations may be masked, but the general trends are revealed.

V20-108 is a long core with a short stratigraphic section. The same sampling interval in this core, as in V21-148, will reveal more detail of shifts in Td values, particularly in the important time interval from 1 m.y. ago to the present. Also, since this core is farther north than V21-148, smaller climatic fluctuations may have affected the diatom flora. The third core, V20-119, was selected because it is the core farthest north in the study.

INTERPRETATION OF Td VALUES

From the bottom to the top of core V21-148, the general trend is toward lower Td values. Three divisions can be made on the basis of the Td values. The oldest division is from 1100 cm to 740 cm (Fig. 5). This division of V21-148 must be interpreted with care as the Td value is based on abundances of cold and warm water taxa determined from a study of core tops. The older floras, such as those from 1100 to 740 cm, may contain different species which

Table 1. Core Locations and Depth in CM to Ranges of Indicator Species

Core No.	Lat	Long	Depth in CM to: *Rhizosolenia* Complex	*A. oculatus*	*A. okhotensis*	*C. mukhinae*
V20-119	47°57′N	168°47′E	from: 346 to: bottom	from: 798 to: 1053	from: 273 to: 485 from: 1001 to: 1098	from: 1127 to: bottom
V20-109	47°19′N	179°39′W	from: 55 to: bottom	from: 375 to: 1365	from: 0 to: 180 from: 1090 to: 1430	
V20-108	45°27′N	179°14′W	from: 290 to: bottom	from: 1060 to: bottom	from: 10 to: 60 from: 365 to: 645	
V21-173	44°22′N	163°33′W	from: 195 to: bottom	from: 516 to: 1029	few at 200 from: 890 to: 1029	from: 1175 to: bottom
V20-107	43°24′N	178°52′W	from: 190 to: bottom	from: 608 to: bottom		
V21-148	42°05′N	160°36′E	from: 185 to: 1100	from: 465 to: 635		from: 690 to: 880
RC10-203	41°42′N	171°57′W	from: 190 to: 950	from: 495 to: 733		from: 799 to: 950

are temperature indicators but are not included in the groups used for the Xw and Xc calculations (Kanaya and Koizumi, 1966). The high Td values, averaging about 90, obtained for this section of V21-148 are the result of an almost total lack of recent cold water indicators from the samples. *C. marginatus* is the only species present from the cold water group. Two of the warm water indicators used in calculating Xw are abundant (*H. cuneiformis, C. nodulifer*), but they are not the same species most common higher in the core (*T. oestrupi*(*i*), *N. marina*).

Six Pliocene species disappear between 760 and 720 cm in V21-148 and suggest a hiatus in this interval. The Td values drop sharply at this point (*see* Fig. 5) due to the appearance in abundance of cool water Pleistocene species (particularly *D. seminae, R. hebetata f. heimalis*). The flora from 720 cm to the top of V21-148 contains all the taxa from both the Xw and the Xc groups, although not every species was present on every slide. The introduction of the cool water species used in compiling group Xc gives more reliability to the interpretation of the Td values. The Td values in V21-148 from 720 cm to 385 cm average 67. The relief of the curve is greater than below 720 cm, and the values range from 32 to 93. The average number of Xw counted is 61 and the average Xc is 30. Both groups

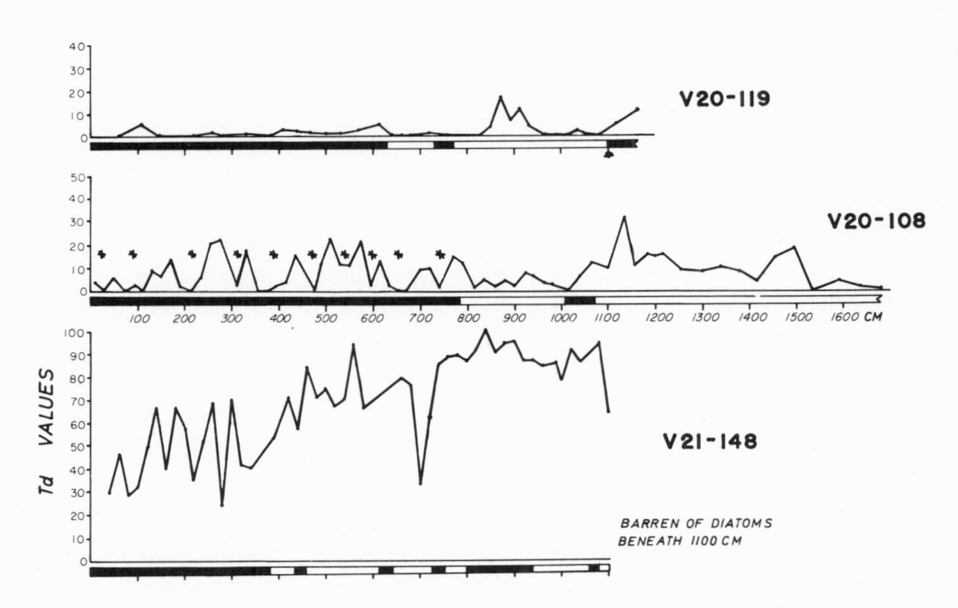

Figure 5. Td values down V21-148, V20-108, and V20-119. Note the three divisions of V21-148 and the general decrease of Td values from core bottom to top. The significant Td minima in V20-108 are marked with an asterisk; a Chi square test for proportions at the 95% confidence level differentiated them from the intervening peaks.

vary but the greater variations occur in the numbers of the Xw while the numbers of Xc are more stable. These shifts in Td values may represent climatic variations.

The third and youngest division of V21-148 is from 385 cm to the top of the core. Here the Td values average 47 and vary from 24 to 68. Compared to the underlying section, the average number of Xw counted has decreased from 61 to 47, while the average number of Xc counted has increased from 30 to 51. In this section, the great relief seen in the Td curve is caused by about equal variations in both the Xw and the Xc groups. The Td values suggest a decrease in average temperature, with greater temperature variations during the last 700,000 years.

The two younger divisions seen in V21-148 are also defined in V20-108. The core section from 1680 cm to about 1000 cm (Fig. 5) shows Td values which average 10, with generally low relief. The section from 1000 cm to 800 cm shows an average Td value of 4 but still does not show sharp fluctuations. This drop in average Td values from 10 to 4 is a result of small shifts in both the Xw and the Xc counts. Xw dropped from an average value of 4 to 2, while Xc increased from 37 to 44.

The average Td value in the younger division of V20-108 (800 to 0 cm) is 7. The average Xw is 3, while the average Xc is 43. The most marked change in this division is the increase in the relief of the curve which may be interpreted as indicating a less stable climate.

Core V20-108 is ideal for examining the fluctuations of Td values seen in the last 700,000 years (800 to 0 cm). It has been suggested by Hays and others (1969) that eight climatic coolings occurred during the last 700,000 years. Examination of the Td curve for V20-108 during that time interval reveals nine minima in which the Td value drops to between 0 and 3 (Fig. 5). These nine minima were selected using a Chi square test for proportions as being significantly different from the intervening peaks at the 95 percent confidence level. There are only six Td minima in this time interval in V21-148 but it is likely that some part of the section is missing.

Td values for core V20-119 remained so low throughout the entire section that divisions or fluctuations were not recognized. A warming is suggested in the lower part of the core (950 to 850 cm).

It is therefore suggested that the changes seen in the diatom temperature curves for V21-148 and V20-108 reflect alternate cooling and warming of the surface water. As a result, the diatom flora in the plankton changes; cool water species increase in abundance with respect to warm water species during drops of temperature.

TAXONOMIC NOTES

Actinocyclus oculatus Jousé 1968, Pl. 2, figs. 6, 7

Valve slightly convex, ranging in diameter from 24 to 65μ (average 34μ). Areolae arranged in curving fascicles with from 10 to 18 (average 12) fascicles on valve face. Each fascicle composed of rows of areolae parallel to a first, longest row. Small spine located at the marginal end of longest row of areolae in each fascicle. These spines difficult to see. Central hyaline area present, usually 1 to 8μ (average 4μ) in longest diameter. Areolae near center of valve may be slightly scattered with hyaline spaces between them; areolae become more closely packed half the radial distance to the margin. Areolae size ranges from 4 to 6 in 10μ for first circle around central hyaline area, 7 to 8 in 10μ for areolae across valve face, to 8 to 9 in 10μ for areolae nearest valve margin. Margin is 1 to 2μ wide and radially striated with 12 to 15 striae in 10μ. One or two concentric cross bars connect each set of radial striae forming brick work pattern. Pseudo-nodule a cluster of several small areolae located just inside margin and often difficult to see.

Geographic Distribution. *A. oculatus* has been observed only in the North Pacific, where it is a cold water form. It is abundant in core V21-156 in the Bering Sea and south to about 42° to 43° north lat. It is not present south of 40° north.

Geologic Range. Middle Miocene (Koizumi, 1968) to middle Pleistocene.

Actinocyclus okhotensis Jousé 1968, Pl. 2, figs. 2-5

Valve flat, ranging in diameter from 21 to 48μ (average 32μ). Areolae hexagonal, ranging in size from 4 to 8 in 10μ (average 7) in valve center to 6 to 8 in 10μ (average 7) at valve margin. Central hyaline area present which contains no areolae, 2.5 to 12μ (average 6.8μ) in diameter; hyaline spaces present between areolae to valve margin. May be irregular hyaline ring around marginal edge of valve. Areolae may be in irregular fascicles with rows parallel to a first, longest row. When pores very scattered in hyaline valve surface, fascicles not seen. Spiral sculpture always seen, most pronounced near valve margin. Three concentric circles of closely spaced, small areolae 8 to 10 in 10μ (average 9), occur on marginal edge of valve. Spines present in outermost circle of small areolae. Margin about 1.5 to 2μ wide, radially striated, 14 to 15 striae in 10μ. Pseudo-nodule present in outer two sub-marginal rows of small areolae.

Geographic Distribution. North Pacific Ocean, north of 45° north. Not seen in core V21-156 in the Bering Sea.

Geologic Range. Middle Miocene (Koizumi, 1968) to Recent.

Remarks. The margin and the outer two rows of small areolae, including the spines and the pseudo-nodule, are very frequently lost from the valve.

Rhizosolenia curvirostris Jousé 1968, Pl. 3, fig. 2 (not figs. 1, 3)

A hollow, curving, cylindrical, siliceous tube, with one end (the apex) closed and terminated by two small spines. Other end always broken; a complete specimen never seen by author. Tube bent at angles varying from 80° to 160°. Size of tube ranges in width between individuals, from 7 to 14μ (average 11μ), and in length from 47 to 79μ (average 66μ). (See fig. 6 for location of measurements, based on 20 specimens.) Thickness of tube wall variable, but two thicknesses predominate; a thick-walled form (2 to 3μ), and a thin-walled form (about 1μ), which may be the result of solution. Wall in cross section displays parallel lineations perpendicular to wall surface. Small, flat triangular spine present on surface of tube on convex side of bend and points toward apex. In some specimens, puncta about 1μ in diameter were observed near the apex of the tube.

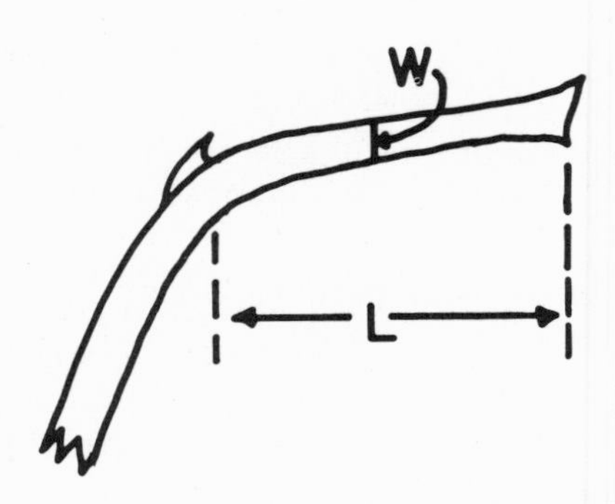

Figure 6. Location of length and width measurements on *R. curvirostris* and *R. curvirostris* var. *inermis*.

Geographic Distribution. *R. curvirostris* found from Bering Sea in core V21-156 to 30° north lat. Most abundant in northern part of its range and becomes rare south of 40° north.

Geologic Range. Lower Pleistocene to upper Pleistocene.

Remarks. These specimens are similar to *R. alata* var. *curvirostris* Gran; since no whole specimen has been seen, the author considers the affinities of this object with the Genus *Rhizosolenia* tentative.

Rhizosolenia curvirostris var. *inermis* Jousé (1968, written commun.)

The form is identical with *R. curvirostris,* except that it lacks the triangular spine and is larger. The form ranges in width between individuals from 9 to 17μ (average 13μ), and in length from 48 to 100μ (average 71μ); measurements are based on 20 specimens.

Geographic Distribution. In the North Pacific, *R. curvirostris* var. *inermis* has the same distribution as the species *R. curvirostris*. However, *R. curvirostris* var. *inermis* is bi-polar as it has been observed in Antarctic deep-sea sediments of Tertiary age (identified as *R. curvirostris* by Donahue, 1967).

Geologic Range. Miocene to upper Pleistocene.

Coscinodiscus mukhinae (Mukhina) Donahue

Thalassiosira convexa Mukhina, Mukhina, V. V., 1965. comb. nov.

Valve strongly convex, ranging in diameter from 27 to 47μ (average 37μ) and valve height from 6 to 15μ (average 10μ). Areolae hexagonal, forming closed network over valve face, ranging in size from 5 to 8 areolae in 10μ (average 6) in valve center to 6 to 8 in 10μ (average 8) at valve margin. In some specimens, aerolae in fascicles, 10 to 12 on valve face, longest row of areolae at center, parallel rows of areolae successively shorter on either side of longest row. In other specimens, areolae appear in radial rows. Margin 1.5 to 3.5μ wide (average 2.5μ) with radial striae 8 to 12 in 10μ (average 9). When fascicles present on valve face, striae also fasciculate. Inner edge of margin set with circlet of small spines projecting up and out, away from valve center. When complete, terminal end of spine spatulate or globular. Spines difficult to see because often broken off.

Geographic Distribution. Widespread in temperate and tropical Pacific Ocean.

Geologic Range. Middle Miocene to uppermost Pliocene.

ACKNOWLEDGMENTS

The author wishes to thank James Hays, Jack Donahue and Andrew McIntyre for critically reviewing the manuscript and suggesting valuable changes. Dr. Hays offered continual encouragement and advice during the preparation of the paper and Dr. Donahue gave invaluable assistance in the statistical analysis of the Td values. The work was supported by NSF Grants GA 861 and GA 1193 and by Office of Naval Research Grant N-00014-67-A-0108-0004.

REFERENCES CITED

Berger, W. H., and Heath, G. R., 1968, Vertical mixing in pelagic sediments: Jour. Marine Research, v. 26, p. 134-143.

Dixon, W. J., and Massey, F. J., Jr., 1957, Introduction to statistical analysis: 2d ed., New York, McGraw-Hill Book Co., Inc., 488 p.

Donahue, J. G., 1967, Diatoms as indicators of Pleistocene climatic fluctuations in the Pacific sector of the Southern Ocean, *in* Sears, M., *Editor*, Progress in Oceanography: Oxford, Pergamon Press, v. 4, p. 133-140.

Hays, J. D., 1970, The stratigraphy and evolutionary trends of Radiolaria in North Pacific deep-sea sediments: Geol. Soc. America Mem. 126, p. 185-218.

Hays, J. D., Saito, T., Opdyke, N., and Burckle, L., 1969, Pliocene-Pleistocene sediments of the equatorial Pacific—their paleomagnetic, biostratigraphic and climatic record: Geol. Soc. America Bull., v. 80, p. 1481-1514.

Jousé, A. P., 1962, Stratigraphic and paleogeographic studies in the northwestern part of the Pacific Ocean: Akad. Nauk. SSSR, Institut Okeanology, Moscow, 258 p. (In Russian with English abstract.)

—— 1968, Species novae Bacillariophytorum in sedimentis fundi Oceani Pacifici et Maris Ochotensis inventae: Akad. Nauk. SSSR, Botanicheskii Institut, Novitates systematicae plantarum non vascularum, p. 12-21.

Kanaya, T., and Koizumi, T., 1966, Interpretation of diatom thanatocoenoses from the North Pacific as applied to a study of core V20-130 (Studies of a deep-sea core V20-130, Part IV): Tohoku Univ. Sci. Repts., Sendai, Japan, 2d. Ser., (Geology), v. 37, p. 89-130.

Koizumi, I., 1968, Tertiary diatom flora of Oga Peninsula, Akita prefecture, northeast Japan: Tohoku Univ. Sci. Repts., Sendai, Japan, 2d. Ser., (Geology), v. 40, p. 171-240.

Kozlova, O. G., and Mukhina, V. V., 1967, Diatoms and silicoflagellates in suspension and floor sediments of the Pacific Ocean: Internat. Geology Rev., v. 9, p. 1322-1342.

Mukhina, V. V., 1965, De specibus diatomearum novis e depositibus partis aequatorialis oceani Pacifici: Akad. Nauk. SSSR, Botanicheskii Institut, Novitates systematicae plantarum non vascularium, p. 22-25.

Opdyke, N. D., and Foster, J. H., 1970, Paleomagnetism of cores from the North Pacific: Geol. Soc. America Mem. 126, p. 83-119.

LAMONT-DOHERTY GEOLOGICAL OBSERVATORY CONTRIBUTION NO. 1486
MANUSCRIPT RECEIVED APRIL 10, 1969
REVISED MANUSCRIPT RECEIVED JUNE 23, 1969

THE GEOLOGICAL SOCIETY OF AMERICA, INC.
MEMOIR 126, 1970

Radiolarian Assemblages in the North Pacific and Their Application to a Study of Quaternary Sediments in Core V20-130

CATHERINE NIGRINI

Box 779, Kanata, Ontario, Canada

ABSTRACT

Distributions of 37 species of Radiolaria are determined from 83 samples of Holocene North Pacific sediments. Recurrent group analysis of these distributions allows recognition of six assemblages which reflect present-day surface water circulation and distribution of upper water masses. Relative abundances of "cold," transitional, and "warm" water recurrent groups in Pleistocene core V20-130 (35° 59′ N., 152° 36′ E.) determine a radiolarian temperature curve which is similar to a previously determined diatom temperature curve. Two temperature maxima and three temperature minima are recognized in the core. One new species is described.

CONTENTS

Figure

Plate

Table

INTRODUCTION

In 1966, Kanaya and Koizumi published a study of modern diatom thanatocoenoses in 88 sediment samples from the North Pacific and in one long Pleistocene core (V20-130, 35° 59′ N., 152° 36′ E.). They found that the distribution of individual diatom species form patterns generally coincident with those of the upper water masses. Analysis of the distribution of 55 of these species by means of recurrent grouping (Fager, 1957; Fager and McGowan, 1963), allowed them to recognize seven distinctive assemblages in North Pacific sediments which reflect present-day surface water circulation and distribution of upper water masses. They then applied their knowledge of the assemblages to a study of a Pleistocene core (V20-130) and inferred temperature variations in surface waters during the Pleistocene. Within this core, three major temperature maxima and four (possibly five) major temperature minima were detected.

A study parallel to that of Kanaya and Koizumi using Radiolaria rather than diatoms is presented herein. Radiolaria are known to be reasonably abundant in some areas of the North Pacific (Kruglikova, 1966), although generally they are not so numerous as in tropical sediments. However, relatively little specific distributional or taxonomic work has been done on polycystine Radiolaria from this area. Bailey (1856) briefly described and illustrated 13 species found in sediments from the Sea of Kamtschatka. In a sediment sample from 20° 52′ N., 151° 50′ W., Ehrenberg (1860) listed 69 species. Nakaseko (1964) described and illustrated 19 species of Liosphaeridae and Collosphaeridae found in a sediment sample from the Japan Trench. The most extensive published taxonomic work is by Dogiel and Reshetnyak (1952) working with plankton hauls. In all, they have described 30 polycystine species from the area of the Sea of Kamtschatka. Dr. Ling, of the University of Washington (Seattle), is currently describing a number of radiolarian species from northeast Pacific sediments (north of 35° N., east of 160° W.), but his work is, as yet, unpublished.

Riedel (1958) suggested that Recent radiolarian assemblages from various latitudes might be distinguished one from another. Hays (1965) showed that there is a marked contrast between assemblages from north and south of the Antarctic Polar Front. He also observed that:

> A number of Antarctic species are confined to the ocean south of the Polar Front or are bipolar, whereas the warm-water fauna is more cosmopolitan. The Antarctic fauna is uniform in its species content throughout the area of its occurrence; warm-water fauna shows increasing diversity northward.

Nigrini (1967) found that low and middle latitude assemblages can be recognized in Indian Ocean sediments.

Although a number of radiolarian species found in sediments are cosmopolitan, most are latitudinally restricted. Their distribution appears to be controlled by surface water circulation, thus confining them to certain upper water masses (Petrushevskaya, 1966; Nigrini, 1968). However, ubiquitous species are apparently able to persist in a submerged water mass (Casey, in press). For instance, *Spongotrochus glacialis* Popofsky is a major component of the subarctic fauna. It decreases in abundance to the south and appears only very rarely in sediment samples from the tropics. However, it is present in every sample examined during the present study. As a result of this ability to dive with or follow a water mass at depth, most of the relatively few species endemic to subarctic regions are present, but in considerably decreased numbers, in tropical sediment samples. The many species endemic to tropical areas are limited in their poleward extent. During the present study, only eight species found in northern and transitional samples were entirely absent from equatorial samples, while 24 species found in equatorial samples were limited in their northward extension.

AREAL DISTRIBUTION OF RADIOLARIAN SPECIES IN THE NORTH PACIFIC

From the North Pacific, 83 samples were prepared and examined (Fig. 1; Table 1). All the samples were provided by the Scripps Institution of Oceanography. As nearly as possible, the sampling of Kanaya and Koizumi was duplicated. The samples were taken from within the upper 12 cm (usually from within the upper 6 cm) of the cores and are thought to represent Holocene sediments. Column 7 of Table 1 shows qualitatively the abundance of radiolarians in the samples relative to the non-calcareous fraction greater than 62μ.

The areal distribution of 92 species (39 Spumellaria and 53 Nassellaria) was studied. Only the presence or absence of a species in each sample was recorded and no estimates of abundance were made. Of these 92 species, 55 were either cosmopolitan over the area represented by the samples, or appeared too infrequently to be of any value in the present study. The remaining 37 species or subspecies are restricted in their distribution (Figs. 2 through 38). From visual inspection of their distribution patterns, it is immediately obvious that there is an assemblage boundary at about 40° N. between subarctic and central water masses (*compare with* Bradshaw, 1959; Fig. 4). It is also clear that an area of mixing exists in the vicinity of the Kuroshio current (35 to 40° N., 145 to 180° E.).

RECURRENT GROUPS OF RADIOLARIA

The theory of recurrent group analysis and the assumptions involved have been discussed elsewhere (Fager, 1957; Fager and McGowan, 1963; Kanaya and Koizumi, 1966). The method employs only the presence or absence of a species, and relative frequencies are not considered. A recurrent group represents the largest possible separate unit, within which all pairs of species show an affinity for each other. An affinity index of 0.5 means that pairs of species are found together in at least 50 percent of their recorded occurrences. In their use of recurrent grouping, both Fager and McGowan (1963) and Kanaya and Koizumi (1966) used an index of affinity 0.5. In order to identify a group in a particular sample, Kanaya and Koizumi found that if the ratio of species present to the total number of species in the group is approximately 0.87, a reasonable distribution pattern is produced; this criterion is followed herein.

In the present study, a recurrent group analysis was made of 37 species in 68 samples and calculations were carried out on a CDC 6400 computer. Using an index of affinity of 0.5, four groups, one of which contains only two species, were determined. When these groups were plotted at each station, a reasonable, though overly dissected pattern resulted. However, undue emphasis was placed on the two species which form one of the groups. Re-analysis using an index of affinity of 0.6 produced three groups which are thought by the author to be more meaningful and which, when plotted, form a more satisfactory and more reliable distribution pattern (Figs. 39 and 41). The groups are:

Group I. (Tropical assemblage): *Spongocore puella**, *Hymeniastrum euclidis**, *Larcospira quadrangula**, *Heliodiscus asteriscus**, *Carpocanium* spp.*, *Theocorythium trachelium trachelium*, *Lamprocyclas maritalis maritalis*, *Eucyrtidium hexagonatum*, *Pterocanium trilobum*, *Polysolenia spinosa*, *Siphonosphaera polysiphonia*, *Panartus tetrathalamus tetrathalamus*, *Polysolenia lappacea*, *Spongaster tetras tetras*, *Euchitonia furcata*, *Anthocyrtidium ophirense*, *Amphirhopalum ypsilon*, *Pterocanium praetextum praetextum*, *Lamprocyclas*

---

*These species range too far north to be meaningful in the analysis of core V20-130.

TABLE 1. CORE LOCATIONS AND INTERVALS STUDIED

No.+	Expedition Abbr. and Core No.	Lat.	Long.	Depth (m)	Interval Sampled (cm from top of core)	Radio-larian Abundance
1	MUK B12G	54°47′N	168°19′W	1960	0-2	rare
* 2	TP 40-1	53°57′N	171°11′E	1382	0-2	absent**
3	MUK B10G	53°15′N	157°02′W	4560	4-6	few
4	CK 8	53°01′N	176°15′W	3660	0-5	few
5	MUK B21G	52°32′N	141°44′W	3792	3-6	few
* 6	USSR 3403	52°14′N	160°56′E	4435	0-8	absent
7	USSR 3359	51°21′N	172°14′E	4880	0-8	rare
8	USSR 3274	51°04′N	162°17′E	5434	0-8	rare
9	CK 11	49°39′N	177°39′W	4850	4-8	few
10	USSR 4151	49°38′N	139°42′W	3970	0-5	few
11	Nth. Hol. 8	48°37′N	157°29′W	5715	0-4	common
12	USSR 4158	46°58′N	144°01′W	4658	0-3	common
13	CK 6	46°57′N	164°49′W	5094	4-8	common
14	USSR 3109	45°55′N	154°48′E	5001	0-7.5	rare
15	CUSP 11G	45°34′N	143°11′W	4700	2-4	common
16	CAS 2	45°02′N	127°13′W	2930	0-5	abundant
17	USSR 3108	44°52′N	155°42′E	4902	0-3	rare
18	CK 13-3	44°45′N	173°02′W	4835	0-4	few
19	CUSP 9G	43°58′N	140°38′W	4450	2-4	common
20	USSR 3163	43°49′N	156°38′E	5441	0-5	few
21	MUK H5G	42°41′N	142°10′W	4290	0-2	common
22	CK 4	42°30′N	162°08′W	5388	5-12	common
23	MUK B31G	42°05′N	125°39′W	2917	3-5	common
*24	Nth. Hol. 9	40°53′N	155°13′W	5343	0-5	absent
*25	Nth. Hol. 1	40°31′N	147°58′W	5370	0-6	absent
26	JYN II 10G	40°30′N	169°48′E	5550	2-4	abundant
27	JYN II 8G	40°29′N	172°33′E	4250	3-5	abundant
28	Nth. Hol. 10	40°14′N	155°55′W	5167	0-7	common
29	TP 71	40°05′N	146°45′E	5050	0-4	common
30	CK 3	39°56′N	158°38′W	5005	0-5	abundant
31	JYN II 12G	39°47′N	160°53′E	5510	0-2	abundant
32	CUSP 4G	39°30′N	125°52′W	3733	4-8	few
*33	MEN 18G	39°30′N	133°05′W	4740	7-9	absent
34	JYN II 14G	39°19′N	156°57′E	5635	0-3	abundant
*35	MEN 27G	39°05′N	139°26′W	5920	4-6	absent
36	Nth. Hol. 11	38°36′N	154°25′W	5590	0-6	common
37	JYN II 17G	38°28′N	153°10′E	5690	1-4	abundant
38	USSR 3449	38°19′N	145°38′E	5286	5-10	common
39	JYN II 6G	37°56′N	178°10′E	5250	0-2	abundant
40	JYN II 19G	37°46′N	149°49′E	5910	0-4	abundant
41	Nth. Hol. 12	37°27′N	154°08′W	5582	0-4	common
42	JYN II 20G	37°04′N	148°14′E	5730	2-4	common
43	CK 16	36°30′N	173°16′W	4195	0-12	few
*44	CK 2	35°09′N	157°17′W	5627	0-4	absent
45	USSR 4084	35°00′N	172°56′W	5971	0-3	common

TABLE 1. CORE LOCATIONS AND INTERVALS STUDIED (CONT.)

No.+	Expedition Abbr. and Core No.	Lat.	Long.	Depth (m)	Interval Sampled (cm from top of core)	Radio-larian Abundance
46	CUSP 24G	34°29′N	126°02′W	4760	0-3	common
47	ZETES III 3G	33°19′N	158°02′E	3032	2-4	abundant
*48	TP 97	32°02′N	139°25′E	1336	0-4	absent
*49	USSR 3478	31°17′N	148°50′E	6177	0-5	absent
*50	CUSP 2G	31°05′N	135°24′W	5160	0-4	absent
*51	CK 1B	29°24′N	153°06′W	5664	0-4	absent
52	JYN IV 4G	29°18′N	150°42′E	5880	0-2	abundant
*53	USSR 3493	29°04′N	142°35′E	5194	0-5	absent
54	ZETES IV 12G	28°33′N	136°17′E	4518	2-4	abundant
55	ZETES IV 11G	27°42′N	146°46′E	5696	2-4	abundant
*56	USSR 3530	27°32′N	131°32′E	4963	7-12	rare
*57	CK 22	26°22′N	168°53′W	4450	0-4	absent
*58	JYN IV 14G	22°01′N	179°51′W	5172	6-8	absent
59	MP 20-2	20°27′N	154°55′W	~5100	2-4	abundant
*60	USSR 4293	19°56′N	134°05′W	5277	0-6	absent
61	MP 24	19°45′N	166°50′W	~5200	0-4	common
62	ZAP 2G	17°51′N	109°31′W	3640	0-2	abundant
63	MSN 155G	15°09′N	137°06′W	4992	2-4	abundant
64	CHUB 4	14°02′N	125°30′W	4505	4-12	common
65	CHUB 5	13°03′N	125°29′W	4448	0-4	abundant
66	CHUB VIg	11°56′N	91°43′W	3566	10-12	abundant
67	MSN 150G	10°59′N	142°37′W	4978	0-2	common
68	CHUB XIg	10°53′N	105°09′W	3255	3-6	abundant
69	ZAP 4P	10°24′N	96°41′W	4020	0-2	abundant
70	RIS 11G	9°45′N	117°37′W	4370	0-4	abundant
71	DWHT 8C	9°03′N	129°00′W	4650	0-2	abundant
72	MSN 147G	8°07′N	145°25′W	5100	0-2	abundant
73	PROA 92PG	7°58′N	164°57′E	5219	0-2	common
74	DWBG 8	7°51′N	130°55′W	5069	8-10	abundant
75	LSDH 90PG	7°19′N	175°28′W	5190	0-3	abundant
76	JYN V 38PG	6°30′N	141°59′W	5018	4-6	abundant
77	MSN 143PG	5°32′N	146°09′W	5100	0-2	abundant
78	RIS 21G	5°32′N	103°09′W	3330	3-5	abundant
79	MSN 142G	5°20′N	146°13′W	5089	0-2	abundant
80	DWBG 149	4°08′N	115°46′W	4160	0-4	abundant
81	MSN 141G	3°32′N	146°51′W	4577	0-2	common
82	DWBG 12	3°12′N	131°31′W	4438	0-2	abundant
83	CAP 2BG1	0°43′N	169°20′E	4320	0-2	abundant

+Numbers correspond to those in Figure 1.
*Denotes samples omitted from recurrent group analysis.
**"Absent" implies absent for all practical purposes; one or two radiolarians may be present.

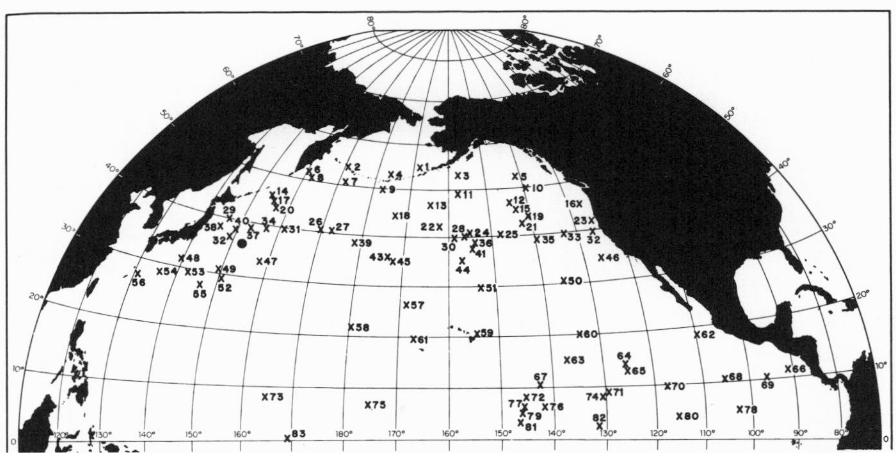

Figure 1. Locations of samples examined. Numbers correspond to the serial numbers in Table 1. Filled circle shows location of core V20-130.

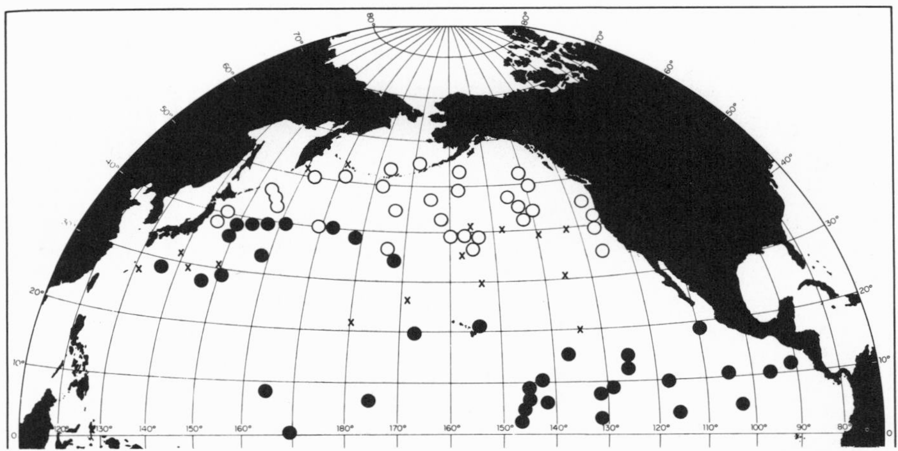

Figure 2. Distribution of *Collosphaera tuberosa*. Filled circles = present; open circles = absent; X = samples barren of Radiolaria.

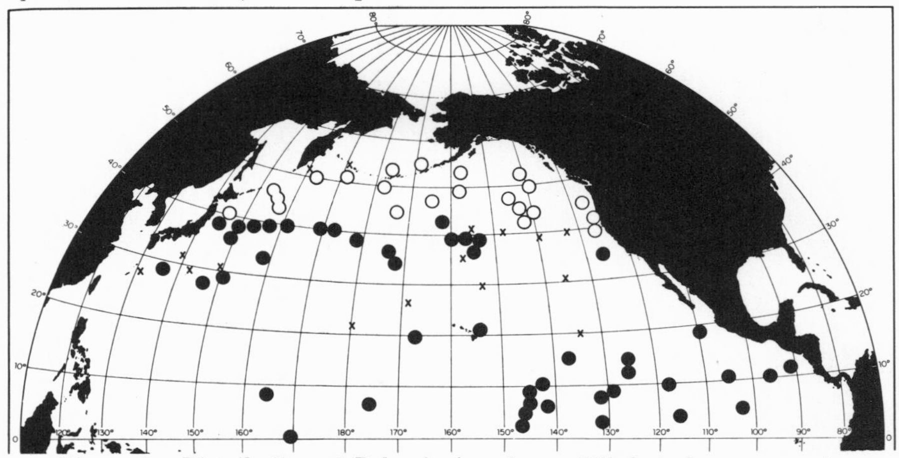

Figure 3. Distribution of *Polysolenia spinosa*. Filled circles = present; open circles = absent; X = samples barren of Radiolaria.

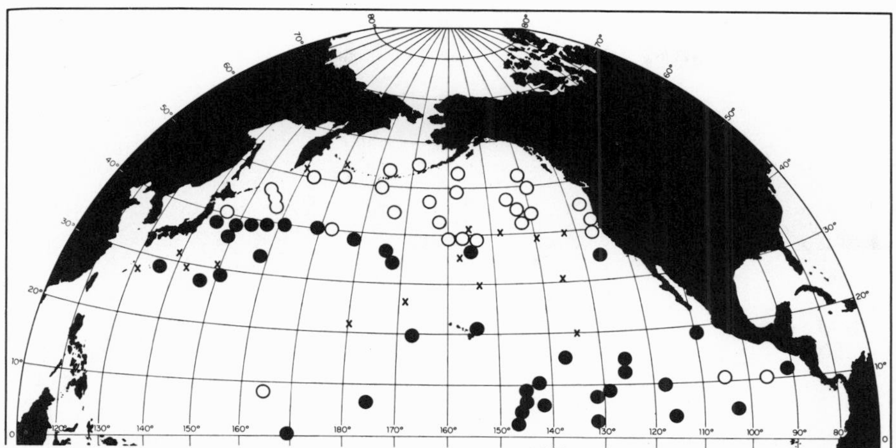

Figure 4. Distribution of *Polysolenia lappacea*. Filled circles = present; open circles = absent; X = samples barren of Radiolaria.

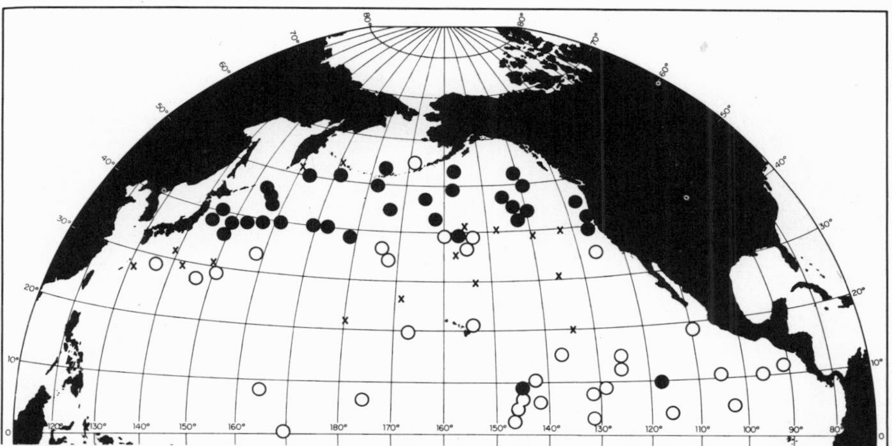

Figure 5. Distribution of *Polysolenia arktios*. Filled circles = present; open circles = absent; X = samples barren of Radiolaria.

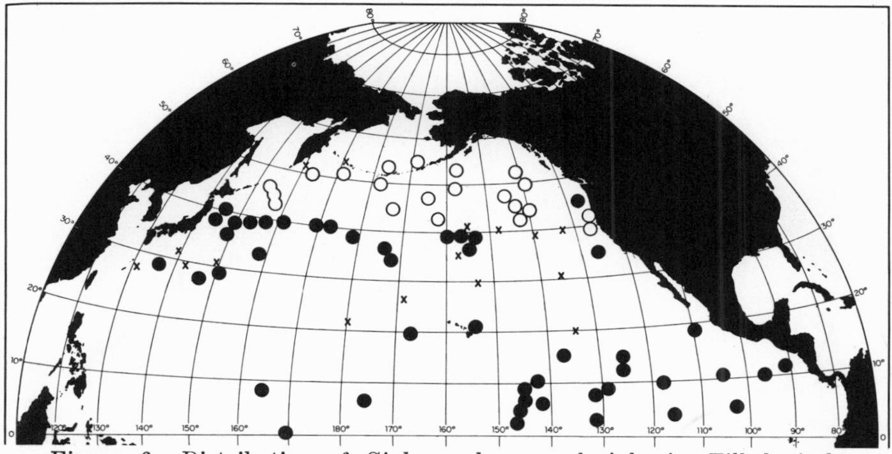

Figure 6. Distribution of *Siphonosphaera polysiphonia*. Filled circles = present; open circles = absent; X = samples barren of Radiolaria.

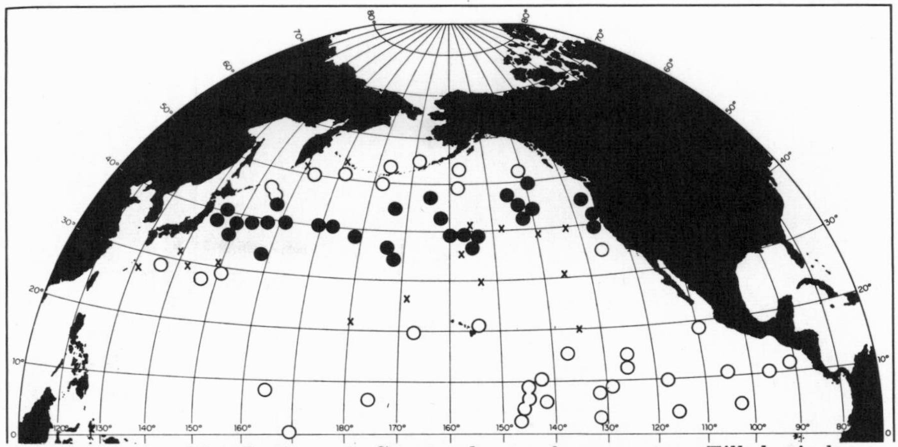

Figure 7. Distribution of *Styptosphaera ? spumacea*. Filled circles = present; open circles = absent; X = samples barren of Radiolaria.

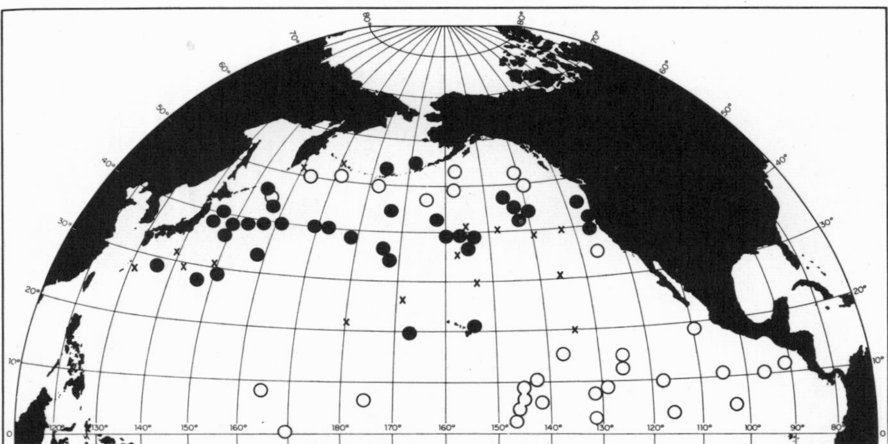

Figure 8. Distribution of *Heteracantha dentata*. Filled circles = present; open circles = absent; X = samples barren of Radiolaria.

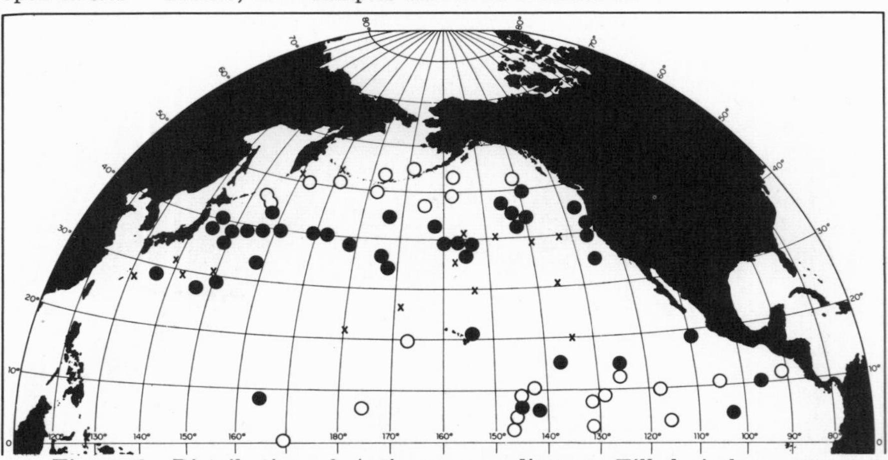

Figure 9. Distribution of *Actinomma medianum*. Filled circles = present; open circles = absent; X = samples barren of Radiolaria.

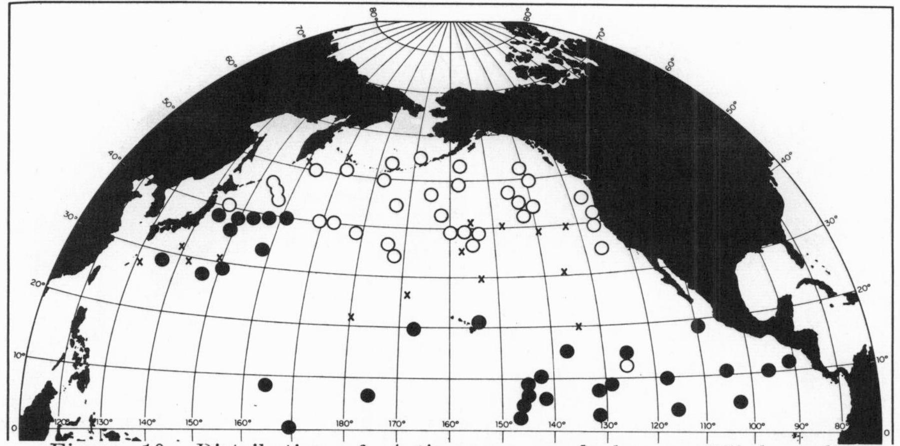

Figure 10. Distribution of *Actinomma arcadophorum*. Filled circles = present; open circles = absent; X = samples barren of Radiolaria.

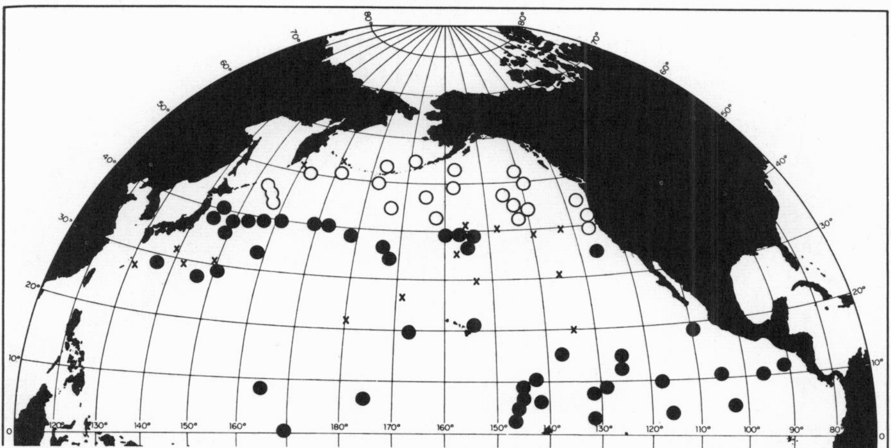

Figure 11. Distribution of *Panartus tetrathalamus tetrathalamus*. Filled circles = present; open circles = absent; X = samples barren of Radiolaria.

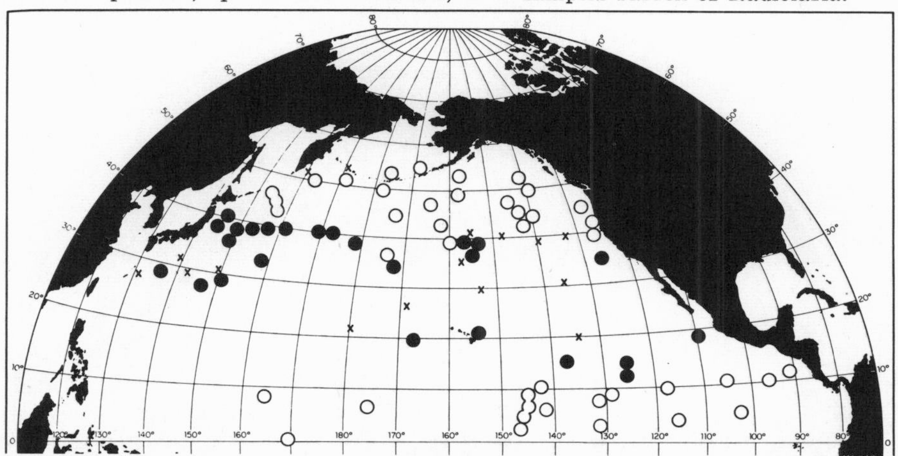

Figure 12. Distribution of *Panartus tetrathalamus coronatus*. Filled circles = present; open circles = absent; X = samples barren of Radiolaria.

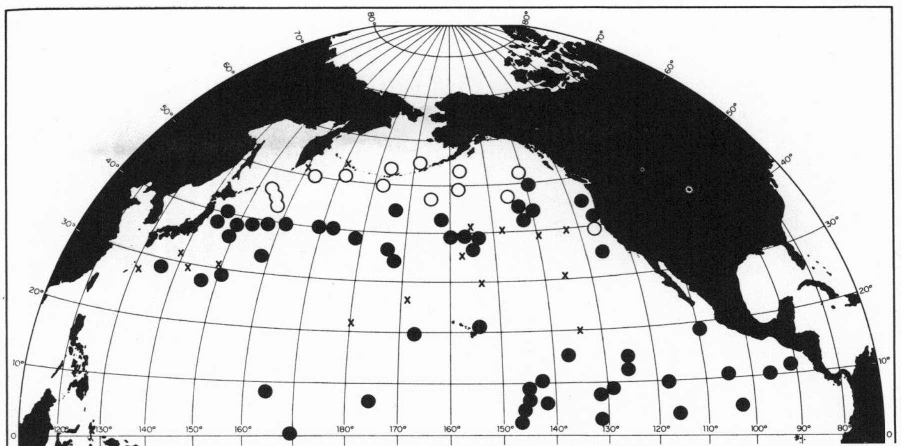

Figure 13. Distribution of *Heliodiscus asteriscus.* Filled circles = present; open circles = absent; X = samples barren of Radiolaria.

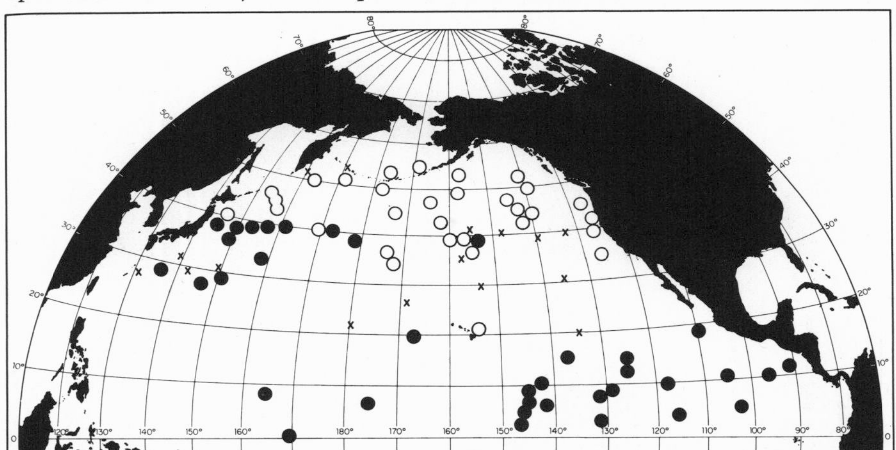

Figure 14. Distribution of *Amphirhopalum ypsilon.* Filled circles = present; open circles = absent; X = samples barren of Radiolaria.

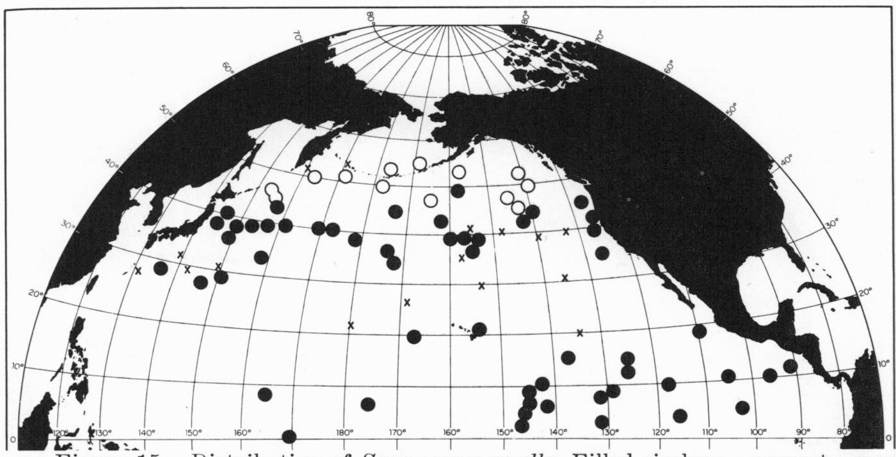

Figure 15. Distribution of *Spongocore puella.* Filled circles = present; open circles = absent; X = samples barren of Radiolaria.

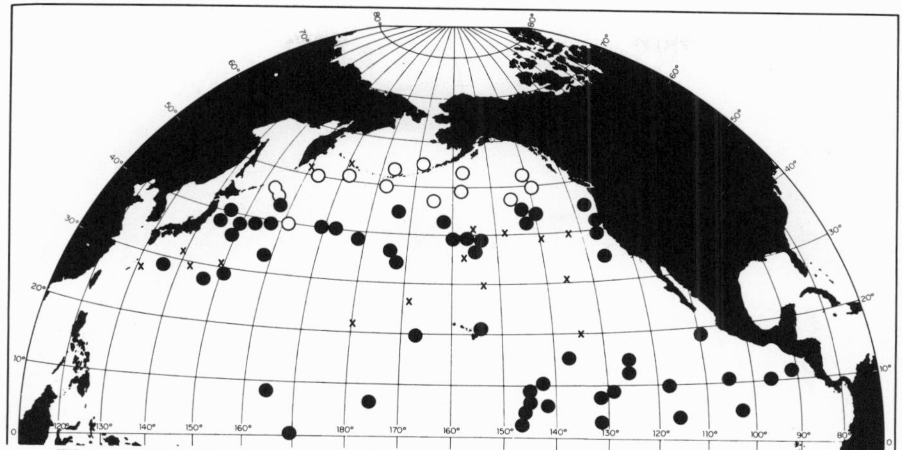

Figure 16. Distribution of *Hymeniastrum euclidis*. Filled circles = present; open circles = absent; X = samples barren of Radiolaria.

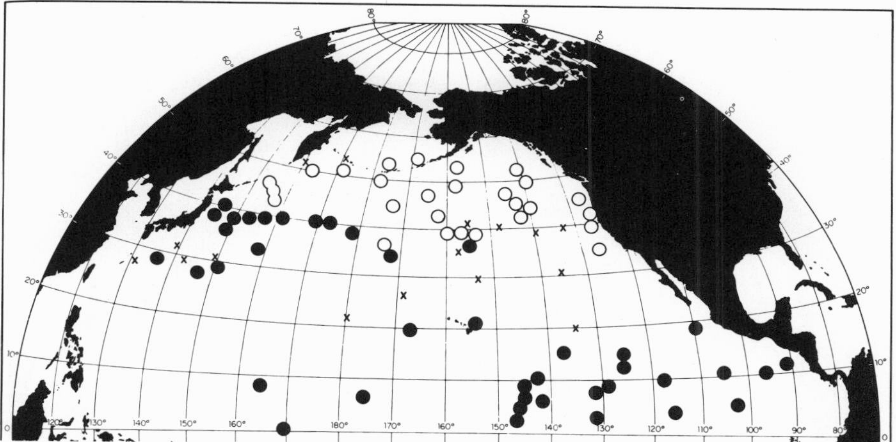

Figure 17. Distribution of *Euchitonia furcata*. Filled circles = present; open circles = absent; X = samples barren of Radiolaria.

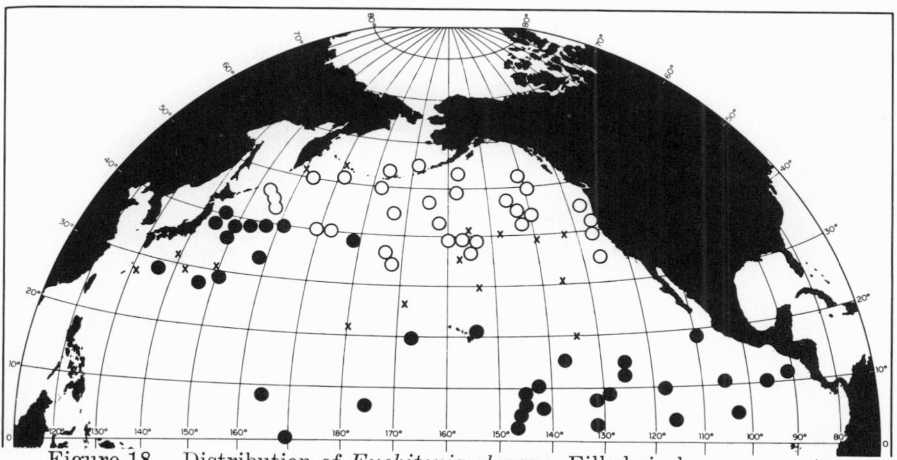

Figure 18. Distribution of *Euchitonia elegans*. Filled circles = present; open circles = absent; X = samples barren of Radiolaria.

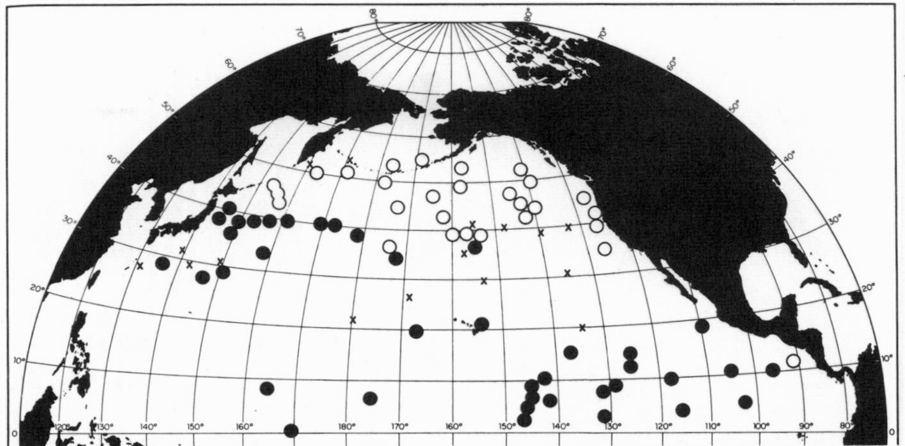

Figure 19. Distribution of *Spongaster tetras tetras*. Filled circles = present; open circles = absent; X = samples barren of Radiolaria.

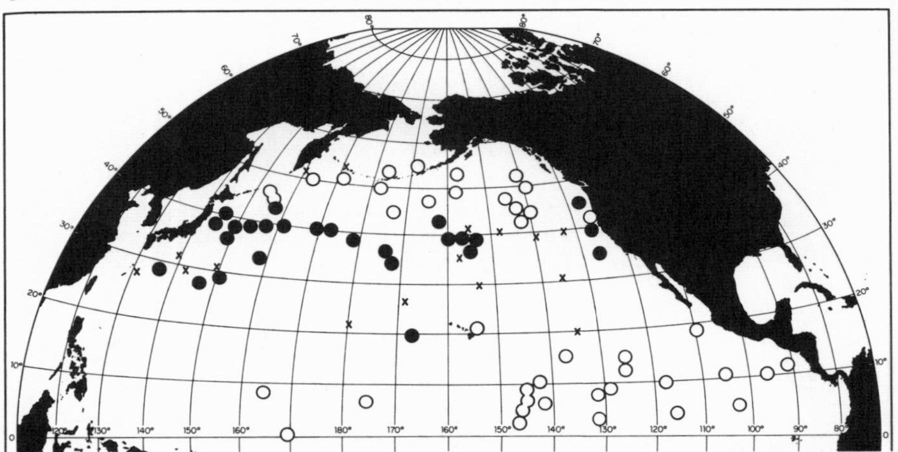

Figure 20. Distribution of *Spongaster tetras irregularis*. Filled circles = present; open circles = absent; X = samples barren of Radiolaria.

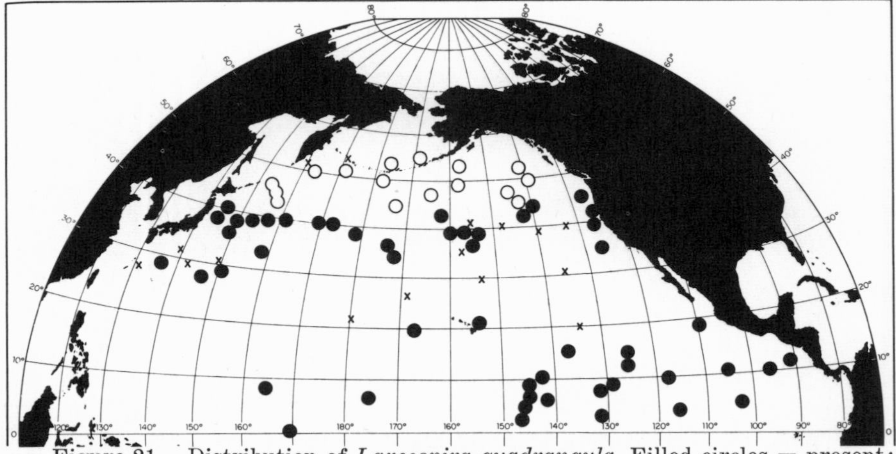

Figure 21. Distribution of *Larcospira quadrangula*. Filled circles = present; open circles = absent; X = samples barren of Radiolaria.

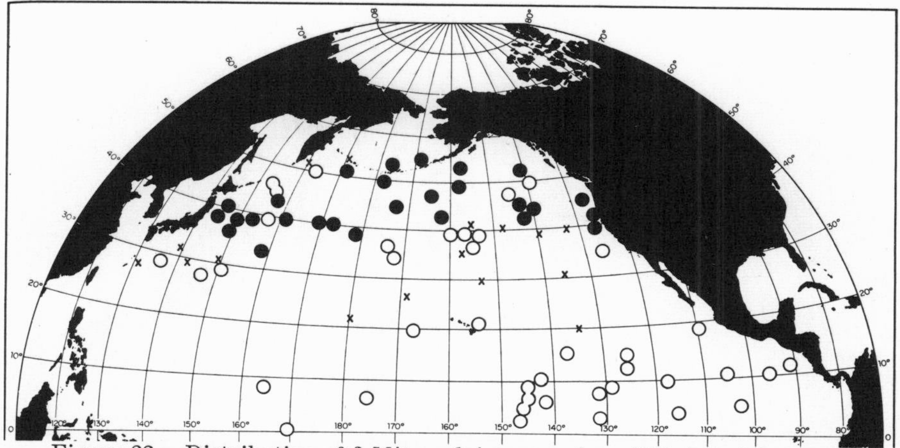

Figure 22. Distribution of ? *Mitrocalpis araneafera*. Filled circles = present; open circles = absent; X = samples barren of Radiolaria.

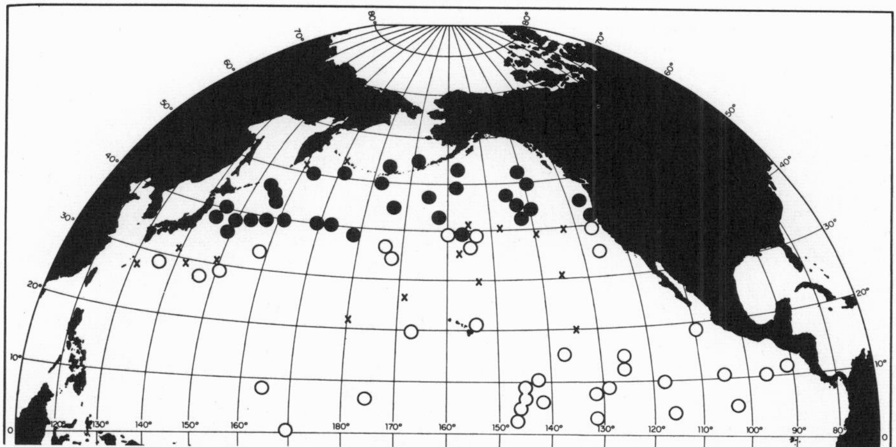

Figure 23. Distribution of *Tristylospyris* sp. Filled circles = present; open circles = absent; X = samples barren of Radiolaria.

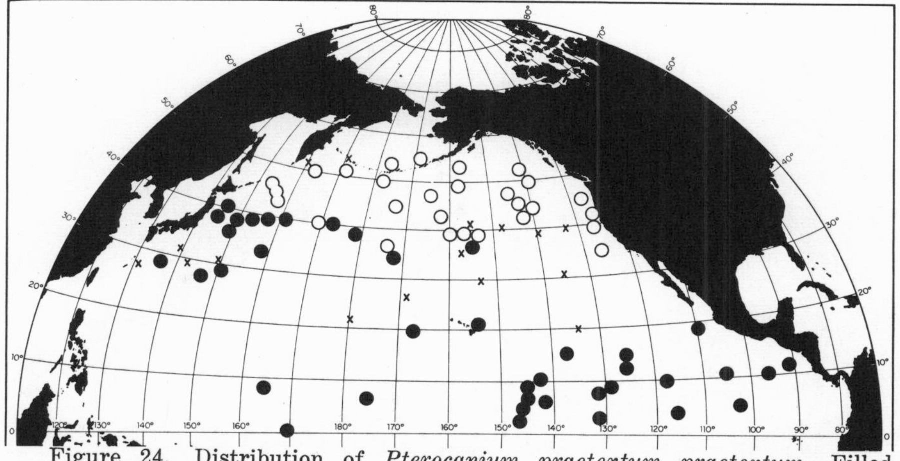

Figure 24. Distribution of *Pterocanium praetextum praetextum*. Filled circles = present; open circles = absent; X = samples barren of Radiolaria.

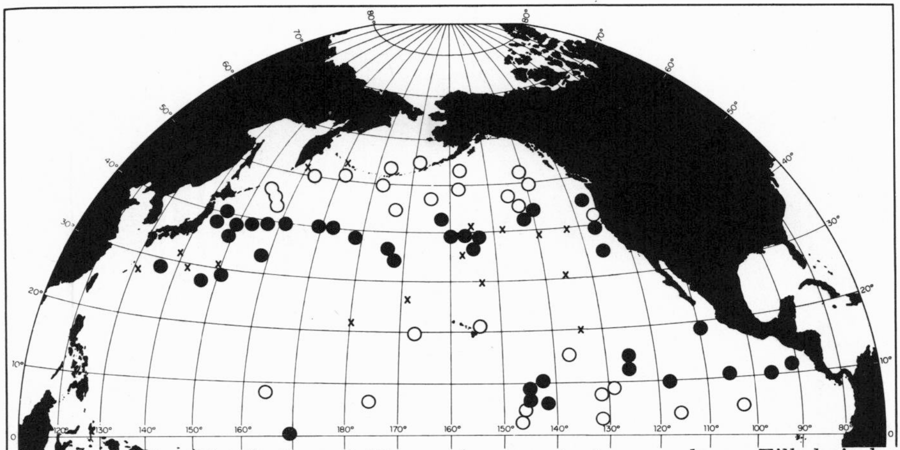

Figure 25. Distribution of *Pterocanium praetextum eucolpum.* Filled circles = present; open circles = absent; X = samples barren of Radiolaria.

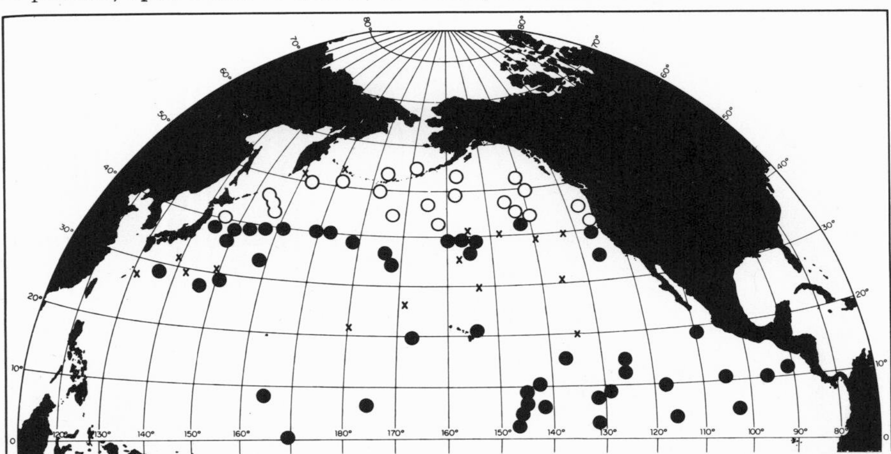

Figure 26. Distribution of *Pterocanium trilobum.* Filled circles = present; open circles = absent; X = samples barren of Radiolaria.

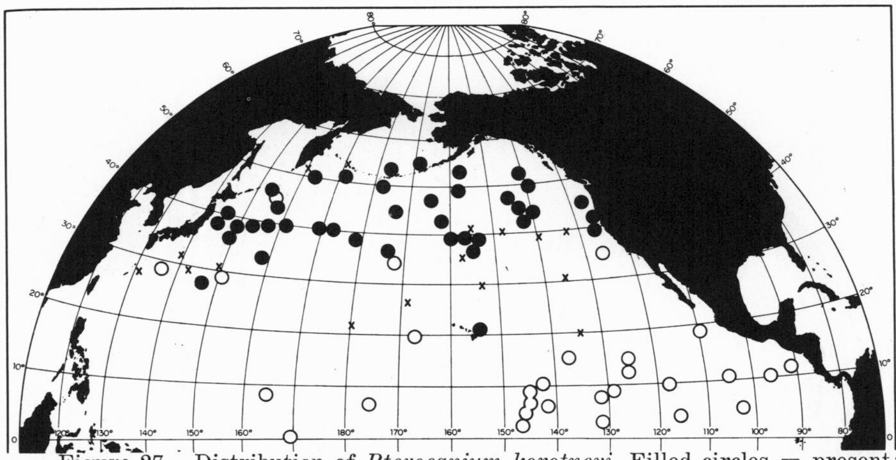

Figure 27. Distribution of *Pterocanium korotnevi.* Filled circles = present; open circles = absent; X = samples barren of Radiolaria.

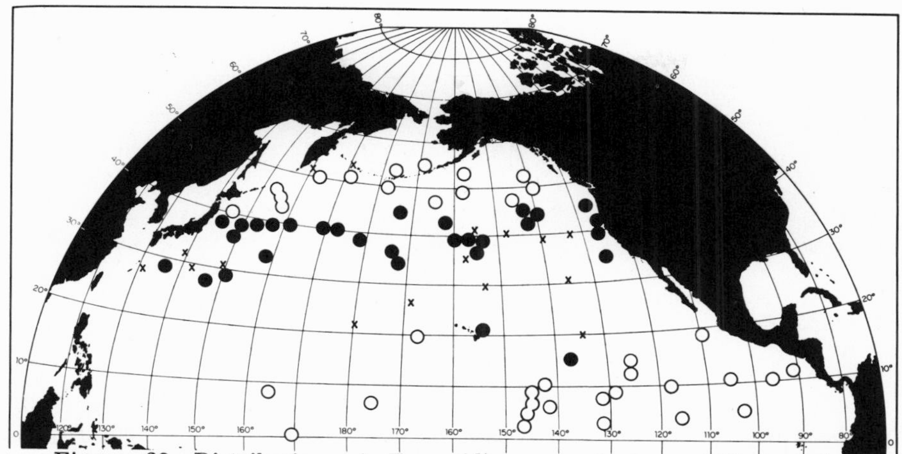

Figure 28. Distribution of *Eucyrtidium acuminatum.* Filled circles = present; open circles = absent; X = samples barren of Radiolaria.

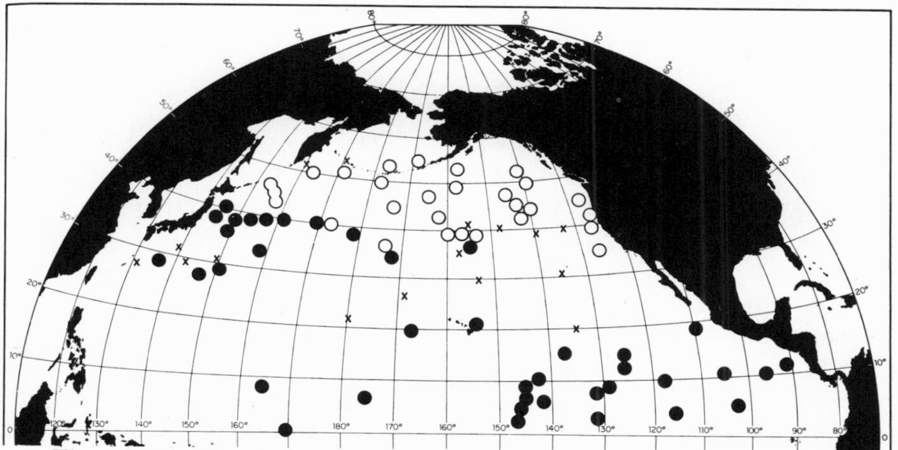

Figure 29. Distribution of *Eucyrtidium hexagonatum.* Filled circles = present; open circles = absent; X = samples barren of Radiolaria.

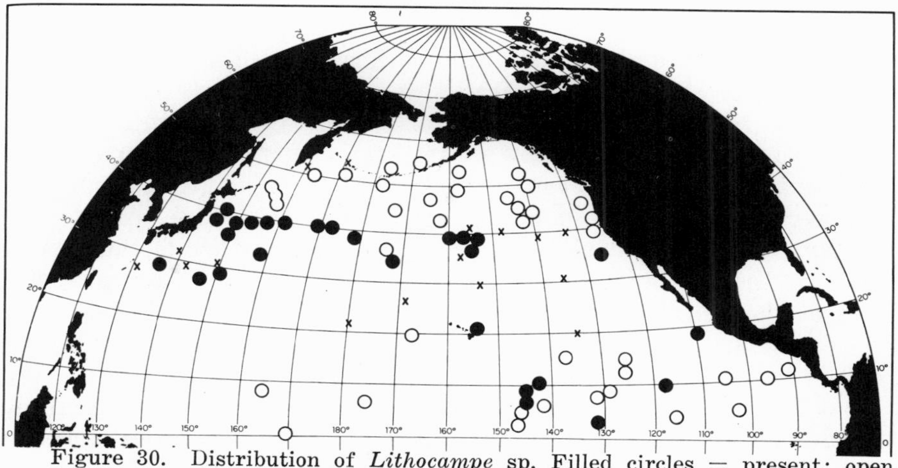

Figure 30. Distribution of *Lithocampe* sp. Filled circles = present; open circles = absent; X = samples barren of Radiolaria.

TABLE 2. RADIOLARIAN RECURRENT GROUP ASSEMBLAGES

Radiolarian Assemblage	Recurrent Groups Identified
Subarctic	Group III only
Subarctic-Transitional	Groups II and III
Transitional	Group II only
Subarctic-Transitional-Tropical	Groups I, II and III
Transitional-Tropical	Groups I and II
Tropical	Group I only

maritalis polypora, Collosphaera tuberosa, Botryocyrtis scutum, Siphocampe corbula, Actinomma arcadophorum, Euchitonia elegans.

Group II (Transitional assemblage): *Actinomma medianum, Eucyrtidium acuminatum, Spongaster tetras irregularis, Panartus tetrathalamus coronatus, Pterocanium praetextum eucolpum, Lithocampe* sp., *Heteracantha dentata.*

Group III (Subarctic assemblage): *Polysolenia arktios, Pterocanium korotnevi, Styptosphaera ? spumacea, Tristylospyris* sp., ? *Mitrocalpis araneafera;* associated: *Saccospyris conithorax.*

Group I is considered to be present in a sample if 20 of its 24 species are present; Group II requires 6 of its 7 species; Group III requires 5 of its 6 species. The areal distribution of these three groups can be used to define six modern radiolarian thanatocoenoses in North Pacific sediments (Figs. 39 and 41; Table 2).

Not unexpectedly, the patterns produced by analyses of modern diatom and radiolarian thanatocoenoses (Figs. 39, 40, 41, and 42) are generally similar to each other and to the known distribution of upper water masses. However, there are some important and probably very useful differences between the diatom and radiolarian patterns. At about 50° N. the diatom analysis is able to distinguish between Subarctic and Northeast Transitional-North assemblages, whereas the radiolarian Subarctic assemblage extends southward to about 40° N. In most of the area of the Northwest Marginal diatom assemblage, no radiolarian groups can be found. Kanaya and Koizumi's Northeast Transitional-South assemblage is somewhat broader than the corresponding Subarctic-Transitional area defined herein, but the radiolarian zone can be traced farther to the west in some of the northern Subarctic-Central Mixed diatom samples. The rest of the "Mixed" diatom samples are also mixed with respect to the Radiolaria. Using radiolarian groups, the northern part of the diatom Central assemblage can be divided into Transitional and Transitional-Tropical. The Tropical radiolarian group is essentially similar in areal extent to the Equatorial diatom group, but embraces two samples containing a Central diatom assemblage.

RADIOLARIA FROM CORE V20-130

Doctors Kanaya and Ujiié gave the author 98 samples from core V20-130 (35° 59′ N., 152° 36′ E.). The samples are splits of those used by Kanaya and Koizumi and are similarly designated, except for D-O which is part of the sample originally set aside for radiolarian work by Dr. Nakaseko of Osaka University, and which represents the top 5 cm of the piston core.

Following the procedure of Kanaya and Koizumi, a count to 200 radiolarian tests was made for each sample. The counts considered three categories of Radiolaria: (1) Xw, "warm" water species (all members of Recurrent Group I except for *Heliodiscus asteriscus, Spongocore puella, Hymeniastrum euclidis, Larcospira quadrangula,* and *Carpocanium* spp., which range too far north to be considered as "warm" water forms, and *Collosphaera tuberosa* which is known [Nigrini, in press] to have some stratigraphic significance); (2) Xt + Xc, transitional plus "cold" water species (all members of Recurrent Groups II and III); and (3) forms not identified in the present study and those species of Group I mentioned above. No additional work on core V20-130 concerning speciation, numbers of particular species, or absolute numbers of radiolarian tests was done by the author. It was, however, verified that the aspect of the radiolarian fauna throughout the core is indeed Quaternary.

In his count to 200 diatom valves, Kanaya found between about 50 and 100 "warm" plus "cold" water specimens, but in the radiolarian counts this number was always less than 50. Therefore, the count on each sample was extended to 300 individuals and, if the number of "warm" plus "cold" water species was still less than 40, the count was further extended to 500 individuals. Kanaya and Koizumi defined a diatom temperature number,

$$\mathrm{Td} = \frac{\mathrm{Xw}}{\mathrm{Xc} + \mathrm{Xw}} \times 100.$$

A corresponding radiolarian temperature number is defined herein as

$$\mathrm{Tr} = \frac{\mathrm{Xw}}{(\mathrm{Xt} + \mathrm{Xc}) + \mathrm{Xw}} \times 100.$$

For each of the 98 samples from V20-130, a count as described above was made, and Tr was calculated. Samples D-95 and D-96 were found to contain too few Radiolaria for a meaningful count and were omitted from the study. For each value of Tr, 95 percent confidence limits were calculated on a CDC 6400 computer using the following equations (Mood and Graybill, 1963, p. 261):

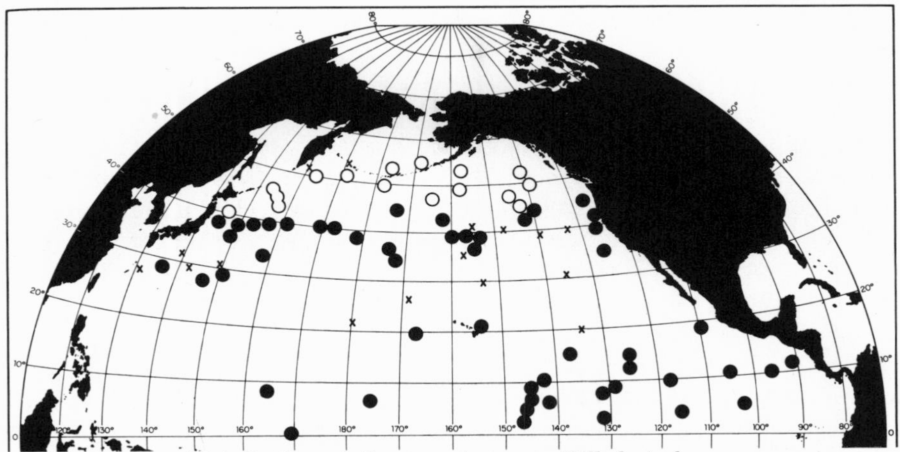

Figure 31. Distribution of *Carpocanium* spp. Filled circles = present; open circles = absent; X = samples barren of Radiolaria.

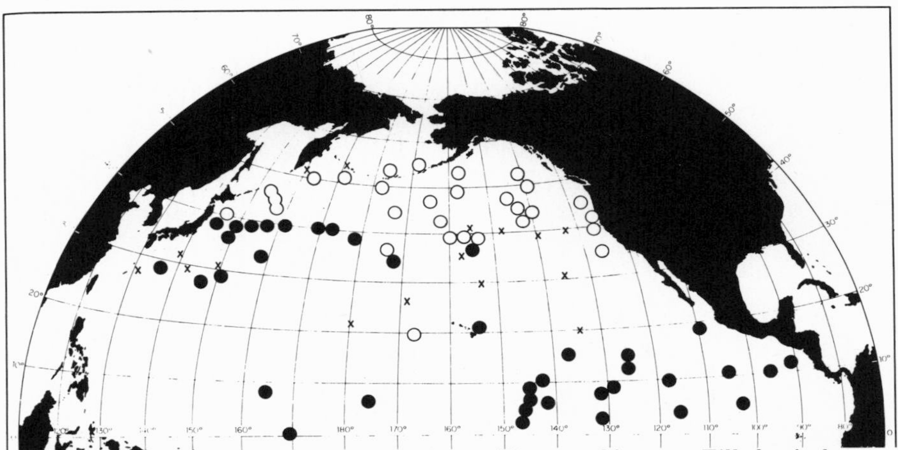

Figure 32. Distribution of *Anthocyrtidium ophirense*. Filled circles = present; open circles = absent; X = samples barren of Radiolaria.

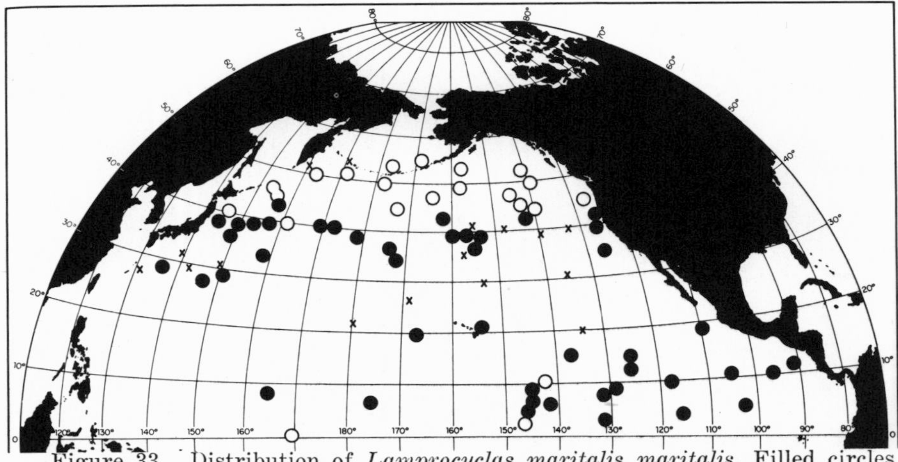

Figure 33. Distribution of *Lamprocyclas maritalis maritalis*. Filled circles = present; open circles = absent; X = samples barren of Radiolaria.

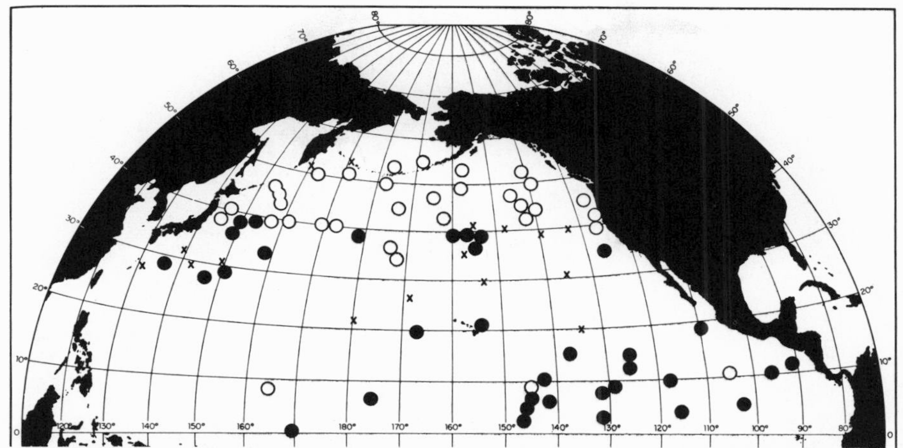

Figure 34. Distribution of *Lamprocyclas maritalis polypora*. Filled circles = present; open circles = absent; X = samples barren of Radiolaria.

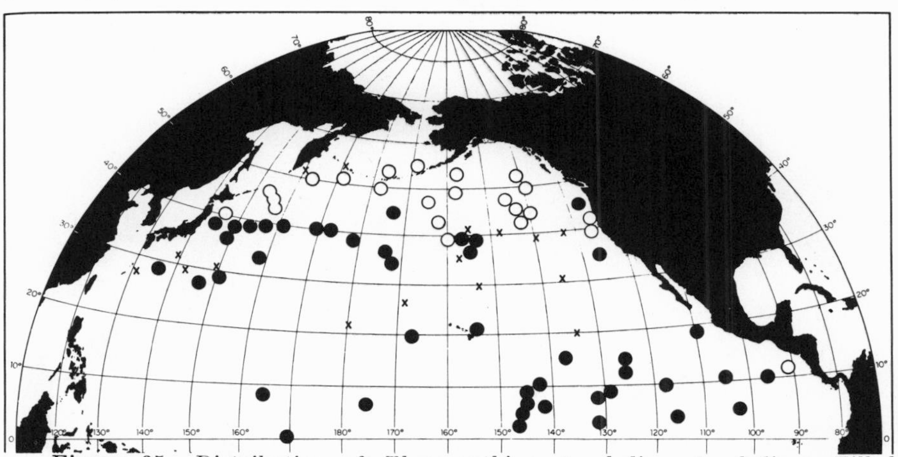

Figure 35. Distribution of *Theocorythium trachelium trachelium*. Filled circles = present; open circles = absent; X = samples barren of Radiolaria.

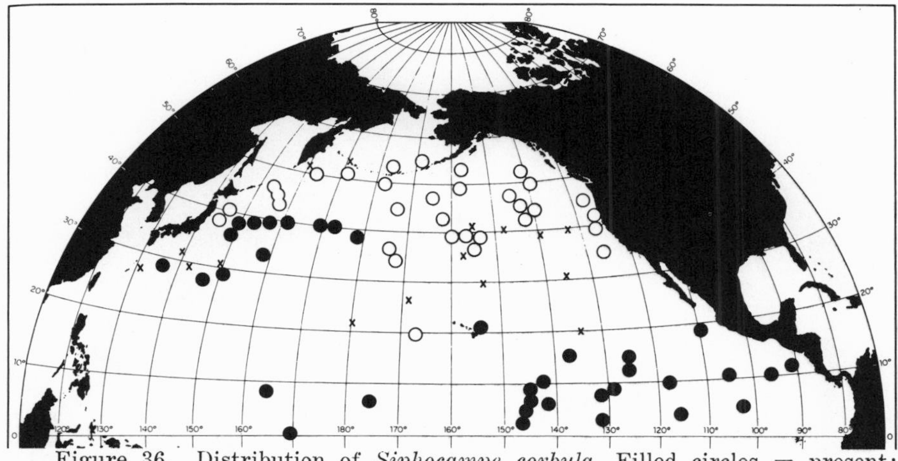

Figure 36. Distribution of *Siphocampe corbula*. Filled circles = present; open circles = absent; X = samples barren of Radiolaria.

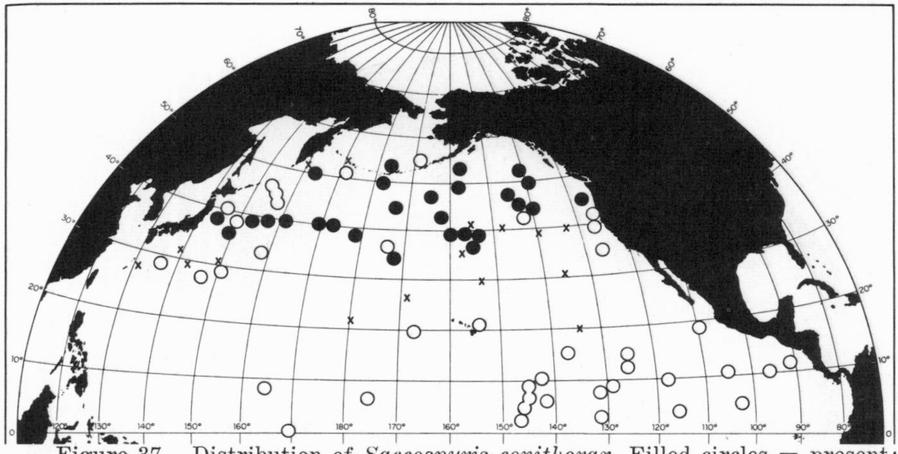

Figure 37. Distribution of *Saccospyris conithorax*. Filled circles = present; open circles = absent; X = samples barren of Radiolaria.

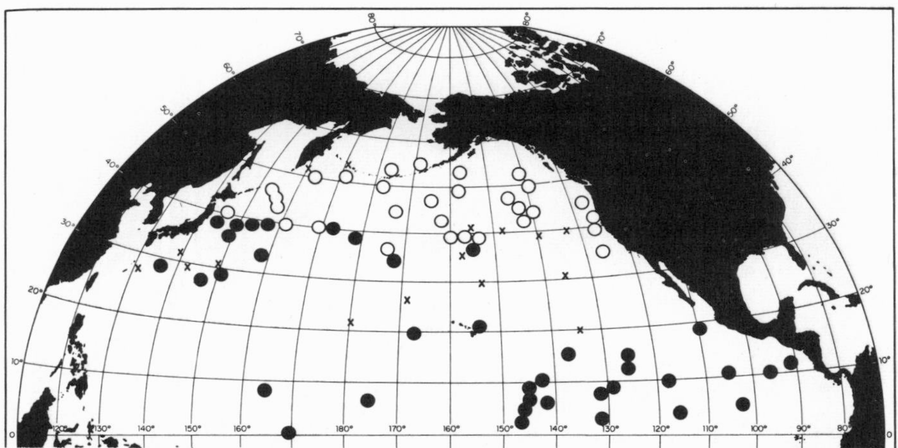

Figure 38. Distribution of *Botryocyrtis scutum*. Filled circles = present; open circles = absent; X = samples barren of Radiolaria.

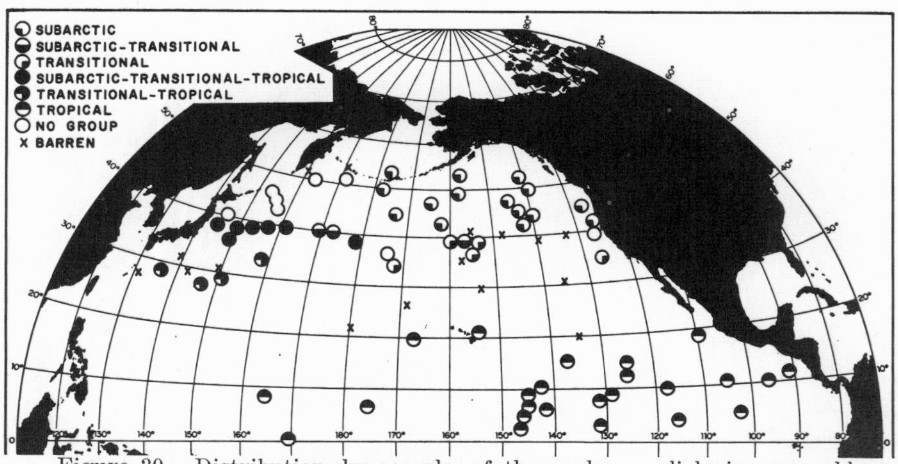

Figure 39. Distribution, by sample, of the modern radiolarian assemblages defined by the areal distribution of three recurrent groups of Radiolaria.

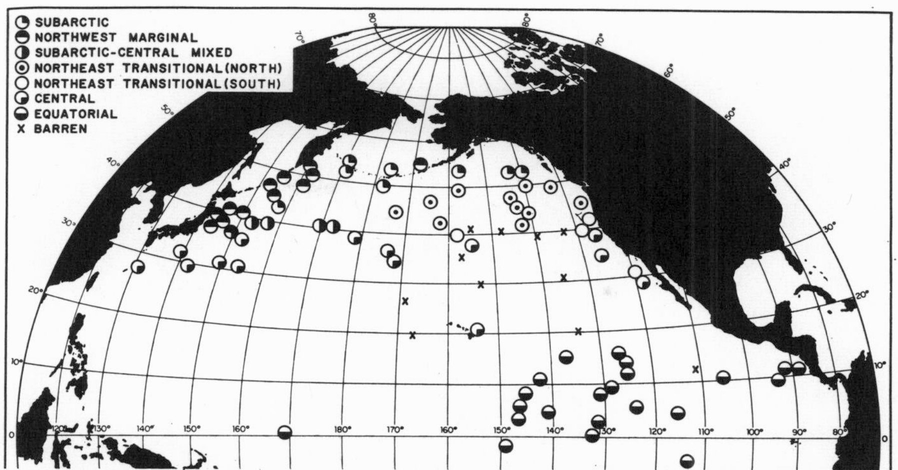

Figure 40. Distribution, by sample, of the modern diatom assemblages defined by the areal distribution of four recurrent groups of diatoms (*after* Kanaya and Koizumi, 1966).

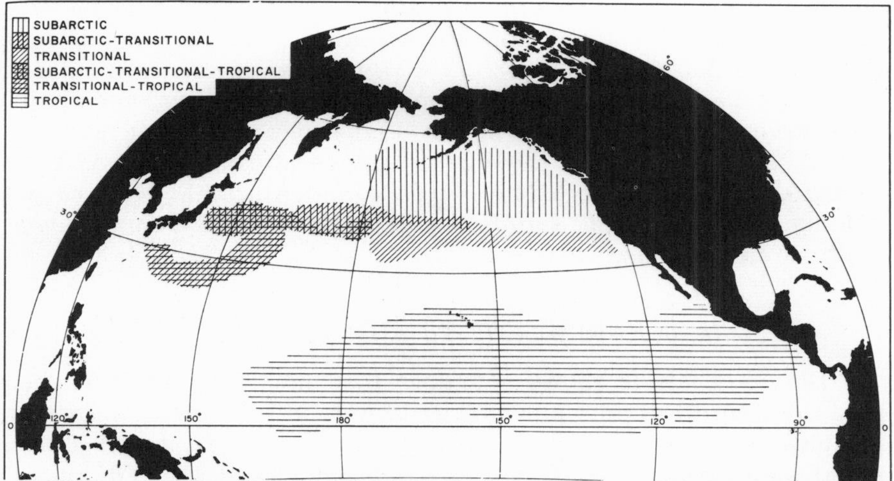

Figure 41. Over-all distribution of the modern radiolarian assemblages defined by the areal distribution of three recurrent groups of Radiolaria.

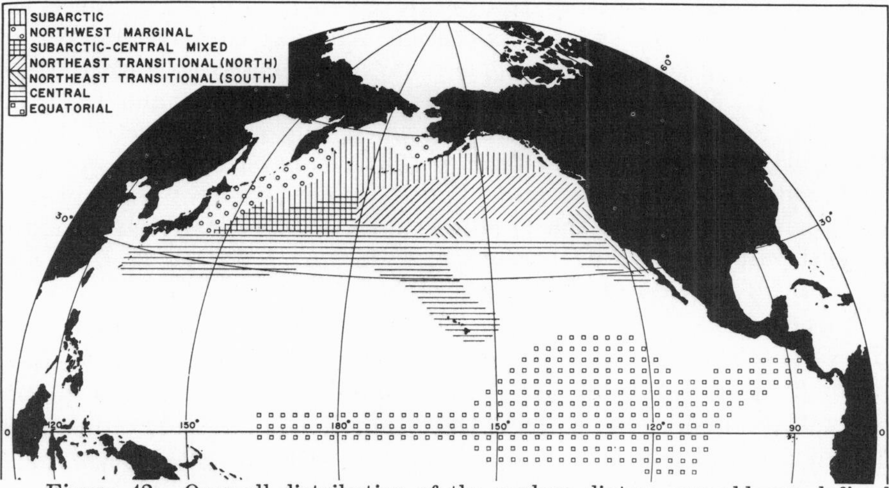

Figure 42. Over-all distribution of the modern diatom assemblages defined by the areal distribution of four recurrent groups of diatoms (*after* Kanaya and Koizumi, 1966).

$$\sum_{y=0}^{k} \binom{n}{y} p^{y} (1-p)^{n-y} = .025$$

$$\sum_{y=k}^{n} \binom{n}{y} p^{y} (1-p)^{n-y} = .025$$

where $n = (Xt + Xc) + Xw$

$k = Xw$

$p =$ lower or upper confidence limit.

At a few critical levels in the core, further counts were made so that the confidence interval could be narrowed. The resultant Tr curve is shown in Figure 43a and can be compared with the Td curve of Kanaya and Koizumi (Fig. 43b). At a number of critical points on the curves, 95 percent confidence limits have been shown, and simplified curves which eliminate minor oscillations have been superimposed. Simplification of the diatom temperature curve was made by the author from the text of Kanaya and Koizumi. Maxima and minima are considered significant if their 95 percent confidence limits do not overlap.

COMPARISON OF Td AND Tr CURVES

Diatom analysis shows a temperature maximum at D-0 where only the Central diatom assemblage is present. A similar maximum is shown by the radiolarian analysis; only the Tropical and Transitional radiolarian assemblages are present. Both the Td and Tr curves decrease from D-0 to significant minima at D-3 where all the radiolarian recurrent groups are present, but where only the Subarctic diatom assemblage is present. Another Td maximum occurs at D-6, but unlike D-0, the sample contains a Subarctic-Central Mixed assemblage. There are also lesser Td maxima at D-11 and D-14. At D-31 there is another minimum in which only the Subarctic diatom assemblage is present. The absence of the Subarctic diatom group at D-22 indicates a slight warming, but this is not shown by the Td value. Between D-3 and D-30, the Tr curve shows no significant maxima; the largest value (Tr = 54) occurs at D-19. At D-14, D-20, D-26 and D-30, the Tropical radiolarian assemblage is absent.

Below D-31 (Td curve) and D-30 (Tr curve), both the Td and Tr values increase, but Td reaches a significant maximum at D-43 while Tr peaks at D-38. At D-43 a Subarctic-Central Mixed diatom assemblage is present; at D-38 the Transitional and Subarctic radiolarian assemblages are present, but the Tropical assemblage must be considered as absent because only 18 of the minimum 20 species required for identification of the group could be found.

Both Td and Tr decrease in value below their respective maxima to minima at D-46 (Subarctic assemblage only) on the Td curve, and D-41 (Tropical, Transitional and Subarctic assemblages all present) on the Tr curve. Below D-46 on the Td curve there is a gradual increase to a significant maximum at D-62 (Central assemblage only). The Tr curve also shows a maximum at D-62 (Tropical, Transitional and Subarctic assemblages), but does not show the gradual increase. There appears instead to be a sharp increase to D-44, followed by generally higher values down to D-53, then a sharp decrease to D-56, and finally an increase to a significant maximum at D-62.

Below D-62 the diatom interpretation becomes less reliable, but there does appear to be a general decrease in Td down to a significant minimum at D-94 (Subarctic assemblage only). The Tr curve shows a pronounced minimum at D-90 with an intermediate low at D-78 and a suggestion of a lesser maximum at D-87.

DISCUSSION AND CONCLUSIONS

It must be noted that shifts in the radiolarian assemblages in V20-130 do not correspond to fluctuations of the Tr value as well as changes in the diatom assemblages correspond to fluctuations of the Td value. For example, the Tropical radiolarian group is absent from a number of samples, including an anomalous absence at D-38, but the Subarctic assemblage is absent from only two samples (D-0 and D-54), one of which (D-54) does not have a particularly large Tr value. This inconsistency between Tr values and the radiolarian assemblages suggests that the assemblages are not sufficiently discriminating at the latitude of V20-130. Analysis of the surface sediments (Figs. 41 and 42) shows that the Subarctic radiolarian group ranges farther south than the Subarctic diatom assemblage. Furthermore, there is an imbalance between the large number of tropical species and relatively few transitional and subarctic forms. Further study of radiolarians in high latitude sediment samples may add a few more species to the Transitional and Subarctic assemblage lists presented herein, but an imbalance will certainly persist.

However, ignoring the actual radiolarian assemblages, a useful comparison between the Tr and Td curves can be made. There is near perfect agreement at levels D-0, D-3 and D-62. The shapes of the curves from D-0 to D-3 are remarkably similar, and the view of Kanaya and Koizumi that the last major temperature minimum of the last major cold stage occurred at time D-3, is fully supported by the present radiolarian analysis. The present study also

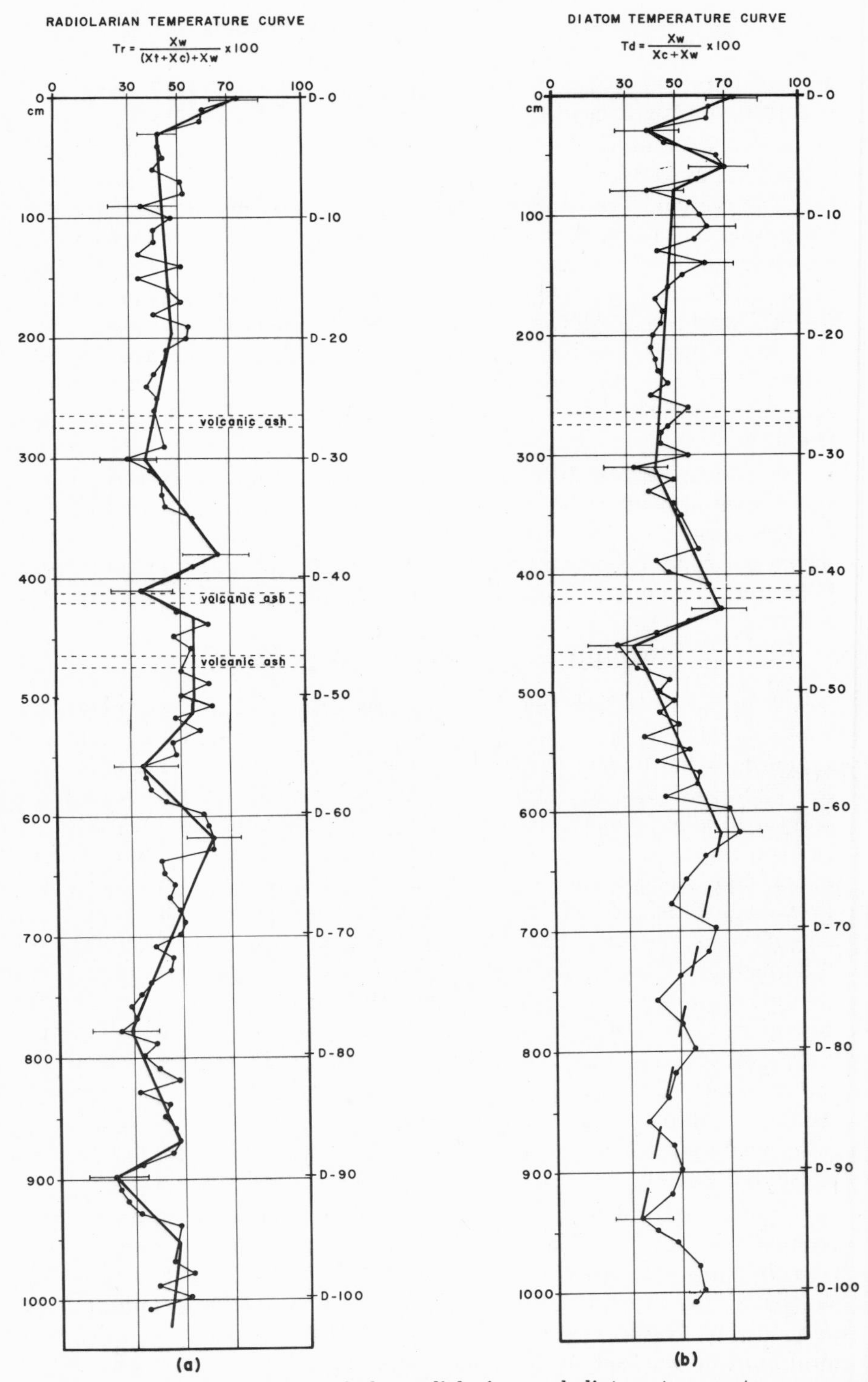

Figure 43. Comparison of the radiolarian and diatom temperature curves for core V20-130 (36°59′N., 152°36′E.; depth, 6547 m; length, 1039 cm).

supports Kanaya and Koizumi's conclusion that at time D-62 there was a major temperature maximum and that conditions then were very like those of the present day.

After D-62 time the Td curve reflects a gradual cooling of the surface waters to D-46 time. However, it would appear from the Tr curve that a more complex cooling, slight warming, and cooling sequence was taking place in the deeper (down to 400 m at least) waters inhabited by radiolarians. These waters reached a minimum temperature at a somewhat later time, D-41, than the surface (upper 200 m) waters. Both curves indicate an increase in temperature after their respective minima at D-46 and D-41, but again there is a time lag with respect to the Tr curve. The Td curve reaches a maximum at D-43 time while the Tr curve peaks at D-38 time. The Td curve shows another cooling period to a minimum at D-31 while the Tr curve lags slightly behind to a minimum at D-30.

Lower down in the core it can also be seen that changes in the Tr curve lag behind those in the Td curve. A Td minimum occurs at D-94 time while the corresponding Tr minimum occurs at D-90 time. If this lag is considered to hold in all cases, then the Td minimum at D-86 may well correspond to the Tr minimum at D-78.

Diatom maxima at D-6, D-11 and D-14 are, perhaps, reflections of minor temperature changes in the surface waters which did not affect the deeper living and possibly more resilient radiolarian fauna. Alternatively, the Tr maximum at D-0 may represent the lag time response to these maxima.

To interpret Pleistocene events from the study of a single faunal group in a single core is, of course, hazardous. Certainly Kanaya and Koizumi were aware of the embryonic nature of their study. However, there now appears to be a considerable similarity between independently conducted diatom and radiolarian analyses of core V20-130. At the very least, it can be said that at about time D-62, conditions in the western North Pacific were similar to those of today, that the last major temperature minimum in the area occurred at time D-3, and that there was a warm period *circa* D-38 to D-43 time. The internal consistency of the diatom analysis and the good agreement between the Tr and Td curves suggest that calculation of such curves is a useful tool in the interpretation of Pleistocene stratigraphy in deep-sea cores, especially where there is strong overlapping of assemblages.

It would be rewarding to pursue studies of the deep-sea Pleistocene stratigraphy of Radiolaria in the North Pacific. A previous study (Nigrini, in press) of tropical Pacific cores showed no variations in the radiolarian fauna that could be attributed to temperature fluctuations. Apparently, if there were any variations in near-surface water temperatures in the tropical Pacific during the Pleistocene, they were not great enough to affect the radiolarian fauna. Therefore, it is suggested that future faunal studies of the Pleistocene stratigraphy of Radiolaria in the Pacific be conducted at higher latitudes. Examination of additional cores in the vicinity of V20-130 would be a useful beginning, and it has been demonstrated that parallel studies involving at least two faunal groups are particularly rewarding. Isotopic age determinations at various levels in such cores and in V20-130 are essential for absolute correlation. Finally, further studies are necessary to determine whether or not the "lag time" suggested by the present comparison does, in fact, exist. If it does, it enables us not only to examine shifts in faunas and, therefore, water masses, but also to study details of such shifts.

TAXONOMIC NOTES

Most of the 37 species or subspecies whose distributions have been determined and used herein have been adequately described elsewhere. All are figured in this paper. They are listed below according to the classification devised by Riedel (1967). References to descriptions are given for each species.

Order: Polycystina Ehrenberg 1838 emend. Riedel 1967
Suborder: Spumellaria Ehrenberg 1875
Family: Collosphaeridae Müller 1858

Collosphaera tuberosa Haeckel, 1887
Pl. 1, fig. 1; Fig. 2

Nigrini, in press.

Polysolenia spinosa (Haeckel), 1862
Pl. 1, fig. 2; Fig. 3

Nigrini, 1967, p. 14, Pl. 1, fig. 1.

Polysolenia lappacea (Haeckel), 1887
Pl. 1, fig. 3; Fig. 4

Nigrini, 1967, p. 16, Pl. 1, figs. 3a, b.

Polysolenia arktios Nigrini, n. sp.
Pl. 1, figs. 4, 5; Fig. 5

Description. Shell smooth, spherical with numerous randomly arranged, subcircular and irregular pores of varying size. Shell very spiny, one or more poreless spines up to one-third the length of the shell diameter project outward from most pore margins. Spines may fork, or 2 or more spines arising from a single pore may meet at a point; they may be needle-like or flat, blunt projections. Rarely, specimens with an extracortical shell have been observed.

Dimensions (based on 20 specimens). Shell diameter 103 to 238 μ. Maximum length of spines 48 μ.

Discussion. This species appears to be closely related to *P. spinosa*, but its restricted distribution in Subarctic sediments (*compare* Figs. 3 and 5) seems to warrant a taxonomic distinction. Intraspecific variations and the relationship between species can, of course, only be fully determined by examination of whole colonies.

Siphonosphaera polysiphonia Haeckel, 1887
Pl. 1, fig. 6; Fig. 6

Nigrini, 1967, p. 18, Pl. 1, figs. 4a, b.

Family: Actinommidae Haeckel 1862 emend. Riedel 1967

Styptosphaera ? *spumacea* Haeckel, 1887
Pl. 1, figs. 7, 8; Fig. 7

? 1887 *Styptosphaera spumacea* Haeckel, "Chall." Rept., p. 87.

Description. Shell spherical, composed entirely of loose, irregular spongy meshwork. Pores are subcircular and of varying size. No central cavity or radial spines. Surface rough, but without thorns.

Dimensions (based on 20 specimens). Diameter of shell 119 to 167 μ.

Discussion. *S. spumacea* was described, but not illustrated, by Haeckel (1887) from "Challenger" station 236 (34° 58′ N., 139° 29′ E.), but no specimen of this general form could be found in topotypic material examined by the author. Haeckel's unillustrated description appears to fit the form found in the North Pacific during this study. However, the shell diameter given by Haeckel is almost twice that of the North Pacific specimens which are, therefore, only tentatively assigned to *S. spumacea*.

Heteracantha dentata Mast, 1910
Pl. 1, fig. 9; Fig. 8

Anomalacantha dentata (Mast) *in* Benson, 1966, p. 170, Pl. 5, figs. 10, 11.

Discussion. According to Benson (1966), this species is widespread in Holocene tropical seas. This was not found to be the case by the author (*compare with* Fig. 8).

Actinomma medianum Nigrini, 1967
Pl. 1, fig. 10; Fig. 9

Nigrini, 1967, p. 27, Pl. 2, figs. 2a, b.

Actinomma arcadophorum Haeckel, 1887
Pl. 1, fig. 11; Fig. 10

Nigrini, 1967, p. 29, Pl. 2, fig. 3.

Panartus tetrathalamus tetrathalamus Haeckel, 1887
Pl. 1, fig. 12; Fig. 11

Nigrini, 1967, p. 30, Pl. 2, figs. 4a-4d. The nominate subspecies is here named to permit a distinction from the following subspecies.

Panartus tetrathalamus coronatus Haeckel, 1887
Pl. 1, figs. 13, 14; Fig. 12

1887 *Cyphinidium coronatum* Haeckel, "Chall." Rept., p. 372.

Description. This subspecies is identical in all respects to *P. tetrathalamus tetrathalamus,* except for the development of a corona of stout, unbranched, 3-bladed spines on the distal ends of the cortical twin-shell and/or on the polar caps. When polar caps are present, the spines commonly extend without interruption beyond the point of junction with the cap. There may be as many as 9 or 10 spines on each end of the cortical twin-shell which is, in general, smaller than that of *P. tetrathalamus tetrathalamus.*

Dimensions (based on 20 specimens). Length of cortical twin-shell 96 to 119 μ. Maximum breadth of cortical twin-shell 64 to 88 μ. Maximum length of polar spines 56 μ.

Discussion. Since in Holocene sediments there is only one species of *Panartus* having strongly developed spines, it seems likely that the form described by Haeckel (1887) is the same as that encountered in North Pacific sediments during the present study. Haeckel described *C. coronatum* from "Challenger" stations 270 through 274, but examination of topotypic material from "Challenger" stations 270, 271 and 272 and MSN 131PG, which lies close to "Challenger" station 274, failed to reveal any spinose specimens of *Panartus.* This is probably due to the extreme rarity, indeed absence for all practical purposes, of such forms in equatorial sediments.

Family: Phacodiscidae Haeckel 1881

Heliodiscus asteriscus Haeckel, 1887
Pl. 2, fig. 1; Fig. 13

Nigrini, 1967, p. 32, Pl. 3, figs. 1a, b.

Family: Spongodiscidae Haeckel 1862 emend. Riedel 1967

Amphirhopalum ypsilon Haeckel, 1887
Pl. 2, fig. 2; Fig. 14

Nigrini, 1967, p. 35, Pl. 3, figs. 3a-d.

Spongocore puella Haeckel, 1887
Pl. 2, fig. 3; Fig. 15

Benson, 1964, Pl. 1, fig. 21; 1966, p. 187, Pl. 8, figs. 1-3.

Hymeniastrum euclidis Haeckel, 1887
Pl. 2, fig. 4; Fig. 16

Popofsky, 1912, p. 136, Fig. 51; Benson, 1966, p. 222, Pl. 12, figs 1-3; Ling and Anikouchine, 1967, p. 1488, Pl. 191, fig. 3, Pl. 192, fig. 3.

Euchitonia furcata Ehrenberg, 1860
Pl. 2, fig. 5; Fig. 17

E. mülleri in Nigrini, 1967, p. 37, Pl. 4, figs. la, b; *E. furcata* in Ling and Anikouchine, 1967, p. 1484, Pl. 189, figs. 5-7, Pl. 190, figs. 5-7.

Euchitonia elegans (Ehrenberg), 1872
Pl. 2, fig. 6; Fig. 18

Nigrini, 1967, p. 39, Pl. 4, figs. 2a, b.

Spongaster tetras tetras Ehrenberg, 1860
Pl. 2, fig. 7; Fig. 19

Nigrini, 1967, p. 41, Pl. 5, figs. 1a, b.

Spongaster tetras irregularis Nigrini, 1967
Pl. 2, fig. 8; Fig. 20

Nigrini, 1967, p. 43, Pl. 5, fig. 2.

Family: Litheliidae Haeckel 1862

Larcospira quadrangula Haeckel, 1887
Pl. 2, fig. 9; Fig. 21

Benson, 1966, p. 266, Pl. 18, figs. 7, 8.

Sub-order: Nassellaria Ehrenberg 1875
Family: Plagoniidae Haeckel 1881 emend. Riedel 1967

? *Mitrocalpis araneafera* Popofsky, 1908
Pl. 3, figs. 1, 2; Fig. 22

Riedel, 1958, p. 232, Pl. 3, figs. 3, 4; Fig. 4.

Discussion. *Mitrocalpis araneafera* was originally described by Popofsky (1908) from Antarctic sediments. It was re-described by Riedel (1958) also from Antarctic sediments. A similar form has been found in North Pacific sediments during the present study. It is generally smaller than the specimens found by Popofsky and Riedel, but Riedel's original material does contain smaller individuals. Certainly the Antarctic and North Pacific forms are closely related, but a detailed comparative study would be necessary to determine their exact relationship. Therefore, the North Pacific form is only tentatively assigned to *M. araneafera.*

Dogiel and Reshetnyak (1952) described from a single specimen a new species, *Arachnocorys dubius* Dogiel. Their illustration bears considerable resemblance to ? *M. araneafera,* although it is inverted. Dogiel and Reshetnyak interpreted their specimen as a 2-segmented form with a poorly visible cephalis. Petrushevskaya (1968, written commun.) searched the original material, but was unable to find the specimen of *A. dubius*. However, considering the location (northwestern Pacific Ocean) at which the specimen was originally found, it seems likely that it also belongs to ? *M. araneafera.*

Family: Acanthodesmiidae Hertwig 1879

Tristylospyris sp.
Pl. 3, figs. 3 to 6; Fig. 23

Discussion. Only one species of Spyroidea is commonly found in subarctic sediments from the Pacific Ocean. It has not previously been described. Within the Haeckelian generic classification of Spyroidea, which is admittedly artificial, the species appears to be most closely related to the genus *Tristylospyris* (Haeckel, 1881, p. 441; 1887, p. 1032) in that it has three basal feet and lacks an apical horn. However, in his doctoral dissertation, Goll (1967) made major revisions, based on phylogenies, of Haeckel's classification. It is now, therefore, more advisable to fit all species of Spyroidea into Goll's classification, and it is further appropriate that such work be left to Dr. Goll as far as possible. Since the purposes of the present study are not primarily taxonomic, the author has chosen simply to illustrate this most useful subarctic species, without providing a morphological description.

Family: Theoperidae Haeckel 1881 emend. Riedel 1967

Pterocanium praetextum praetextum (Ehrenberg), 1872
Pl. 3, fig. 7; Fig. 24

Nigrini, 1967, p. 68, Pl. 7, fig. 1.

Pterocanium praetextum eucolpum Haeckel, 1887
Pl. 3, fig. 8; Fig. 25

Nigrini, 1967, p. 70, Pl. 7, fig. 2.

Pterocanium trilobum Haeckel, 1861
Pl. 3, fig. 9; Fig. 26

Nigrini, 1967, p. 71, Pl. 7, figs. 3a, b.

Pterocanium korotnevi (Dogiel), 1952
Pl. 3, figs. 10, 11; Fig. 27

1952 *Pterocorys korotnevi* Dogiel, Dogiel and Reshetnyak, Issledovanya Dalnevostochnykh Morei SSSR, no. 3, p. 17; Fig. 11.

Discussion. Dogiel and Reshetnyak (1952) placed this species in the genus *Pterocorys.* Haeckel's written description of this genus warrants such a generic assignment, but the type species, *Pterocorys campanula* Haeckel, 1887, is a quite different 3-segmented form having a lobed cephalis. Therefore, the species is reassigned herein to the genus *Pterocanium* (type species = *Pterocanium proserpinae* Ehrenberg, 1858).

Eucyrtidium acuminatum (Ehrenberg), 1844
Pl. 4, fig. 1; Fig. 28

Nigrini, 1967, p. 81, Pl. 8, figs. 3a, b.

Eucyrtidium hexagonatum Haeckel, 1887
Pl. 4, fig. 2; Fig. 29

Nigrini, 1967, p. 83, Pl. 8, figs. 4a, b.

Lithocampe sp.
Pl. 4, fig. 3; Fig. 30

Nigrini, 1967, p. 87, Pl. 8, figs. 6a, b.

Family: Carpocaniidae Haeckel 1881 emend. Riedel 1967

Carpocanium spp.
Pl. 4, figs. 4 to 6; Fig. 31

C. petalospyris in Benson, 1966, p. 434, Pl. 29, figs. 9, 10; Fig. 25.

Discussion. Haeckel (1887) described a number of species belonging to the genus *Carpocanium.* It is thought by the author that many of these are conspecific, and that some variations in the shape of the thorax and the nature of the peristomal teeth are intraspecific. Benson (1966) is in agreement with this opinion. Furthermore, Haeckel also described a number of species belonging to the genus *Carpocanistrum,* which he described as being similar to *Carpocanium,* but lacking a rudimentary cephalis. It seems probable that *Carpocanium* and *Carpocanistrum* are congeneric and that the cephalis is simply more difficult to detect in some specimens than in others.

For the purposes of the present study, all specimens of the *Carpocanium-Carpocanistrum* form were considered together except for the distinctive *Carpocanium* sp. A. (Nigrini, 1968, p. 55, Pl. 1, fig. 4).

Family: Pterocoryidae Haeckel 1881 emend. Riedel 1967

Anthocyrtidium ophirense (Ehrenberg), 1872
Pl. 4, fig. 7; Fig. 32

Nigrini, 1967, p. 56, Pl. 6, fig. 3.

Lamprocyclas maritalis maritalis Haeckel, 1887
Pl. 4, fig. 8; Fig. 33

Nigrini, 1967, p. 74, Pl. 7, fig. 5.

Lamprocyclas maritalis polypora Nigrini, 1967
Pl. 4, fig. 9; Fig. 34

Nigrini, 1967, p. 76, Pl. 7, fig. 6.

Theocorythium trachelium trachelium (Ehrenberg), 1872
Pl. 4, fig. 10; Fig. 35

Nigrini, 1967, p. 79, Pl. 8, fig. 2, Pl. 9, fig. 2.

Family: Artostrobiidae Riedel 1967

Siphocampe corbula (Harting), 1863
Pl. 4, fig. 11; Fig. 36

Nigrini, 1967, p. 85, Pl. 8, fig. 5, Pl. 9, fig. 3.

Family: Cannobotryidae Haeckel 1881 emend. Riedel 1967

Saccospyris conithorax Petrushevskaya, 1965
Pl. 4, fig. 12; Fig. 37

Petrushevskaya, 1965, p. 98, Fig. 11; 1967, p. 152, Fig. 85-I.

Discussion: Forms encountered in North Pacific sediments are very similar to a specimen figured by Petrushevskaya (1965, Fig. 11-VI). Petrushevskaya (1968, written commun.) has suggested that neither the North Pacific form nor her 1965 Figure 11-VI belong to *S. conithorax,* but constitute a separate species distinguishable from *S. conithorax* by its flatly truncated thorax. However, until a detailed comparative study of Antarctic and North Pacific specimens can be made, these forms are being included in *S. conithorax.* North Pacific specimens are generally smaller and somewhat less robust than Antarctic ones.

Botryocyrtis scutum (Harting), 1863
Pl. 4, fig. 13; Fig. 38

Nigrini, 1967, p. 52, Pl. 6, figs. 1a to 1c.

ACKNOWLEDGMENTS

Samples used in the present study were given to the author by the Scripps Institution of Oceanography, Dr. Kanaya of Tohoku University, and Dr. Ujiié of the National Science Museum, Tokyo. Their generosity is indeed appreciated. I would also like to thank Dr. E. W. Fager who lent me a computer program for recurrent group analysis; Mr. W. R. Riedel who, as usual, was kind enough to offer constructive criticism of the manuscript; and Miss P. B. Helms whose assistance with the photographs was invaluable. Financial support for the project, which was carried out in the Department of Geology, Northwestern University, came from National Science Foundation Grant GA-635.

EXPLANATION OF PLATES

Photographed specimens are located according to sample number, slide letter (in the case of more than one slide from a particular location), and England Finder coordinates. Type and figured specimens will be deposited in the U. S. National Museum, Washington, D.C., and the USNM numbers used herein are from their Cenozoic Catalogue No. 132.

RADIOLARIAN SPECIES FROM NORTH PACIFIC HOLOCENE SEDIMENTS

All figures × 175
Figure

1. *Collosphaera tuberosa* Haeckel; ZETES III 3G, A-J50/0; USNM No. 651199.
2. *Polysolenia spinosa* (Haeckel); JYN II 17G, A-X34/1; USNM No. 651200.
3. *Polysolenia lappacea* (Haeckel); ZETES III 3G, A-L46/1; USNM No. 651201.
4. *Polysolenia arktios* Nigrini, n. sp.; CK 11, A-W33/2; HOLOTYPE; USNM No. 651202.
5. *Polysolenia arktios* Nigrini, n. sp.; CK 11, A-Q31/3; PARATYPE; USNM No. 651203.
6. *Siphonosphaera polysiphonia* Haeckel; ZETES III 3G, A-L49/2; USNM No. 651204.
7. *Styptosphaera ? spumacea* Haeckel; CAS 2, A-M25/0; USNM No. 651205.
8. *Styptosphaera ? spumacea* Haeckel; CK 3, A-Y12/2; USNM No. 651206.
9. *Heteracantha dentata* Mast; × 160; JYN II 10G, A-W47/0; USNM No. 651207.
10. *Actinomma medianum* Nigrini; CAS 2, A-Q51/0; USNM No. 651208.
11. *Actinomma arcadophorum* Haeckel; JYN II 19G, A-Q37/0; USNM No. 651209.
12. *Panartus tetrathalamus tetrathalamus* Haeckel; ZETES III 3G, A-T39/0; USNM No. 651210.
13. *Panartus tetrathalamus coronatus* Haeckel; JYN II 19G, A-Y25/2; USNM No. 651211.
14. *Panartus tetrathalamus coronatus* Haeckel; ZETES III 3G, A-N20/0; USNM No. 651212.

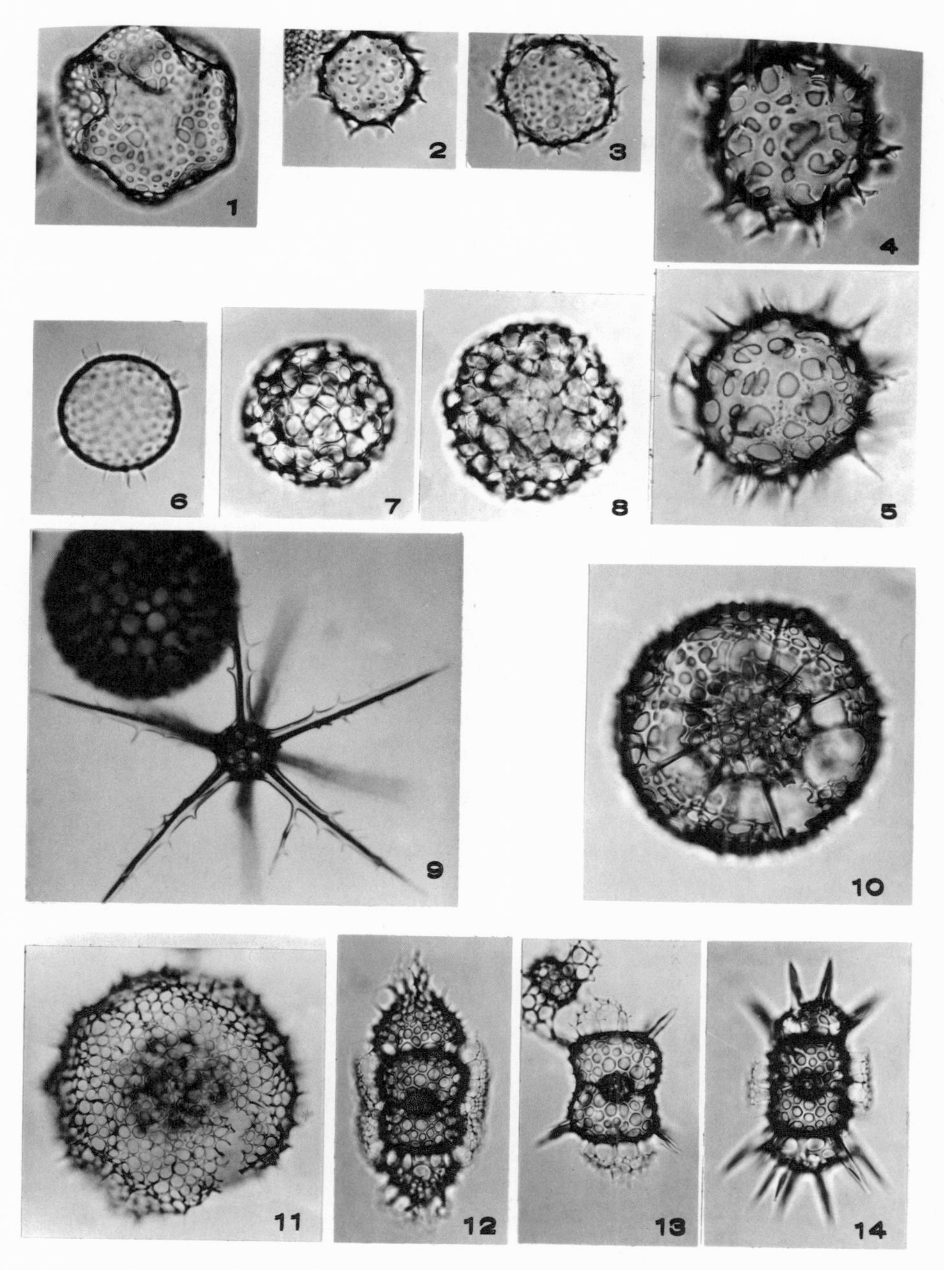

RADIOLARIAN SPECIES FROM NORTH PACIFIC HOLOCENE SEDIMENTS

NIGRINI, PLATE 1

RADIOLARIAN SPECIES FROM NORTH PACIFIC HOLOCENE SEDIMENTS

All figures × 175
Figure

1. *Heliodiscus asteriscus* Haeckel; ZETES III 3G, A-T43/0; USNM No. 651213.
2. *Amphirhopalum ypsilon* Haeckel; JYN II 17G, A-N53/2; USNM No. 651214.
3. *Spongocore puella* Haeckel; CAS 2, A-L25/3; USNM No. 651215.
4. *Hymeniastrum euclidis* Haeckel; CAS 2, A-Y44/0; USNM No. 651216.
5. *Euchitonia furcata* (Ehrenberg); JYN II 19G, B-D37/0; USNM No. 651217.
6. *Euchitonia elegans* (Ehrenberg); JYN II 17G, A-055/3; USNM No. 651218.
7. *Spongaster tetras tetras* Ehrenberg; JYN II 20G, A-L23/2; USNM No. 651219.
8. *Spongaster tetras irregularis* Nigrini; CK 4, A-T38/2; USNM No. 651220.
9. *Larcospira quadrangula* Haeckel; CAS 2, A-G36/4; USNM No. 651221.

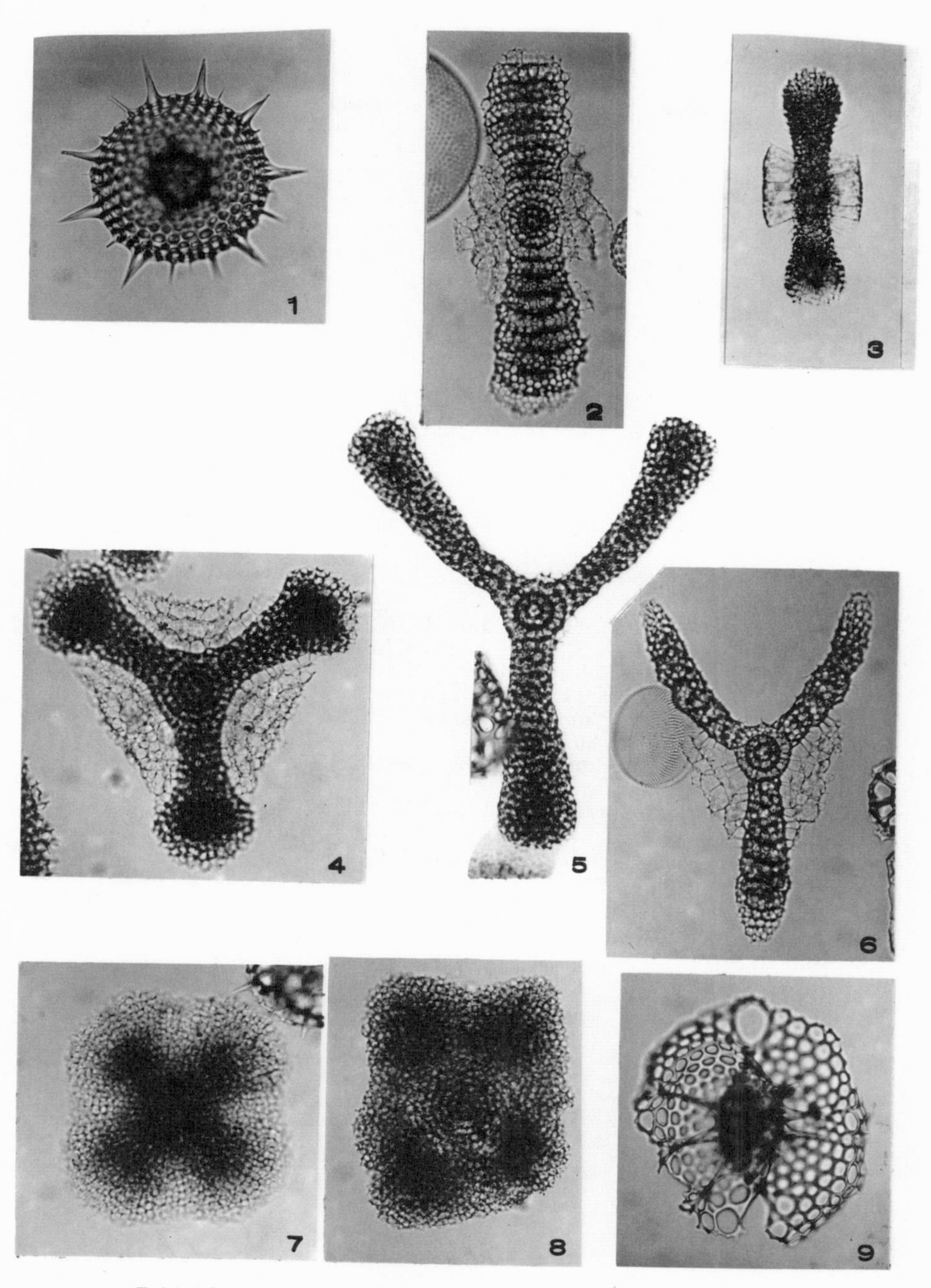

RADIOLARIAN SPECIES FROM NORTH PACIFIC
HOLOCENE SEDIMENTS

NIGRINI, PLATE 2
Geological Society of America Memoir 126

RADIOLARIAN SPECIES FROM NORTH PACIFIC HOLOCENE SEDIMENTS

All figures × 175
Figure

1. ? *Mitrocalpis araneafera* Popofsky; MUK B31G, B-W32/2; USNM No. 651222.
2. ?*Mitrocalpis araneafera* Popofsky; JYN II 8G, A-C50/3; USNM No. 651223.
3. *Tristylospyris* sp.; CK 11, A-J51/3; USNM No. 651224.
4. *Tristylospyris* sp.; CK 8, A-A16/3; USNM No. 651225.
5. *Tristylospyris* sp.; CK 8, A-U43/2; USNM No. 651226.
6. *Tristylospyris* sp.; CK 11, A-P22/1; USNM No. 651227.
7. *Pterocanium praetextum praetextum* (Ehrenberg); JYN II 19G, B-X19/0; USNM No. 651228.
8. *Pterocanium praetextum eucolpum* Haeckel; ZETES III 3G, A-T38/2; USNM No. 651229.
9. *Pterocanium trilobum* (Haeckel); JYN II 17G, A-K30/4; USNM No. 651230.
10. *Pterocanium korotnevi* (Dogiel); CK 8, A-U32/1; USNM No. 651231.
11. *Pterocanium korotnevi* (Dogiel); MUK B21G, A-028/2; USNM No. 651232.

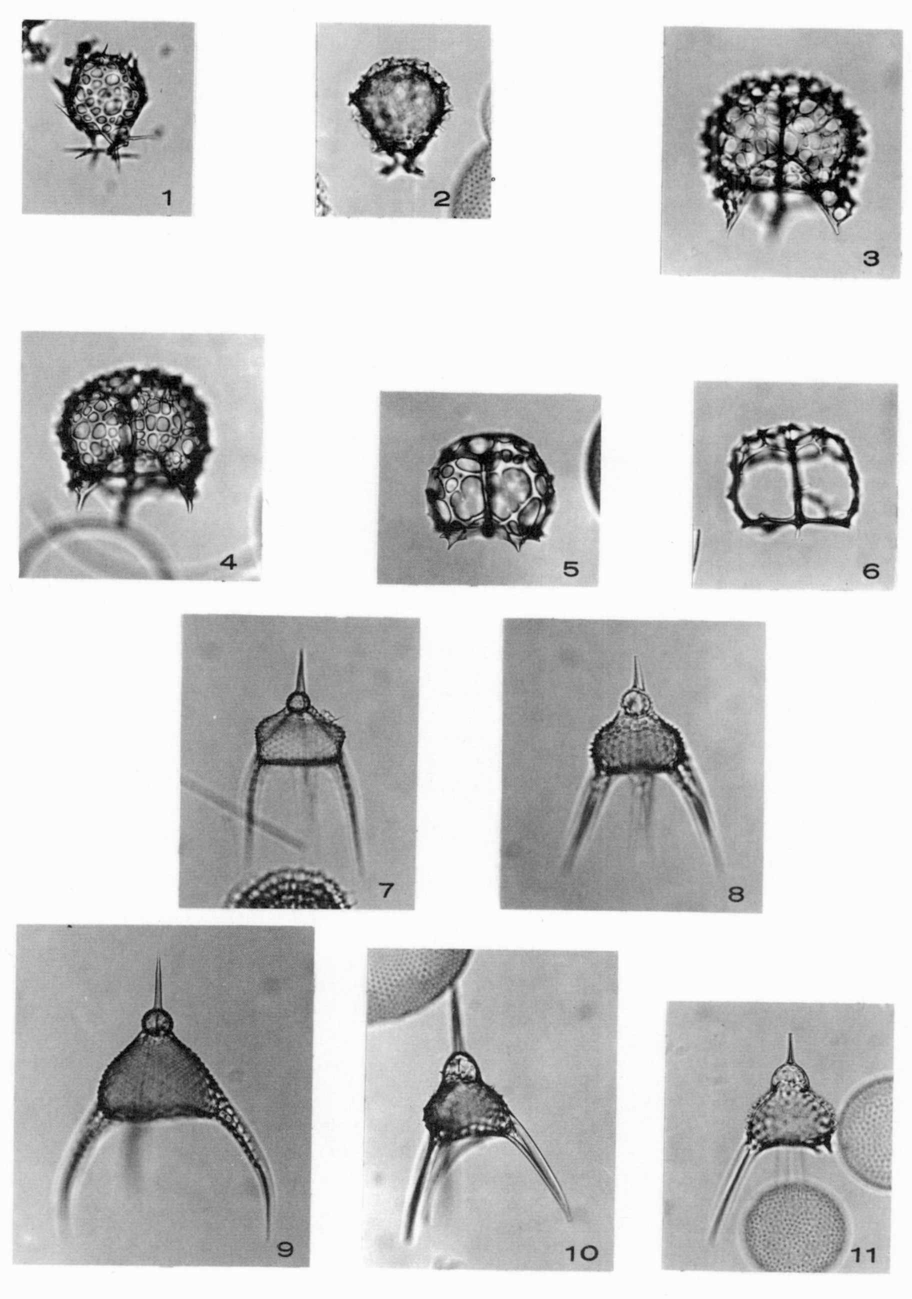

RADIOLARIAN SPECIES FROM NORTH PACIFIC
HOLOCENE SEDIMENTS

RADIOLARIAN SPECIES FROM NORTH PACIFIC HOLOCENE SEDIMENTS

All figures × 175
Figure

1. *Eucyrtidium acuminatum* (Ehrenberg); JYN II 19G, A-W36/0; USNM No. 651233.
2. *Eucyrtidium hexagonatum* Haeckel; JYN II 19G, B-D23/2; USNM No. 651234.
3. *Lithocampe* sp.; JYN II 19G, B-E15/1; USNM No. 651235.
4. *Carpocanium* spp.; MSN 155G, A-X51/2; USNM No. 651236.
5. *Carpocanium* spp.; MUK B31G, A-R31/3; USNM No. 651237.
6. *Carpocanium* spp.; JYN II 19G, B-C36/3; USNM No. 651238.
7. *Anthocyrtidium ophirense* (Ehrenberg); JYN II 19G, A-U27/1; USNM No. 651239.
8. *Lamprocyclas maritalis maritalis* Haeckel; JYN II 19G, A-U28/1; USNM No. 651240.
9. *Lamprocyclas maritalis polypora* Nigrini; ZETES III 3G, A-K40/4; USNM No. 651241.
10. *Theocorythium trachelium trachelium* (Ehrenberg); ZETES III 3G, A-N43/0; USNM No. 651242.
11. *Siphocampe corbula* (Harting); JYN II 19G, A-S43/2; USNM No. 651243.
12. *Saccospyris conithorax* Petrushevskaya; CK 11, A-H25/0; USNM No. 651244.
13. *Botryocyrtis scutum* (Harting); JYN II 19G, B-T21/4; USNM No. 651245.

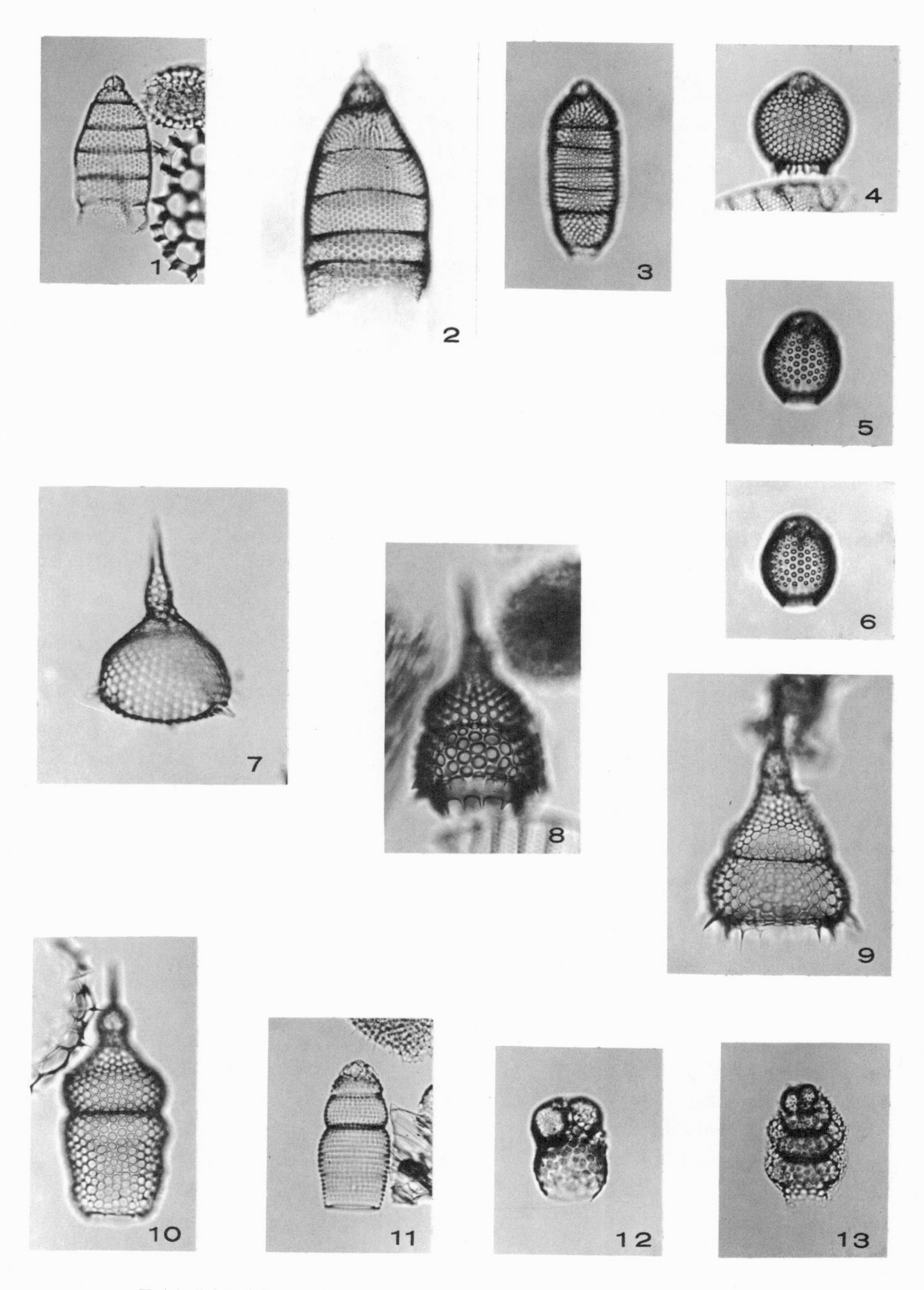

RADIOLARIAN SPECIES FROM NORTH PACIFIC
HOLOCENE SEDIMENTS

REFERENCES CITED

Bailey, J. W., 1856, Notice of microscopic forms found in the soundings of the Sea of Kamtschatka: Am. Jour. Sci., Ser. 2, v. 22, p. 1-6.

Benson, R. N., 1964, Preliminary report on Radiolaria in Recent sediments of the Gulf of California, *in* Marine Geology of the Gulf of California —A Symposium: Am. Assoc. Petroleum Geologists Mem. 3, p. 398-400.

—— 1966, Recent Radiolaria from the Gulf of California: Ph.D. thesis, Minnesota Univ., 577 p.

Bradshaw, J., 1959, Ecology of living planktonic Foraminifera in the north and equatorial Pacific Ocean: Cushman Found. Foram. Research Contr., v. 10, pt. 2, p. 25-64.

Casey, R., 1970, Radiolarians as indicators of past and present water masses—a series of investigations: SCOR Symp. Micropal. Marine Bottom Sediments, Cambridge, England, Sept., 1967 (in press).

Dogiel, V. A., and Reshetnyak, V. V., 1952, Materialy po radiolyariyam severozapadnoi chasti tikhogo okeana: Issledovanya Dalnevostochnykh Morei SSSR, no. 3, p. 5-36.

Ehrenberg, C. G., 1860, Über den Tiefgrund des stillen Oceans zwischen Californien und den Sandwich-Inseln aus bis 15,600′ Tiefe nach Lieut. Brooke. Monatsber. Kgl.: Preuss. Akad. Wiss. Berlin, Jahrg. 1860, p. 819-833.

Fager, E. W., 1957, Determination and analysis of recurrent groups: Ecology, v. 38, no. 4, p. 586-595.

Fager, E. W., and McGowan, J. A., 1963, Zooplankton species groups in the North Pacific: Science, v. 140, no. 3566, p. 453-460.

Goll, R. M., 1967, Classification and phylogeny of Cenozoic Trissocyclidae (Radiolaria) in the Pacific and Caribbean basins: Ph.D. thesis, Ohio State Univ., 147 p.

Haeckel, E., 1881, Entwurf eines Radiolarien—Systems auf Grund von Studien der Challenger-Radiolarien: Naturw. Jenaische Zeitschr., v. 15 (new ser., v. 8), no. 3, p. 418-472.

—— 1887, Report on the Radiolaria collected by H. M. S. *Challenger* during the years 1873-76: Rept. Voyage *Challenger*, Zool., v. 18, clxxxviii + 1803 p.

Hays, J. D., 1965, Radiolaria and late Tertiary and Quaternary history of Antarctic seas: Am. Geophys. Union, Biol. Antarctic Seas, v. 2, Antarctic Research Ser. 5, p. 125-184.

Kanaya, T., and Koizumi, I., 1966, Interpretation of diatom thanatocoenoses from the North Pacific applied to a study of core V20-130 (studies of deep-sea core V20-130, part IV): Tohoku Univ. Sci. Repts., Sendai, 2d Ser. (geol.), v. 37, no. 2, p. 89-130.

Kruglikova, S. B., 1966, Kolichestvennoe raspredelenie radiolyarii v poverkhnostnom sloe donnykh osadkov severnoi poloviny Tikhogo okeana (Quantitative distribution of radiolarians in the surface layer of sediments in the northern half of the Pacific Ocean): p. 246-261 *in* Geokhimiya Kremnezema: Izdatelstvo Nauka, Moscow, 424 p.

Ling, H. -Y., and Anikouchine, W. A., 1967, Some Spumellarian Radiolaria from the Java, Philippine and Mariana trenches: Jour. Paleontology, v. 41, no. 6, p. 1481-1491.

Mood, A. M., and Graybill, F. A., 1963, Introduction to the theory of statistics: 2d. ed., McGraw-Hill, 443 p.

Nakaseko, Kojiro, 1964, Liosphaeridae and Collosphaeridae (Radiolaria) from the sediment of the Japan Trench (on Radiolaria from the sediment of the Japan Trench 1): Osaka Univ. Sci. Repts., v. 13, no. 1, 39-57.

Nigrini, C., 1967, Radiolaria in pelagic sediments from the Indian and Atlantic Oceans: Scripps Inst. Oceanog. Bull., v. 11, 125 p.

—— 1968, Radiolaria from eastern tropical Pacific sediments: Micropaleontology, v. 14, no. 1, p. 51-63.

—— 1970, Radiolarian zones for the Quaternary of the equatorial Pacific Ocean: SCOR symp. Micropal. Marine Bottom Sediments, Cambridge, England, Sept. 1967 (in press).

Petrushevskaya, M. G., 1965, Osobennosti konstruktsii skeleta radiolyarii Botryoidae (otr. Nassellaria): Trudy Zoologicheskogo Inst., Akad. Nauk SSSR, v. 35, p. 79-118.

—— 1966, Radiolyarii v planktone i v donnykh osadkakh (Radiolaria in the plankton and bottom sediments), p. 219-245 *in* Geokhimiya Kremnezema: Izdatelstvo Nauka, Moscow, 424 p.

—— 1967, Radiolyarii otryadov Spumellaria i Nassellaria antarkticheskoi oblasti (Antarctic spumelline and nasselline radiolarians): Issled. Fauny Morei 4 (12), Rez. biol. Issled. sov. antarkt. Eksped. 1955-58, 3, p. 5-186.

Popofsky, A., 1908, Die Radiolarien der Antarktis (mit Ausnahme der Tripyleen): Deutsche Südpolar-Exped. 1901-1903, v. 10, (Zool. v. 2), no. 3, p. 183-305.

—— 1912, Die Sphaerellarien des Warmwassergebietes: Deutsche Südpolar-Exped. 1901-1903, v. 13, (Zool. v. 5), no. 2, p. 73-159.

Riedel, W. R., 1958, Radiolaria in Antarctic sediments: Repts. B.A.N.Z. Antarctic Research Exped., Ser. B, v. 6, pt. 10, p. 217-255.

——1967, Protozoa, class Actinopoda, p. 291-298 *in* The Fossil Record, a Symposium with Documentation: Geol. Soc. London Pub., 827 p.

MANUSCRIPT RECEIVED APRIL 10, 1969
REVISED MANUSCRIPT RECEIVED JULY 7, 1969

THE GEOLOGICAL SOCIETY OF AMERICA, INC.
MEMOIR 126, 1970

Stratigraphy and Evolutionary Trends of Radiolaria in North Pacific Deep-Sea Sediments

JAMES D. HAYS

Lamont-Doherty Geological Observatory
of Columbia University
Palisades, N.Y.

ABSTRACT

The abrupt, widespread, and simultaneous extinction of radiolarian species makes them ideal stratigraphic markers. Five species became extinct in the North Pacific during the last 3 m.y. Four of these are used to define four stratigraphic zones. The boundary between the oldest two zones correlates with the Pliocene/Pleistocene boundary, as defined in southern Italy. These zones can be related through paleomagnetic stratigraphy to previously established radiolarian and foraminiferal zonations.

One species (*Eucyrtidium matuyamai*) evolved and became extinct during the last 2.5 m.y. Its evolution can be related to the invasion of and adaptation to a new habitat. The extinction of *E. matuyamai* shows a striking correlation with the magnetic reversal at the base of the Jaramillo event. The rapid evolution of this species probably reduced the genetic variability of the population, making it more vulnerable to extinction than other less rapidly evolving species. The environmental change that caused the extinction is unclear; however, there is suggestive evidence that it is in some way related to a reversal of the earth's field. If reversals cause abrupt environmental changes, they may have played an important selective role through geologic time.

CONTENTS

INTRODUCTION

In this volume we present two papers treating different aspects of North Pacific Radiolaria. The paper by Nigrini is the first distributional and paleoecological study of Radiolaria over a broad area of the North Pacific. My paper has two purposes. The first is to study stratigraphically significant Radiolaria and relate their ranges through the paleomagnetic stratigraphy, as determined by Ninkovich and others (1966) and by Opdyke and Foster (1970, this volume), to the ranges of radiolarian species in the equatorial Pacific and Antarctic. The second is to follow in some detail the evolution and dispersion of a single species and its ultimate extinction near a reversal of the earth's magnetic field. These are efforts to determine if the evolutionary responses of the species may shed some light on the cause of its extinction.

Previous investigations of Radiolaria in North Pacific sediments north of 30°N. have been purely taxonomic. Nigrini (1970, this volume) has reviewed the few earlier studies.

OCEANIC CIRCULATION AND SEDIMENTS

The physical characteristics and circulation pattern of the North Pacific surface waters have been thoroughly covered by Dodimeed and others (1963). It will suffice here to present only a brief review. The surface waters in the area of study (Fig. 1) are divided into two major water masses; the subarctic water north of about 40°N. and the subtropic water of the central North Pacific to the south. The subarctic water is characterized by a salinity minimum at the surface and a permanent halocline between 100 and 200 m, above which a thermocline develops in summer but vanishes in winter. Near its southern limit, a marked north-south temperature gradient persists, called the Polar Front. By contrast, in subtropic waters the salinity is a maximum at the surface and decreases to a minimum below 500 m. The stability of the water column depends solely on the temperature structure. The magnitude and depth of the thermocline varies throughout the year but does not vanish.

The boundary between these two water masses is designated the subarctic boundary. It is characterized by an isohaline structure that extends from the surface to depths of 200 to 400 m and crosses the North Pacific at about 40°N. lat. The subarctic boundary then marks a transition, separating cold, low salinity surface waters to the north, from warmer, higher salinity surface waters to the south.

The mean circulation is wind-driven; the movement of the subarctic water is cyclonic, and the subtropical water is anticyclonic (Fig. 2). Stommel (1957), in reviewing theories of such circulation, has shown that within the wind-driven anticyclonic gyres surface waters converge and sinking occurs, while within cyclonic gyres surface waters diverge, causing upwelling. Under such conditions of upwelling, nutrient-rich waters are brought to the surface and productivity is increased. Ried (1962) has given data for the North Pacific that show higher values for phosphate-phosphorus and zooplankton volume in the subarctic water than in the subtropic water.

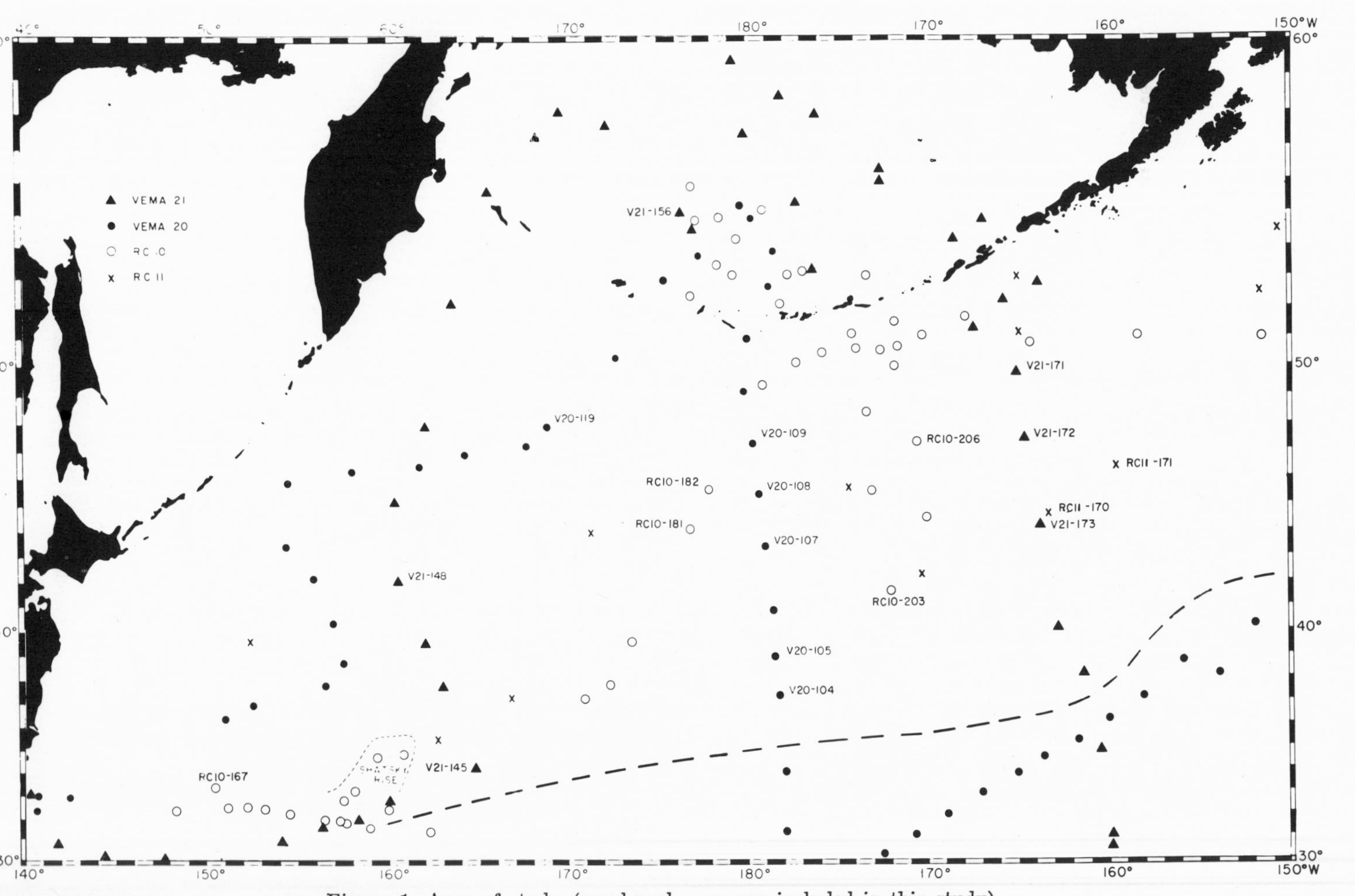

Figure 1. Area of study (numbered cores are included in this study).

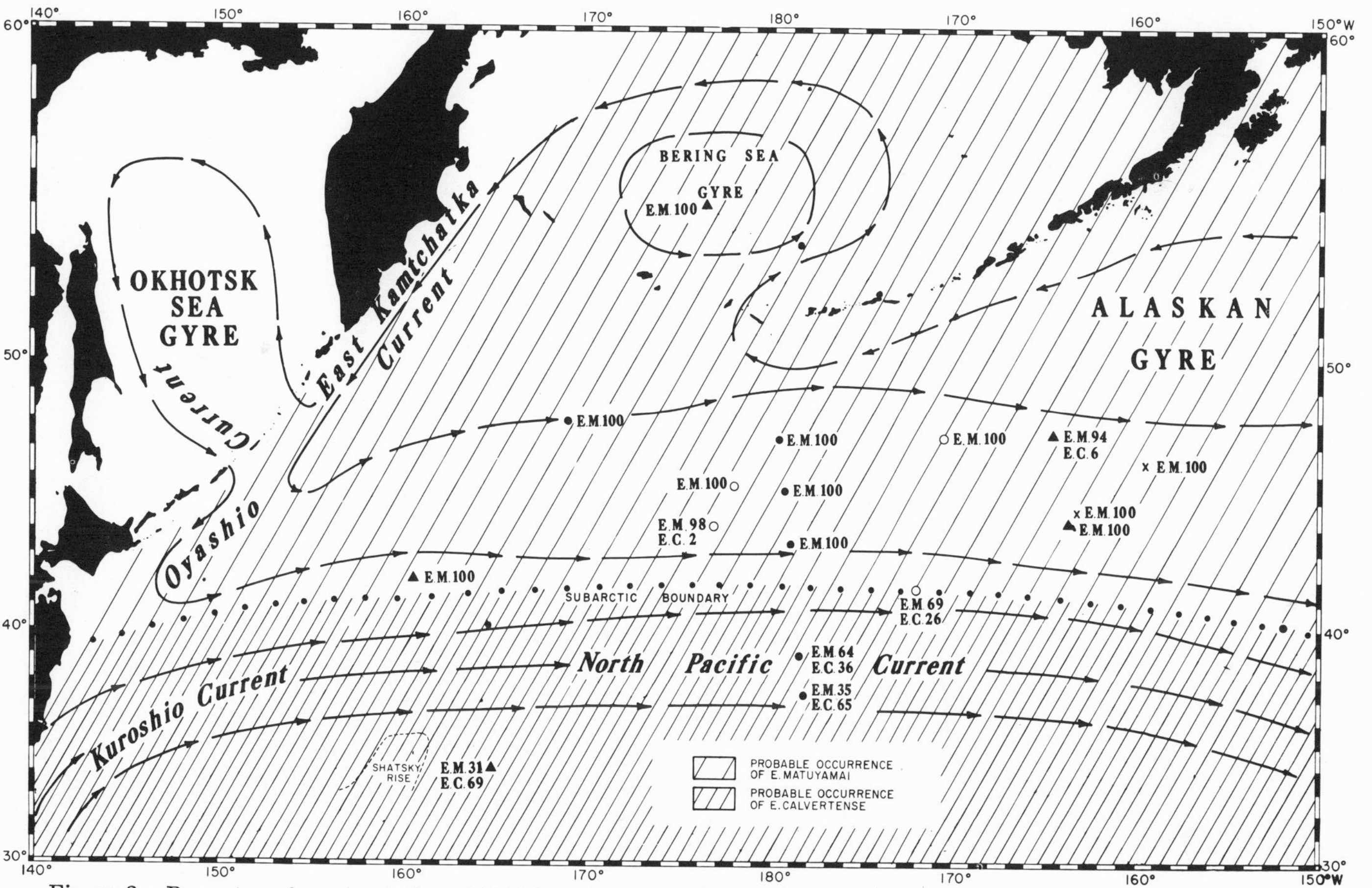

Figure 2. Present surface circulation of the North Pacific and geographical distribution of *Eucyrtidium matuyamai* (E.M.) and *Eucyrtidium calvertense* (E.C.) population at 1.0 m.y.B.P. Numbers next to core locations give relative abundance of E.C. to E.M. as percentage of the total of both.

The higher productivity of the subarctic water, as compared with the subtropical water, is reflected in the concentration of biogenic opaline silica in the underlying sediments. South of the dashed line in Figure 1, the sediments contain few to no radiolarian or diatom skeletons. To the north of this line the concentration increases, reaching maximum values south of the Aleutians and in parts of the Bering Sea. This line approximately corresponds to the boundary between the central North Pacific sedimentary province and the transitional areas to the north (Horn and others, 1970, this volume). The concentration also increases from east to west. Cores to the east of the study area contain generally few Radiolaria.

Although there are local variations, there is a regional trend of increasing rates of accumulation from east to west and from south to north (Opdyke and Foster, 1970, this volume), which roughly corresponds to the increase in biogenic silica.

Cores that represent the longest time intervals and contain Radiolaria, generally lie north of the dashed line in Figure 1, south of 45° N., and east of 175° E. Exceptions to this, such as V20-119 and V21-148, probably owe their long stratigraphies to multiple hiatuses.

RADIOLARIAN STRATIGRAPHY

Stratigraphic Index Species

Riedel, over a period of years, has provided much useful stratigraphic information on equatorial Radiolaria (Riedel, 1957, 1959; Riedel and others, 1963; Riedel and Funnell, 1964; Friend and Riedel, 1967). Recently Riedel (1969) has proposed a stratigraphic zonation of Neogene equatorial Pacific sediments based on Radiolaria. Hays (1965), and Hays and Opdyke (1967), have published radiolarian zonations for Pliocene and Quaternary deep-sea sediments in high southern latitudes. No stratigraphic work has yet been published on the Radiolaria of the North Pacific and it is the purpose of this section to establish a Pliocene and Quaternary radiolarian zonation for this area and relate it to the previously published work from other areas. The paper by Donahue (1970, this volume) is a parallel study that establishes a diatom zonation.

The tops and bottoms of all Lamont-Doherty cores in the North Pacific were sampled and examined for their radiolarian content. Cores selected for detailed study (Fig. 1) were chosen because they contain Radiolaria and span long intervals of time, at least back to the last reversal of the earth's magnetic field.

Nineteen cores were selected for detailed study (Fig. 1; Table 1). Radiolaria were separated on a 62μ screen from samples taken at 40 cm intervals. Additional samples were taken at 10 cm intervals to determine the upper limits of stratigraphically useful species.

Five North Pacific Radiolaria have restricted ranges within the Pleistocene and upper Pliocene, and can be used as stratigraphic indicators. These species, in order of increasing upward range, are: *Eucyrtidium elongatum peregrinum* Riedel, *Lamprocyclas heteroporos* Hays, *Eucyrtidium matuyamai* n. sp., *Stylatractus universus* n. sp. and *Druppatractus acquilonius* n. sp. (Fig. 3 and Pl. 1).

All were found in the oldest sediments studied (about 4 m.y. B.P.), except *E. matuyamai,* which evolved about 2 m.y. B.P., and all have continuous ranges until their extinction. The ranges of these species are shown in Figure 3 for three cores, and in Figure 2 in Hays and Ninkovich (1970, this volume) for 15 cores.

Druppatractus acquilonius becomes extinct at about 310,000 years B. P. (Figs. 3 and 4; also Fig. 3 *in* Hays and Ninkovich, 1970, this volume). During its range it is most abundant beneath the subarctic water mass where it represents 2 to 5 percent of the radiolarian assemblage. It is apparently restricted to the North Pacific for it has not been seen in either the Antarctic or the equatorial Pacific.

Stylatractus universus n. sp., which represents about one percent of the North Pacific radiolarian fauna during its range, was first described from Antarctic sediments (Hays, 1965). Its upper limit is used in the Antarctic to mark the boundary between the Ψ and Ω zones. The age of this boundary is estimated to be 400,000 years B.P., both by extrapolation from paleomagnetic reversals (Opdyke and others, 1966) and measurement of rates of sedimenta-

TABLE 1. LOCATION OF CORES INCLUDED IN THIS STUDY

Core No.	Lat.	Long.	Depth in Corr. m	Length in cm
V20-104	37°18′N	178°10′W	5449	1165
V20-105	39°00′N	178°17′W	5336	1237
V20-107	43°24′N	178°52′W	5872	1282
V20-108	45°27′N	179°14.5′W	5625	1671
V20-109	47°19′N	179°39′W	5629	1452
V20-119	47°57′N	168°47′E	2739	1170
V21-145	34°03′N	164°50′E	6088	1225
V21-148	42°05′N	160°36′E	5477	1448
V21-156	55°05′N	176°20′E	3418	1660
V21-171	49°53′N	164°57′W	5013	850
V21-172	47°40′N	164°21′W	5198	1081
V21-173	44°22′N	163°33′W	5493	1218
RC10-167	33°24′N	150°23′E	6092	1777
RC10-181	44°05′N	176°50′E	5698	1161
RC10-182	45°37′N	177°52′E	5561	1130
RC10-203	41°42′N	171°57′W	5883	1130
RC10-206	47°13′N	170°26′W	5497	1152
RC11-170	44°29.4′N	163°21.1′W	5451	1002
RC11-171	46°36.2′N	159°39.7′W	5167	1161

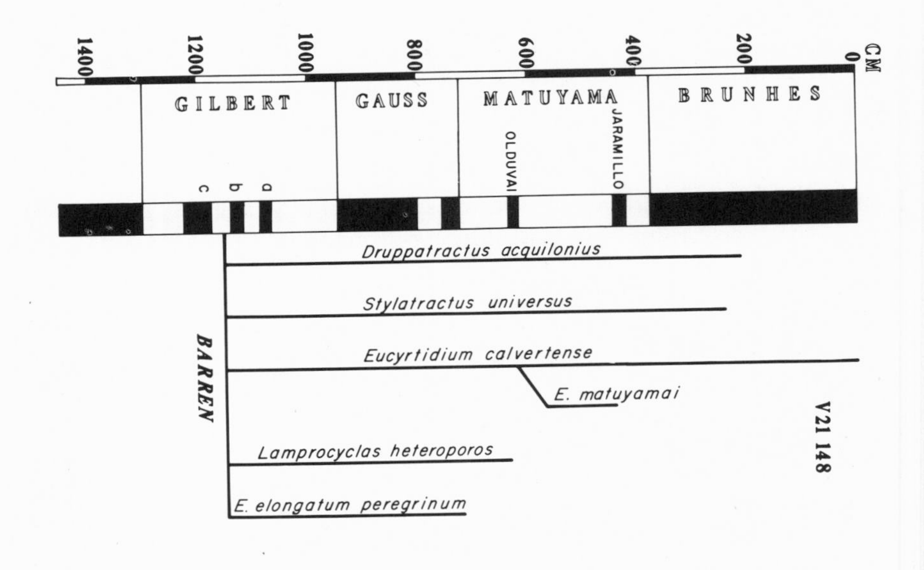

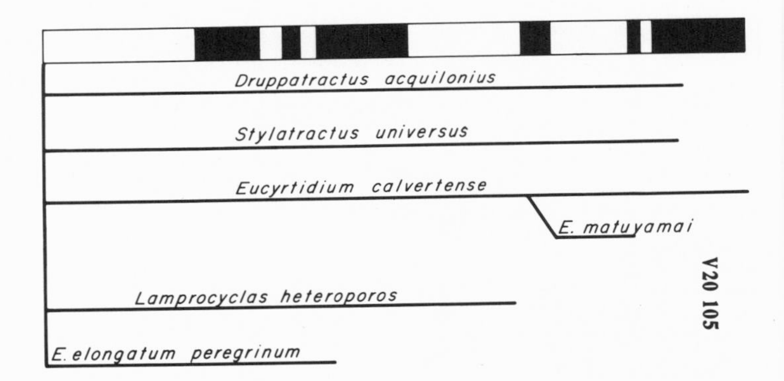

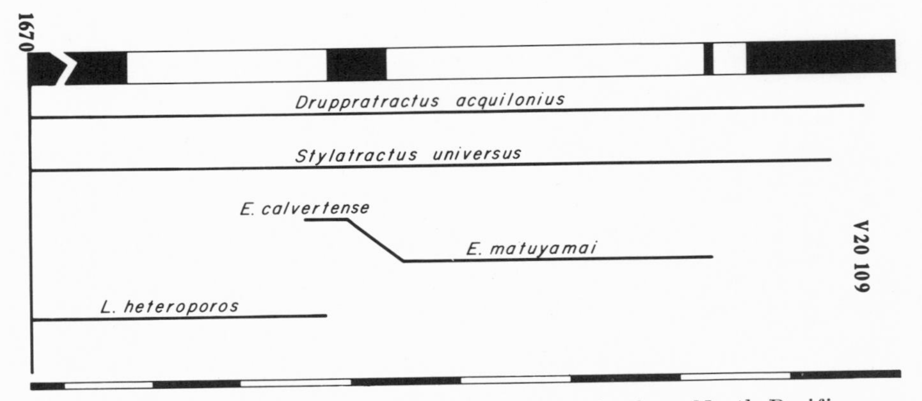

Figure 3. Stratigraphic ranges of indicator species in three North Pacific cores.

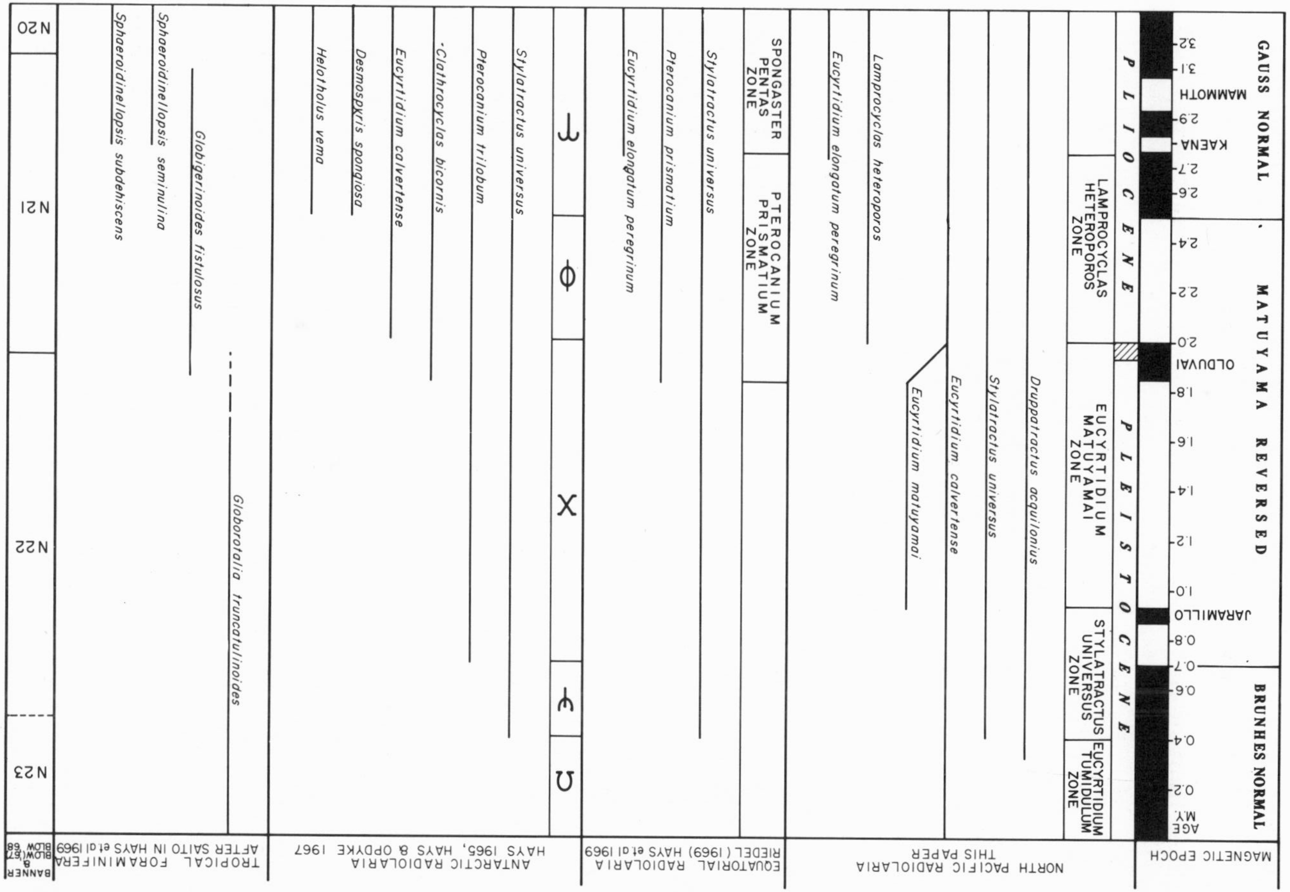

GAUSS NORMAL | MATUYAMA REVERSED | BRUNHES NORMAL — MAGNETIC EPOCH

MAMMOTH, KAENA, OLDUVAI, JARAMILLO

AGE M.Y.: 3.2, 3.1, 2.9, 2.7, 2.6, 2.4, 2.2, 2.0, 1.8, 1.6, 1.4, 1.2, 1.0, 0.8, 0.7, 0.6, 0.4, 0.2

PLIOCENE | *PLEISTOCENE*

NORTH PACIFIC RADIOLARIA THIS PAPER

LAMPROCYCLAS HETEROPOROS ZONE | EUCYRTIDIUM MATUYAMAI ZONE | STYLATRACTUS UNIVERSUS ZONE | EUCYRTIDIUM TUMIDULUM ZONE

Druppatractus acquilonius

Stylatractus universus

Eucyrtidium calvertense

Eucyrtidium matuyamai

Lamprocyclas heteroporos

Eucyrtidium elongatum peregrinum

EQUATORIAL RADIOLARIA RIEDEL (1969) HAYS et al 1969

SPONGASTER PENTAS ZONE | PTEROCANIUM PRISMATIUM ZONE

Stylatractus universus

Pterocanium prismatium

Eucyrtidium elongatum peregrinum

ANTARCTIC RADIOLARIA HAYS 1965, HAYS & OPDYKE 1967

Stylatractus universus

Pterocanium trilobum

Clathrocyclas bicornis

Eucyrtidium calvertense

Desmospyris spongiosa

Helotholus vema

TROPICAL FORAMINIFERA AFTER SAITO IN HAYS et al 1969

Globorotalia truncatulinoides

Globigerinoides fistulosus

Sphaeroidinellopsis seminulina

Sphaeroidinellopsis subdehiscens

BANNER & BLOW '67 BLOW '68

N20 | N21 | N22 | N23

Figure 4. Worldwide correlation of late Neogene radiolarian zones and their relationship to foraminiferal zonation.

tion derived from measuring excess Th^{230} (Ku and Broecker, 1966). In the North Pacific, based on the analysis of 15 cores and on extrapolation from the Brunhes/Matuyama boundary, the mean age of the upper limit of this species is about 400,000 years B.P. (*see* Fig. 3 *in* Hays and Ninkovich, 1970, this volume). Hays and others (1969) determined the upper limit of this species in four equatorial Pacific cores and estimated an average age for its disappearance of 341,000 years B.P. Although this is somewhat younger than the age in Antarctic and North Pacific sediments, the reality of this difference can only be ascertained by the analysis of more cores. In any case, the extinction of this cosmopolitan species occurred about the same time in the Antarctic and North Pacific. This simultaneous extinction effectively rules out local temperature changes and fluctuations of nutrients and points toward some global change in the environment.

Eucyrtidium matuyamai evolved from *Eucyrtidium calvertense* near the base of the Olduvai event and steadily increased in size until its extinction at the base of the Jaramillo event (Fig. 3). A detailed discussion of the evolution of this species is given later in this paper. It is interesting to note here that *Eucyrtidium calvertense* becomes extinct in the Antarctic near the base of the Olduvai event (Opdyke and others, 1966) at about the same time as it gives rise to *Eucyrtidium matuyamai* in the North Pacific. *Eucyrtidium calvertense* is found in Recent sediments south of 40° N., but is not present in Recent sediments north of this latitude.

Lamprocyclas heteroporos becomes extinct near the base of the Olduvai event (Fig. 3). It is possible that *L. heteroporos* evolves into *Androcyclas gamphonycha* Jorgensen. The two are sufficiently similar as to suggest a close relationship. In the Antarctic, *L. heteroporos* is found in the late Pliocene of cores taken north of the Antarctic convergence. In this area its disappearance seems to precede its last occurrence in the North Pacific. Hays (1965) showed the upper limit of *L. heteroporos* in V16-66 at a little more than 9 m. In treating the paleomagnetics of V16-66, Opdyke and others (1966) showed what appears to be the top of the Gauss normal polarity series at a little less than 9 m in this core. More Antarctic cores north of the Polar Front will have to be studied to be certain, but, on the basis of current data, *L. heteroporos* seems to have disappeared from Antarctic sediments 0.5 m.y. before its final occurrence in North Pacific sediments.

Eucyrtidium elongatum peregrinum was seen in only three cores (V21-148, V20-105 and RC11-203). In all of these, it occurs last in the middle of the Gauss normal polarity series. In V20-105, the core having the best preserved magnetic record, this species disappears just above the upper (Kaena) reversed event in the Gauss series. *E. elongatum peregrinum* also disappears just above the Kaena event in two equatorial Pacific cores (V24-59 and V24-62) taken some 35 degrees to the south (Hays and others, 1969). Thus, as with *Stylatractus universus,* the extinction of this species is, within the limits of sampling, nearly isochronous over broad areas and appears to be independent of latitude.

Radiolarian Zones

The species discussed above can be used to define four zones which can then be related to previously established radiolarian and foraminiferal zones in other parts of the world (Fig. 4).

Zone of *Eucyrtidium tumidulum*. Defined by the occurrence of *Eucyrtidium tumidulum* Bailey subsequent to the extinction of *Stylatractus universus. Eucyrtidium tumidulum* was originally described by Bailey (1956) from the Sea of Kamchatka. It is a common species in the North Pacific and ranges continuously from the youngest to the oldest sediments included in this study. The upper limit of *Druppatractus aquilonius* falls just above the base of this zone. This zone spans the last 400,000 years in North Pacific sediments and is equivalent to the Ω zone (Hays, 1965) in Antarctic sediments.

Zone of *Stylatractus universus*. Defined by the range of *S. universus* subsequent to the extinction of *Eucyrtidium matuyamai.* This zone roughly corresponds to the Ψ zone in Antarctic sediments, ranging in time from about 1 m.y. B.P. to 400,000 years B.P.

Zone of *Eucyrtidium matuyamai*. Defined by the occurrence of *Eucyrtidium matuyamai* subsequent to the last occurrence of *Lamprocyclas heteroporos.* This zone contains all of the range of *Eucyrtidium matuyamai,* corresponding approximately to the X zone in Antarctic sediments spanning the interval between about 2 m.y. B.P. and 1 m.y. B.P. The base of this zone can be used to define the base of the Pleistocene in North Pacific sediments.

Zone of *Lamprocyclas heteroporos*. Defined by the range of *Lamprocyclas heteroporos* subsequent to the extinction of *Eucyrtidium elongatum peregrinum.* This is the uppermost Pliocene zone falling within the zone of *Pterocanium prismatium,* as defined by Riedel (1969) in equatorial Pacific cores. It has a similar upper boundary but extends below the lower boundary of the Antarctic Φ zone. The *L. heteroporos* zone spans the time between 2.8 and 2.0 m.y. B.P. (Fig. 4).

The oldest sediments included in this study are those at the bottom of V21-148 (Fig. 3). This core becomes barren between the Gilbert "b" and "c" events in the range of *Eucyrtidium elongatum peregrinum.* Equatorial Pacific cores that passed through the Gilbert "c" event (Hays and others, 1969) did not penetrate below the range of the Foraminifera species *Sphaeroidinella dehiscens,* the first appearance of which is a marker for the base of the Pliocene (Banner and Blow, 1965). Therefore, the zones defined in this study are within the Pliocene and Pleistocene.

The Pliocene/Pleistocene Boundary in the North Pacific and Worldwide Stratigraphic Correlation

Since the advent of extensive paleomagnetic work on deep-sea cores a few years ago, a tightly controlled, late Neogene worldwide chronostratigraphy has rapidly developed (Opdyke and others, 1966; Berggren and others, 1967; Hays and Opdyke, 1967; Glass and others, 1967; Ericson and Wollin, 1968; Hays and Berggren, 1971; Hays and others, 1969).

Banner and Blow (1965) showed that the evolutionary appearance of the planktonic foraminiferal species *Globorotalia truncatulinoides,* from its immediate ancestor *Globorotalia tosaensis,* occurs near the base of the holostratotype Calabrian at Santa Maria di Catanzaro, in southern Italy. Berggren and others (1967) found the evolutionary transition of these species in a North Atlantic deep-sea core within the Olduvai magnetic event, and thus established the age of the Pliocene/Pleistocene boundary at about 1.8 m.y. ago.

The level of the boundary established by Berggren and others (1967) was similar to that obtained by the application of criteria recommended previously by Ericson and others (1963) for the North Atlantic, Riedel and others (1963) for the Equatorial Pacific, and Hays (1965) for the Antarctic.

Since the Olduvai event occurs in a number of North Pacific cores, and faunal changes are associated with it, it is possible to establish faunal criteria for recognizing the base of the Pleistocene in this area. Hays and Berggren (1971), in reviewing the status of the Pliocene/Pleistocene boundary, recommended that the first evolutionary appearance of *E. matuyamai* be used as a criterion for recognizing the boundary in North Pacific sediments. Since this transition extends over a considerable period of time and the distinction between *E. calvertense* and *E. matuyamai* at any horizon must be arbitrary, the upper limit of *L. heteroporos* may be a more practical criteria. There is no evidence of any major climatic change across this boundary in the North Pacific.

Figure 4 compares various radiolarian zonations and shows the relationship of these zones to the ranges of certain Foraminifera from the Equatorial Pacific. Hays and others (1969) have related the magnetic stratigraphy of the last 4.5 m.y. to the previous foraminiferal zonations of Bandy (1963), Bolli (1966), Bandy and Wade (1967), Berggren and others (1967), Parker (1967), Banner and

Blow (1967), and Blow (1968). It is possible then through this paper to relate North Pacific radiolarian zones to the foraminiferal zonations of these workers.

EVOLUTION AND EXTINCTION OF *Eucyrtidium Matuyamai*

The evolutionary development of *Eucyrtidium matuyamai* from its immediate ancestor *Eucyrtidium calvertense* near the base of the Olduvai event, and its ultimate extinction near the base of the Jaramillo event, affords an opportunity to closely examine the entire evolutionary development of a species (Fig. 5). Since the species lived for only a million years and no other radiolarian species in the area became extinct at the base of the Jaramillo event, it is possible that a close examination of the evolutionary development and dispersion of this species will shed some light on the cause of its extinction.

Evolution and Dispersion of *Eucyrtidium matuyamai*

The main difference between *E. matuyamai* and its ancestor, *E. calvertense,* is a difference in size. Although measurements of a number of parameters would demonstrate quantitatively this increase in size, the one that demonstrates it best is the maximum width of the test. The change in size through time of this species (Fig. 5) is delineated in this section through extensive measurements of the width of the fourth segment (almost invariably equivalent to the maximum width of the test).

Eucyrtidium calvertense is found in the surface sediments of the North Pacific south of 40° N. and the living population is presumably restricted to the central water masses south of the subarctic boundary. *E. calvertense* has lived in this area for at least the last 4 m.y., judging from its presence throughout the length of core V20-105.

About 2 m.y. ago, *E. calvertense* invaded the subarctic water masses, for just below the Olduvai event in cores V20-109, V21-173, RC11-170 and RC11-171 (Fig. 1), specimens of *E. calvertense* are found that are similar in morphology to specimens found at comparable levels in cores to the south.

Shortly after its arrival in the north, the size and abundance of individuals in the population began to increase. In general, the increase in size continues upward to the base of the Jaramillo event; however, it is more rapid during the first 400,000 years (2.0 m.y. to 1.6 m.y. B.P.) of its residence in the north. During this time, the mean width of individuals (based on 20-60 measurements per sample) in northern cores (V20-109, V21-173) increases from about

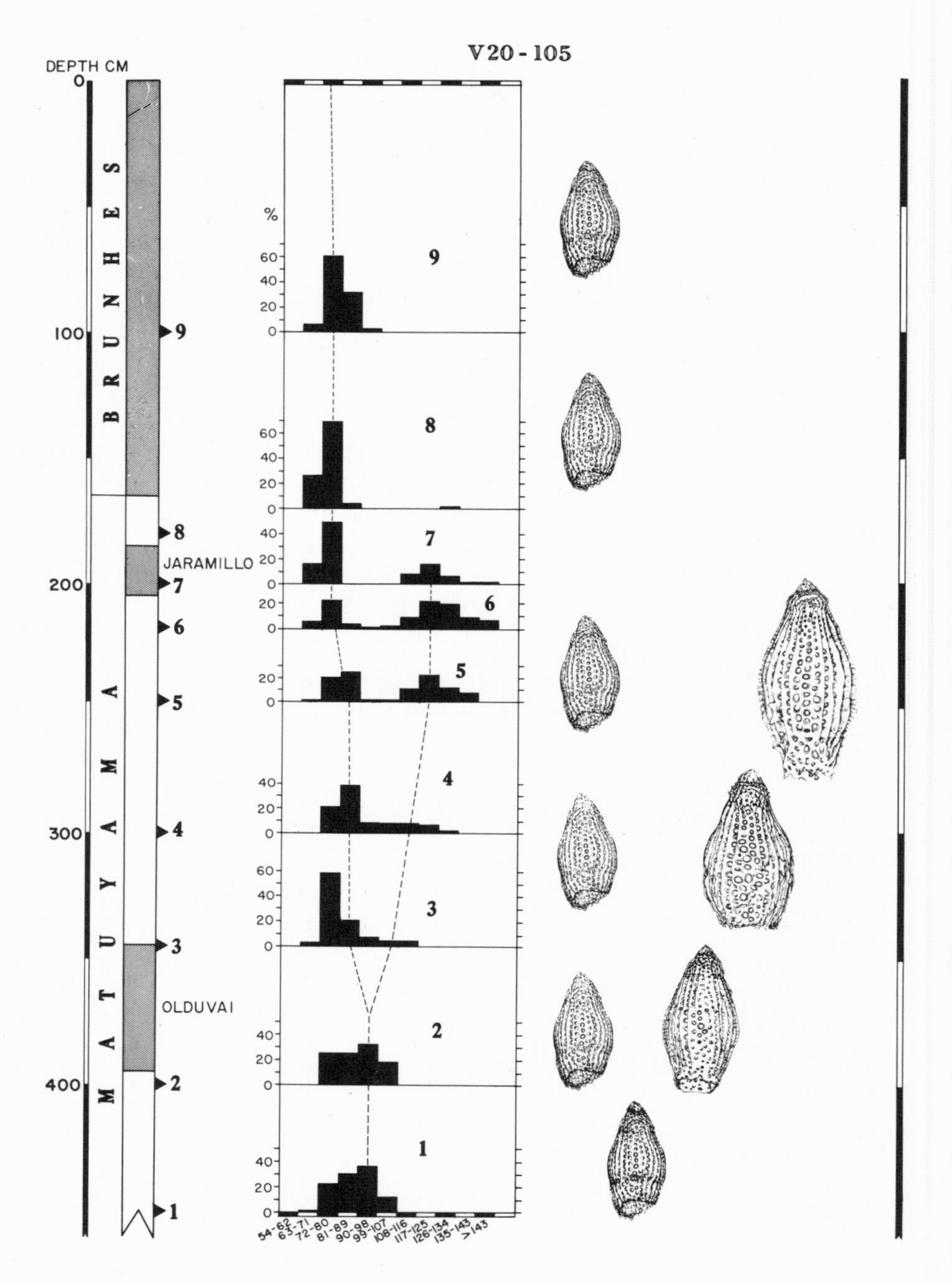

Figure 5. Morphological development of *Eucyrtidium matuyamai* from *Eucyrtidium calvertense*.

98μ to over 120μ for an increase in this dimension of about 20 percent (Fig. 6). Subsequent to 1.6 m.y. B.P., the increase in size is at a significantly slower rate. In V21-173, the mean maximum diameter increases from 120 microns at 1.6 m.y., to 132 microns at 0.9 m.y., an increase of 10 percent in 700,000 years. This is about one-quarter the rate of the first 400,000 years. In V20-109, the increase in size is about 12 percent between 1.6 m.y. and 1.2 m.y.; however, the species then decreases in size until its extinction, at which time it is about the same size as at 1.6 m.y. As will be documented below, the level at which the evolutionary rate decreases (1.6 m.y. B. P.) occurs at about the time the large northern individuals invade the subtropical water to the south and co-exist with the smaller southern individuals. Because the geographic ranges of the large and small populations begin to overlap at this point, it seems an appropriate place to separate the large species (*Eucyrtidium matuyamai*) from the smaller one (*Eucyrtidium calvertense*). By the end of its range, *E. matuyamai* was widely distributed in the subtropical water mass (Fig. 2).

The appearance and evolutionary development of *Eucyrtidium matuyamai* in the south follows a significantly different pattern from its development beneath the subarctic water. During the time of rapid evolution in the north, between 2 m.y. and 1.6 m.y. ago, there was no increase in the size of individuals in southern cores (V20-105 and V21-145; Fig. 6).

In order to closely examine the simultaneous evolutionary development of *Eucyrtidium matuyamai* within the subarctic and subtropical water masses, cores V20-109 and V21-145 were selected for detailed study (Fig. 7). After determining the absolute ages of various levels in the two cores by means of the magnetic stratigraphy (assuming an age of 2.0 m.y. B.P. for the base of the Olduvai event, and an age of 0.9 m.y. for the base of the Jaramillo event), samples were selected from the two cores so that pairs of samples represent similar ages in each core. The pairs of samples were spaced through the cores so as to closely cover portions of the cores representing the time of rapid evolution in the north (between 2 m.y. and 1.4 m.y. ago) and the period just prior to the extinction of *E. matuyamai*. The mean widths of at least 50 individuals from each sample were measured and the data grouped into 8 micron intervals and expressed as histograms in Figure 7.

Between 2.0 and 1.8 m.y. a strong trend toward increasing size is evident in the northern core (V20-109). The mode of the distribution shifts from 90-98μ to 117-125μ. During this time there is no striking change in the size distribution of the population in the southern core (V21-145). However, in the south between 1.7 m.y. and 1.6 m.y., the population rapidly becomes bimodal with the center of the coarser mode at 117-125μ. This size is identical to the mode of the northern population at this time (V20-109, Fig. 7).

It could be argued that this change to a bimodal population in the south (V21-145 and V20-105) signifies the evolution in the central water masses of *E. matuyamai*, and so it would appear without the additional information from V20-109. An alternate, and preferable, interpretation is that the sudden (within 150,000 years) appearance in the south of specimens having a similar size to those already living in the north represents a migration of the subarctic population into subtropical waters. Since there is constant mixing of subarctic and subtropical waters along the subarctic boundary,

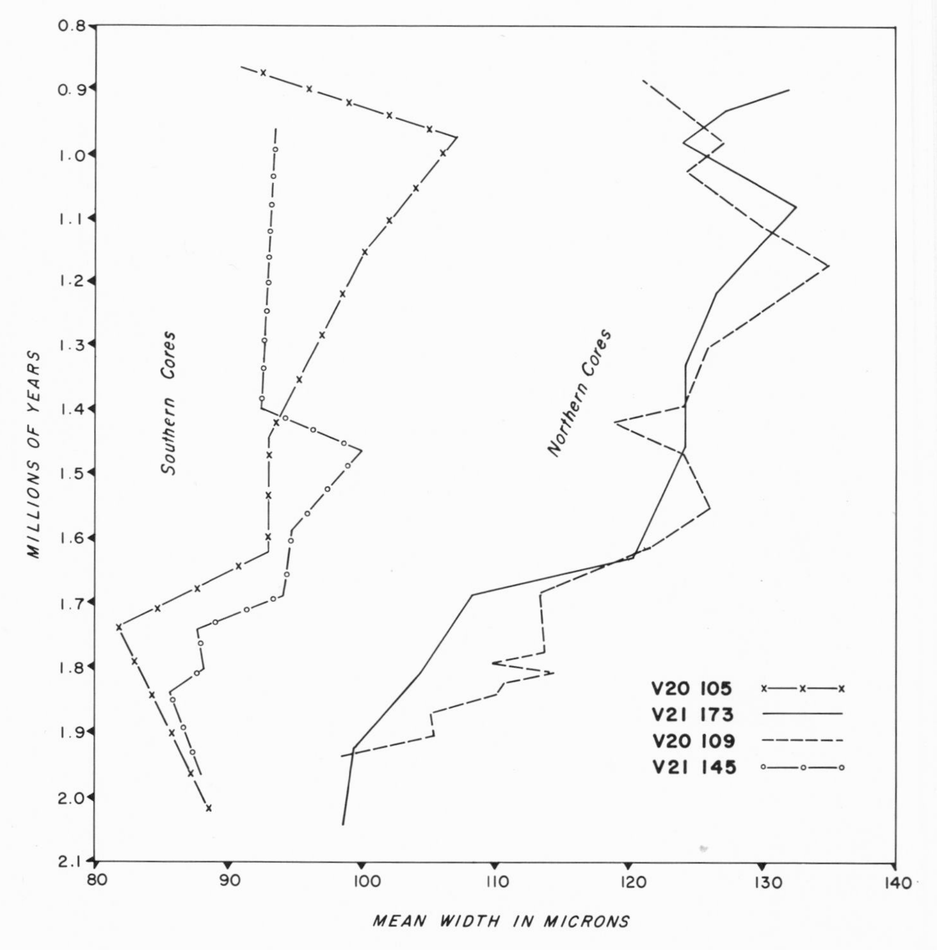

Figure 6. Change in mean width through time of the *Eucyrtidium calvertense*—*Eucyrtidium matuyamai* lineage in four cores.

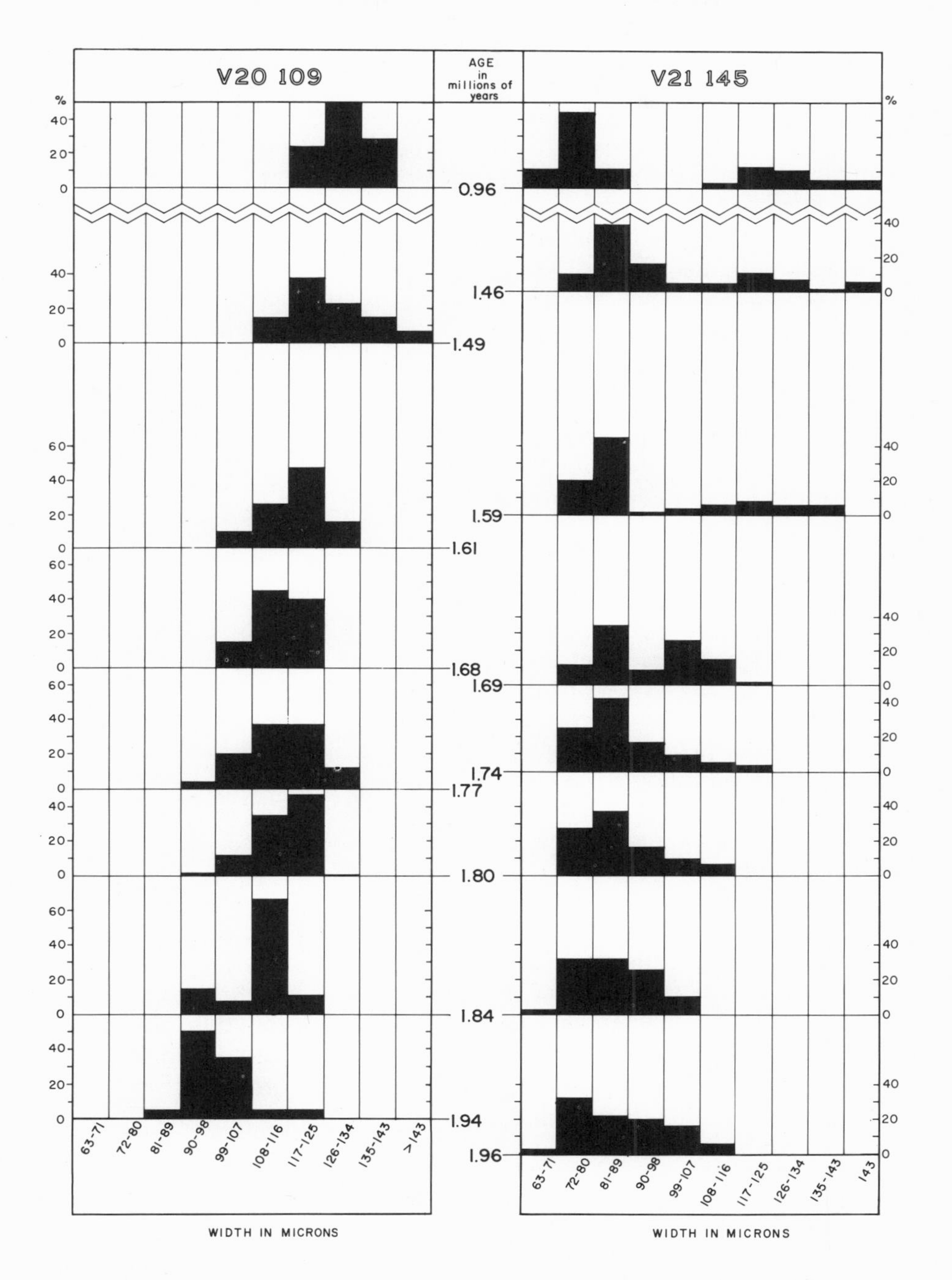

Figure 7. Size distribution of maximum shell width of *Eucyrtidium matuyamai* at selected ages in a northern core V20-109 and a southern core V21-145.

individuals of the northern population were certainly transported into subtropical waters prior to 1.7 m.y. B.P. and very possibly could live there. However, because of insufficient differences between the two populations, it is probable that prior to 1.7 m.y. individuals that were washed into subtropical waters from the larger sized northern population were not able to maintain their identity and were absorbed through interbreeding into the southern population. By 1.6 m.y. B.P., the northern population (*E. matuyamai*) must have established sufficient genetic difference from the southern population (*E. calvertense*) so that they no longer interbred, and *E. matuyamai* could maintain itself in the south as a separate population from *E. calvertense*. Using this line of reasoning, I place the lower limit of *E. matuyamai* at between 1.7 m.y. and 1.6 m.y., for it was at this time that it became a new species, separate from its ancestor *E. calvertense*.

At any given time during its range, *Eucyrtidium matuyamai* is most abundant and the individuals are largest in the sediments underlying subarctic waters (for example, V20-119 and V20-109). The relative abundance decreases southward, as does the size of individuals. Samples were selected at the 1 m.y. level in 13 cores from beneath the subarctic and subtropical water masses, and the mean width of the *E. matuyamai* assemblage was determined. The mean widths of assemblages in southern cores are significantly smaller than those in more northerly cores (Fig. 8). This ecophenotypic variation is possibly a result of the same selective pressures that caused the evolutionary change in the northern population. Of the various ecological parameters measured in this area, for example, phosphate-phosphorus, oxygen, salinity, and temperature, temperature is the strongest and most consistently latitude dependent. Nutrients, trace elements, depth of thermocline, and so on, cannot be ruled out as being important influences on this ecophenotypic variation. After an extensive study of the zooplankton in this area, Fager and McGowan (1963) concluded that zooplankton abundances are either controlled by a complex of factors, or the usual hydrographic measurements are not the critical biological parameters.

The possibility that *E. matuyamai* was washed into the southern area and never lived there, or at least never reproduced there, seems very unlikely for the following reasons: first, it is quite abundant at some levels in V21-145, which is well south and upcurrent from the subarctic boundary; second, the mean size of the southern population (in cores V20-104, V20-105 and V21-145) at about 1.0 m.y. B.P., are significantly different from the northern populations (Fig. 8).

In summary, then, the history of this species seems to have been as follows: About 2 m.y. ago, due to a mutation or series of

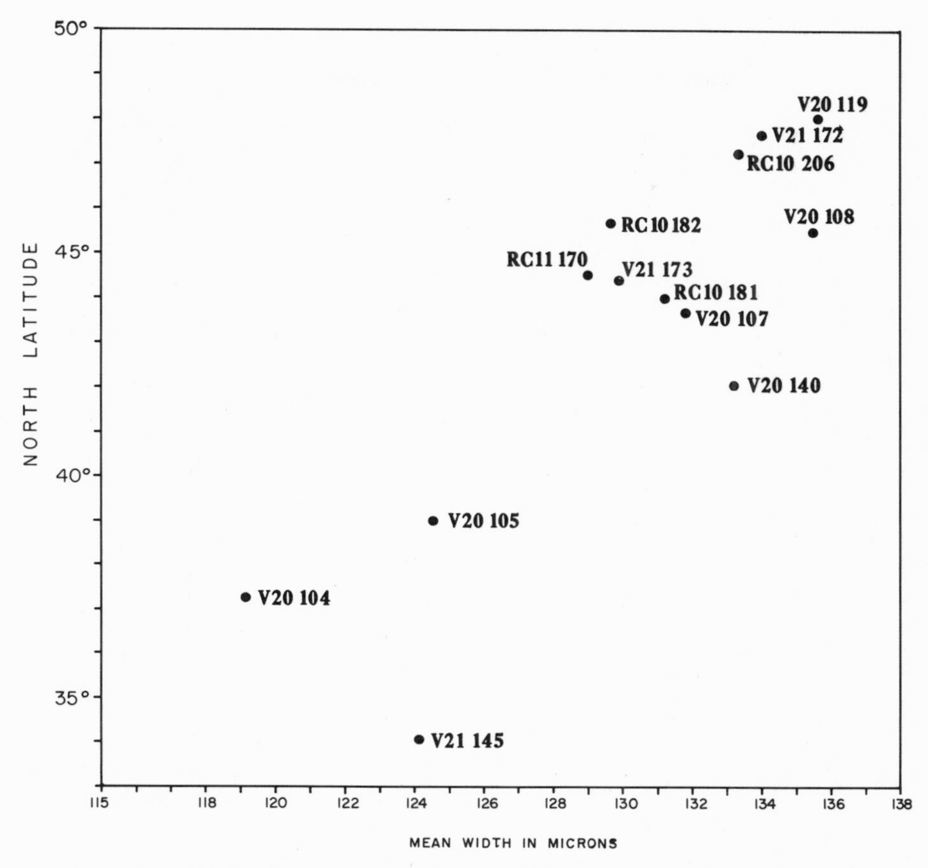

Figure 8. Relation between mean shell width of *Eucyrtidium matuyamai* and latitude at 1.0 m.y. B.P. in 11 cores.

mutations, a part of the *E. calvertense* population inhabiting the subtropical waters of the central Pacific successfully invaded the subarctic water to the north and established itself. In the subarctic water the population responded to the new environmental stresses and rapidly evolved, demonstrating this evolution morphologically by an increase in size. The evolution was most rapid between 2.0 m.y. and about 1.6 m.y. ago. During this time, the size of individuals in the *Eucyrtidium calvertense* population south of the subarctic boundary did not change. After 1.6 m.y. ago, individuals appear in the south that are as large as those living at the same time in the north, indicating: (1) an invasion of the subtropical water by the large-sized subarctic population, and (2) the subarctic population had evolved sufficiently so that individuals of this population could live in the same area with the ancestral population and maintain their identity. A new species, *Eucyrtidium matuyamai,* had evolved.

This species lived on in both the subarctic and subtropical waters until the beginning of the Jaramillo event when it became extinct, abruptly and isochronously, throughout its geographic range.

The Extinction of *Eucyrtidium matuyamai*.

Judging by the relationship in 15 cores between the upper limit of the range of *E. matuyamai* and the base of the Jaramillo event, there is apparently an excellent correlation between the extinction of *E. matuyamai* and the reversal that began the Jaramillo event (Fig. 9). However, the degree to which the sedimentary record accurately represents the true time relationship between these two events is dependent upon several factors: (1) the amount of upward mixing of the species, (2) depression of the record of the reversal by burrowing organisms, and (3) any lag in time of permanent magnetization of the sediments. Each of these factors will be considered separately.

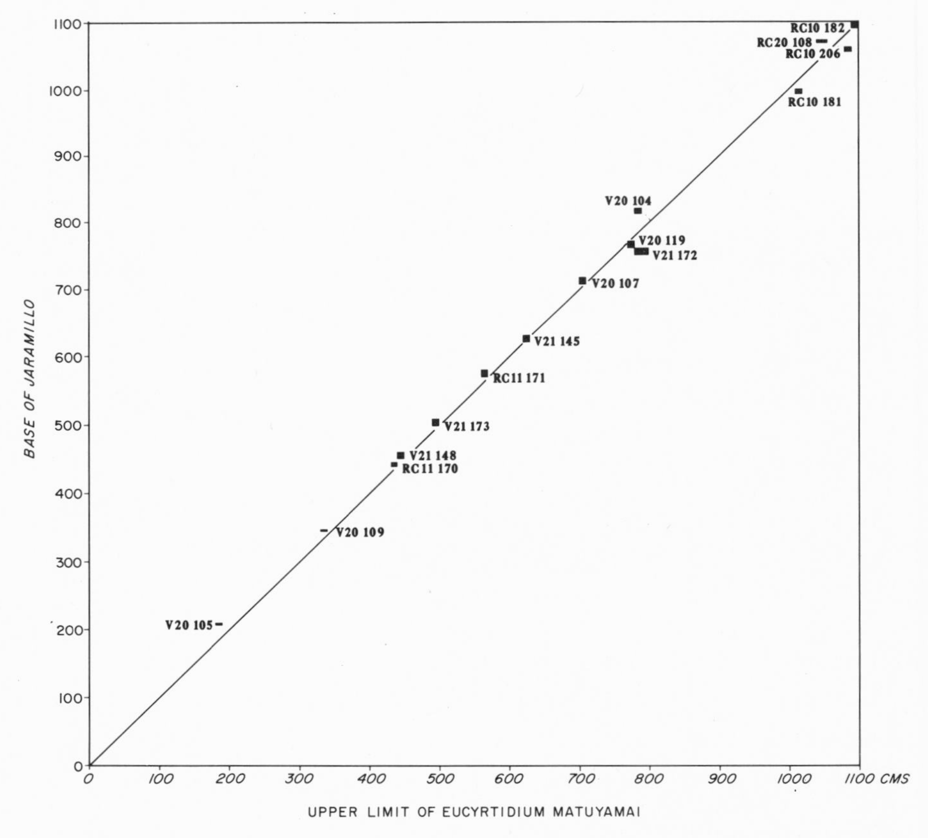

Figure 9. Relationship between base of Jaramillo event and extinction of *Eucyrtidium matuyamai* in 16 cores.

Upward Mixing of Species. In order to determine the degree of upward displacement of specimens due to mixing, counts were made of over 1000 Radiolaria in 11 cores at intervals through the last occurrence of *E. matuyamai.* These data are expressed graphically (Fig. 10) as numbers of individuals of *E. matuyamai* per 1000 Radiolaria counted. The decline in abundance just before disappearance is rapid in most cores, ranging from 10 cm to a maximum of 80 cm (Table 2). The rate of decline of the fossil population is probably primarily a result of upward mixing; however, it may to some extent reflect a declining living population. If this is so, then picking the level of extinction relative to the reversal that marks the base of the Jaramillo event is made more difficult.

In cores having high rates of accumulation (RC10-182 to V21-172, Fig. 10) the beginning of the decline in abundance always occurs below the base of the Jaramillo event, while the absolute upper limit is above the base. The point of extinction must fall between these two points. Berger and Heath (1968) have shown theoretically that the time of extinction will be represented by the level at which the abundance of individuals falls to 37 percent of its normal abundance, if one assumes the decline in abundance is due solely to upward mixing. There is sufficient variation in abundance in most of these cores so as to make any firm statement of normal abundance unwise. However, V20-119, which has the greatest relative abundance of any of the high sedimentation rate cores, shows a rather constant abundance (14 specimens per 1000) up to 780 cm, where it drops sharply to 5 specimens per 1000 at 770 cm and then declines steadily to 0 at 740 cm. The 37 percent of normal level falls at 770 cm, which is coincident with the reversal within the sampling interval (Table 2).

Assuming a spread of normal abundances for the other high sedimentation rate cores, and assuming the decline in abundance is due solely to upward mixing, the calculated extinction point either falls at the reversal or slightly below it (maximum 35 cm.; Table 2). In cores having slower rates of accumulation (7.6-3.9 mm/1000 yrs, V20-107 to V20-119, Fig. 10), the drop in abundance begins above the base of the Jaramillo event and so the extinction point is always above it (Table 2).

Depression of Level of Reversal by Burrowing Organisms

In the highly fluid upper layer of ocean sediments it is conceivable that immediately following a reversal disturbance of some thickness of this layer by burrowing organisms may erase the previous direction of remnant magnetism and cause these sediments to align themselves in the direction of the new field. This would cause a depression of the reversal boundary in the sediments. Presumably,

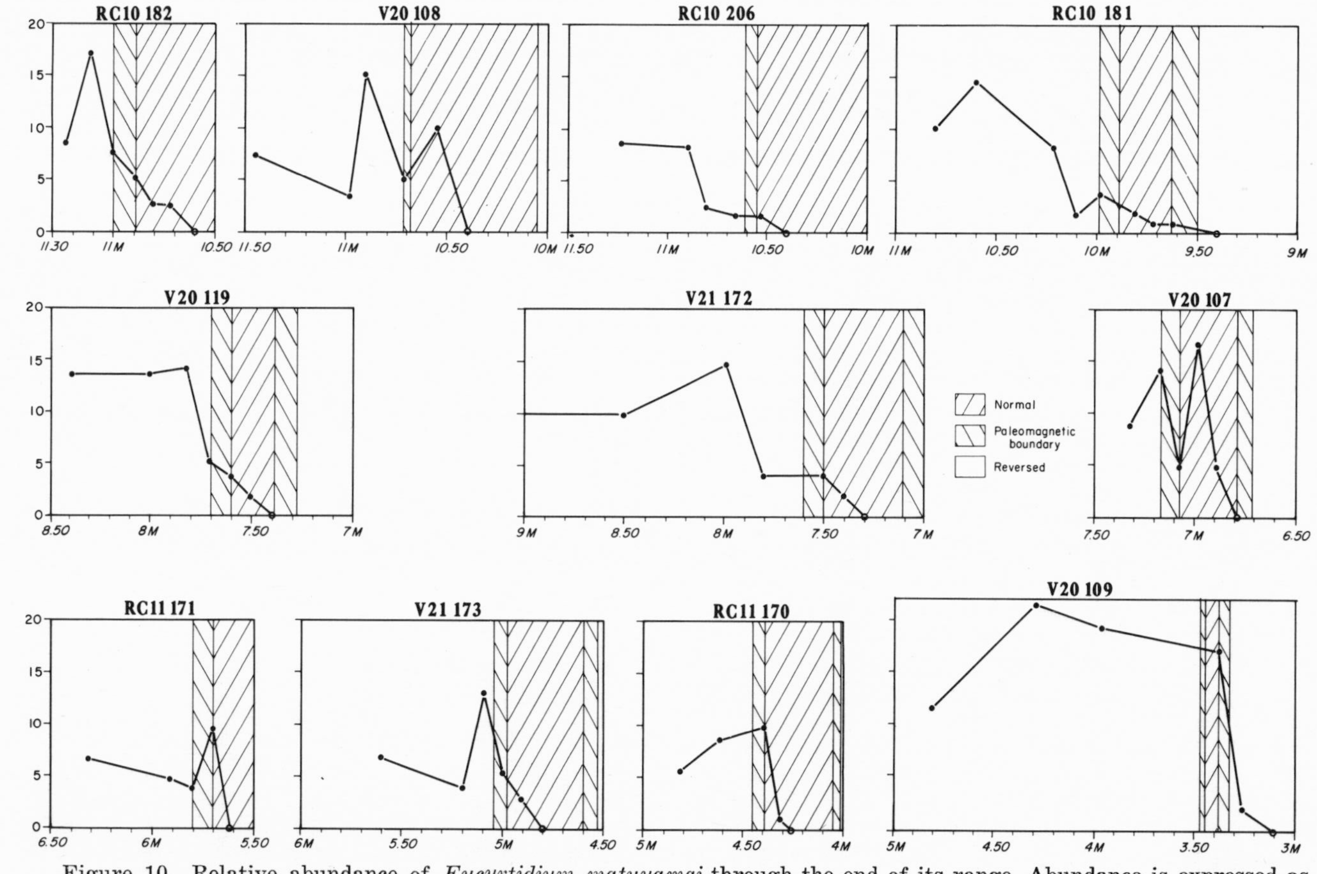

Figure 10. Relative abundance of *Eucyrtidium matuyamai* through the end of its range. Abundance is expressed as numbers of *Eucyrtidium matuyamai* per 1000 Radiolaria. Right diagonal hachures represent normal magnetization. Left diagonal hachures indicate the zone within which the reversal must occur.

TABLE 2. DATA RELATING TO CORRELATION BETWEEN EXTINCTION POINT OF *E. matuyamai* AND BASE OF JARAMILLO EVENT

Core No.	Sedimentation Rate mm/1000 yrs	Interval of Decline in cm	Normal Abundance Numbers/1000	37% of Normal Numbers/1000	Level of 37% Normal	Level of Reversal	Min. and Max. possible distance between extinction point and reversal in cm. Extinction point above reversal (+); below reversal (−).
RC10-182	11.8	1100-1060 (40 cm)	9-16	3-6	1082-1092	1090-1100	0 to −18
V20-108	11.5	?					
RC10-206	11.2	1090-1040 (50 cm)	8	3	1080	1050-1060	−20 to −30
RC10-181	11.1	1020-940 (80 cm)	8-14	3-5	990-1015	990-1000	0 to −25
V20-119	9.2	780-740 (40 cm)	14	5	770	760-770	0 to −10
V21-172	8.2	800-730 (70 cm)	10-15	3-6	745-785	750-760	−35 to +15
V20-107	7.6	700-680 (20 cm)	5-15	2-6	685-695	710-720	+15 to +35
RC11-171	6.3	560-570 (10 cm)	5-10	2-4	560-570	570-580	0 to +20
V21-173	5.5	510-480 (30 cm)	4-13	1-5	490-500	495-510	0 to +20
RC11-170	5.0	440-425 (15 cm)	5-10	2-4	430-440	437-445	0 to +15
V20-109	3.9	340-310 (30 cm)	12-21	4-8	330-335	345-347	+10 to +17

the degree of depression would be a function of the thickness of the mixed layer. There seems to be no way to raise the reversal boundary by mixing or any other process. The thickness of the mixed layer will vary with such variables as rates of sedimentation, sediment type, and intensity of burrowing.

The difference between the position of the extinction point of *E. matuyamai* in the high sedimentation rate cores and in the low sedimentation rate cores, suggests such a depression of the reversal. The fact that the calculated extinction point consistently falls above the base of the Jaramillo event in the low sedimentation rate cores, suggests that a small depression of the reversal is possible (assuming a co-occurrence of extinction and reversal). Since the extinction point ranges from as much as 35 cm below the Jaramillo event in cores with high accumulation rates, to as much as 20 cm above the base of this event in cores with low accumulation rates, the maximum possible depression of the reversal by burrowing is 55 cm. However, I would consider this number to be too large and a better estimate for these cores would be about 20 to 30 cm.

Delay in Time of Acquisition of Permanent Magnetization

It has been suggested recently (Dymond, 1969) that there may be a considerable lag between the time of sediment deposition and the time when the sediment acquires a permanent direction of remnant magnetism. If this is so, the reversal recorded in the sediment might be considerably older than the actual time of reversal. The amount of lag would be a function of time if some diagenetic process was responsible, or depth if compaction was important. In the former case the amount of depression would vary with sedimentation rate, and in the latter case it would probably vary with water content and sediment type. The very close correlation between the extinction point of *E. matuyamai* and the lower boundary of the Jaramillo in cores that have widely varying rates of accumulation and different lithologies (calcareous ooze V20-119, diatom ooze V20-109, radiolarian clay V21-145), rules out the possibility of any long lag. A delay amounting to a few tens of centimeters is possible.

The close relationship between the extinction point of *E. matuyamai* and the base of the Jaramillo event is strong evidence that the sediments contain an accurate record of the time of reversal of the earth's magnetic field, and when upward mixing is taken into account they also provide an accurate record of the time of extinction of planktonic organisms.

The strong correlation between the extinction of this species and the base of the Jaramillo event (Fig. 9), therefore indicates a close correlation between the time of the reversal and the time of the extinction of *E. matuyamai*.

Cause of Extinction of *Eucyrtidium matuyamai*

Although there is a widely accepted general theory of evolution, there is no general theory to explain extinctions. In fact, documenting the cause of any extinction is very difficult and so well-documented cases are rare.

Certain facts about the extinction of some Radiolaria are now becoming clear and these can be used at least to limit the possible causes of extinction.

First, the extinctions are rapid. This is not only true for species that are not abundant, such as *Stylatractus universus* and *Eucyrtidium matuyamai* in this study, but also for *Clathrocyclas bicornis* Hays in the Antarctic, where it may comprise 10 to 15 percent of the radiolarian fauna during its range (Hays and Opdyke, 1967).

Second, the extinction of some species occurs simultaneously throughout the geographic range of the population.

Third, it is rare for more than one species to become extinct at a time. This, incidentally, is probably good evidence against dimorphism in Radiolaria.

From these facts we can infer that the cause of extinction is rapid and wide-spread, possibly global. The nature of the cause is such that it only affects one species at a time.

The simultaneous extinction of the entire population of *E. matuyamai* at a time that coincides or nearly coincides with the base of the Jaramillo event, strongly suggests that environmental factors connected with the reversal caused the extinction of this species. The relationship between extinctions of planktonic organisms and reversals of the earth's magnetic field has been given much attention recently (Opdyke and others, 1966; Watkins and Goodell, 1967; Hays and Opdyke, 1967; Hays and others, 1969). These investigations have raised the interest of a number of workers (Waddington, 1967; Black, 1967; Harrison, 1968) who have examined the theoretical amount of increase of radiation expected at the earth's surface at the time of a field reversal, assuming the dipole field goes to zero (Bullard, 1955). They have concluded that the increased radiation would be insufficient to cause any large-scale increase in the mutation rate, or wholesale extinction. These studies cast doubt on Uffen's (1963) hypothesis that extinction at the time of a reversal may be caused by increased radiation. Nevertheless, the correlation between extinction of *E. matuyamai* and a reversal is striking, and when combined with similar correlations in the Antarctic (Opdyke and others, 1966; Hays and Opdyke, 1967) and equatorial Pacific (Hays and others, 1969), indicates that a connection between reversals and extinctions must be considered.

Since *E. matuyamai* is the only species to become extinct at the base of the Jaramillo event, it must have had unique qualities not shared by other Radiolaria living at the time that made it vulnerable to some small environmental change.

Throughout its geologic range, *E. matuyamai* was continuously evolving, indicating instability and either poor adaptation to its environment or great sensitivity to small changes of its environment. This sensitivity and specialization to the environment is further borne out by the fact that the size of the species varies with latitude, indicating specialization to some latitude-dependent property of the water masses in which it lived (possibly temperature).

Its degree of specialization and vulnerability to extinction may have been products of the way it evolved. Experiments with the fruit fly *Drosophila robusta* (Carson, 1961) have shown that laboratory populations derived from the center of the species geographic range show greater genetic variability and over-all fitness than the structurally homozygous populations derived from the margin of the species range. Carson (1961) concludes that his findings provide support for the hypothesis that chromosomal polymorphism, as found in central populations, provides heterozygotes which are superior to their corresponding homozygotes in all facies of the environment. In contrast to this, marginal populations appear to be more narrowly specialized and exploit the environment through fixation of genes or gene combinations in the homozygous state.

Eucyrtidium matuyamai evolved from individuals of the *E. calvertense* population living at the northern edge of the range of this latter species. Therefore, these individuals probably had less genetic variability than individuals living farther south. In addition, the rapid evolution of the individuals that invaded the subarctic water probably further limited the genetic variability. As a consequence, in spite of the fact that *E. matuyamai* was able to reinvade the subtropical water, it was probably in many ways far more rigid than the *E. calvertense* population. This would explain the greater vulnerability of *E. matuyamai* to extinction.

The vulnerability of *E. matuyamai* is only one aspect of the problem of its extinction. There must have also been some external environmental pressure that exploited this vulnerability in order to cause extinction.

Sufficient data is not yet in hand to be able to say with any certainty what the environmental change was that caused the extinction of *E. matuyamai.* The evidence suggests that the environmental change was connected in some way with the reversal of the earth's magnetic field.

Increased radiation at the time of the field reversal (Uffen, 1963) remains a possibility, but the concensus at this time is against its having any effect on marine planktonic populations.

The apparently strong sensitivity and specialization of *E. matuyamai* to some latitude-dependent parameter, possibly temperature of the water mass in which it lived, suggests that it might be vulnerable to an abrupt change in this parameter. There has been some speculation that a reversal might have a climatic effect (Harrison, 1968) and there is some evidence that reversals are associated with temperature changes (Hays and Opdyke, 1967; Hays and others, 1969) but much more data is needed to convincingly document this.

CONCLUSIONS

The extinctions of radiolarian species make excellent stratigraphic markers and can be used to define biostratigraphic zones which are correlative over broad areas of the ocean floor. When combined with the record of geomagnetic reversals these zones can be correlated on a global scale and related to the type sections on land.

The record of reversals provides a near absolute time scale for deep-sea sediments that can be used to measure rates of evolution and extinction. One species (*Eucyrtidium matuyamai*) evolved and became extinct during the time spanned by the deep-sea cores included in this study. *E. matuyamai's* evolution is initially rapid and can be related to the invasion of and adaptation to a new habitat. The evolution of this species is significantly slower during the latter half of its range than during the first half. The extinction of *E. matuyamai* is abrupt and shows a striking correlation with the magnetic reversal at the base of the Jaramillo event. No other radiolarian species preserved in the sediments became extinct at this time. The rapid evolution of this species probably reduced the genetic variability of the population, making it more vulnerable to environmental changes than co-existing species. The environmental change that caused the extinction is unknown; however, there is suggestive evidence that it is in some way related to a reversal of the earth's field. The absence of multiple extinctions at magnetic reversals suggests that any simultaneous environmental effect must be small. It may be that the abruptness of such an environmental change is more important than its magnitude. If reversals, as the evidence indicates, cause abrupt environmental changes, they may have played an important selective role through geologic time.

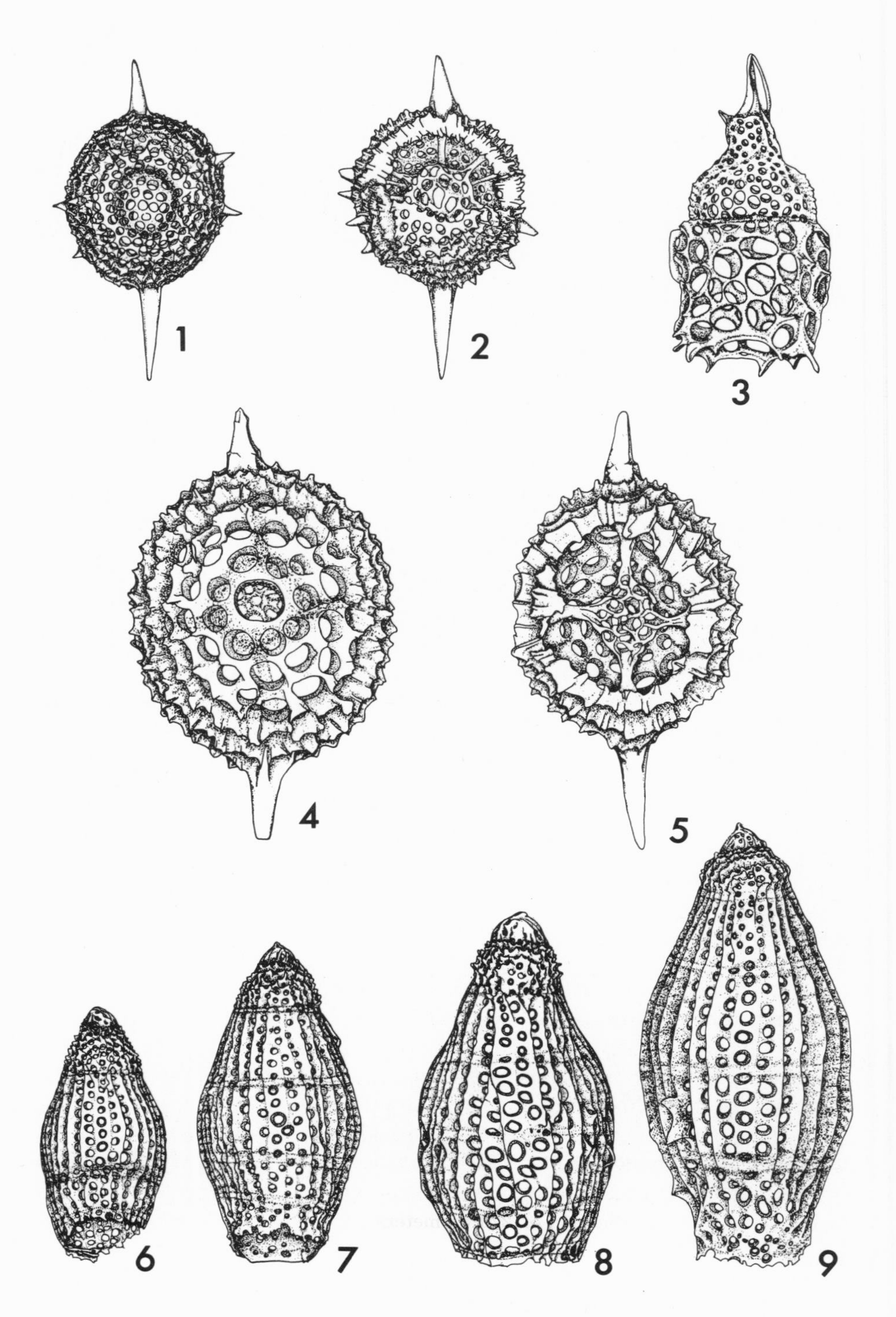

ILLUSTRATIONS OF SELECTED SPECIES

HAYS, PLATE 1
Geological Society of America Memoir 126

TAXONOMIC NOTES

In the following descriptions all measurements are in microns. Measurements were made on at least 20 specimens of each species. The illustrations were made using a Wild M-20 microscope with a drawing tube and all are at a magnification of 270 diameters. Holotypes of new species are deposited in the United States National Museum, Washington, D.C.

Eucyrtidium matuyamai n. sp.
Pl. 1, figs. 7, 8, 9

Cephalis spherical to hemispherical bearing short thorn-like apical spine. Pores either absent or very small scattered over surface. Thorax usually inflated, heavy with rough surface, pores circular arranged in longitudinal rows bordered by raised hexagonal frames. Lumbar stricture distinct. Abdomen conical, pores circular regularly increasing in size distally, arranged in longitudinal rows set in shallow furrows. Fourth segment usually inflated, shell has greatest width at this segment, pores circular of uniform size arranged in longitudinal rows set in shallow furrows. Fifth segment tapering distally, pores circular arranged in longitudinal rows set in shallow furrows decreasing in size distally. Sixth segment short, usually terminal segment in complete specimens, thinner than other segments, pores irregular in size and arrangement.

E. matuyamai ranges from near the base of the Olduvai event (about 2 m.y. B.P.) to the base of the Jaramillo event (about 0.95 m.y. B.P.). During this time it increases in size reaching its maximum dimensions near the end of its range. The measurements below are made on mature specimens near the end of the range of the species.

Length of cephalis 15-23, of thorax 17-30, of 3rd segment 35-50, of 4th segment 36-68, of 5th segment 30-70, of 6th segment 23-60. Width of cephalis 20-35, of 4th segment 118-150 (near end of range), 72-110 (near beginning of range). Length of apical horn 3-6. Diameter of pores, thorax 2-6, 3rd segment 3-9, 4th segment 6-12, 5th segment 6-12, 6th segment 3-12.

This species ranged from the Bering Sea to 34° N. It has not been seen in either equatorial Pacific sediments or Antarctic sediments.

Eucyrtidium calvertense Martin
Pl. 1, fig. 6

Eucyrtidium calvertense Martin (1904, p. 450, Pl. 130, fig. 5)

← ILLUSTRATIONS OF SELECTED SPECIES

Illustrations were made using a Wild M-20 microscope with a drawing tube and all are at a magnification of 270 diameters.

Figure
1, 2. *Stylatractus universus* n. sp.
3. *Lamprocyclas heteroporos?* Hays
4, 5. *Druppatractus acquilonius* n. sp.
6. *Eucyrtidium calvertense* Martin
7, 8, 9. *Eucyrtidium matuyamai* n. sp.

This species in the North Pacific agrees well with the description given by Hays (1965) for the Antarctic form. It differs slightly in that the thorax is usually inflated in the North Pacific while it is not in the Antarctic. The longitudinal furrows are often not as pronounced on the North Pacific individuals as the Antarctic forms.

In general, the dimensions of the Antarctic and North Pacific forms fall within the same range and are probably in the same species.

E. calvertense is still living in the North Pacific; it is probably confined to the central water masses because it is only very rarely encountered in equatorial sediments. It disappeared from the Antarctic 2 m.y. ago.

Druppatractus acquilonius n. sp.
Pl. 1, figs. 4, 5

Cortical shell, ellipsoidal, usually thick-walled, but showing considerable variation in thickness, pores evenly spaced circular to oval, with raised hexagonal borders, 6-7 across minor axis, short thorn-like projections arising from nodes. In some thick-walled individuals, the distal ends of these projections are connected. Shell bears two polar spines unequal in length, circular in cross section, distally sharpened, weakly three-bladed at base. Medullary shell single, ellipsoidal, composed of loose meshwork, pores large, irregular in shape, supported by 8-10 stout beams, 6-8 approximately in the equatorial plane, two along main axis being internal extensions of polar spines.

Length of major axis cortical shell 164-185, width 132-162, pore diameter 6-21 (usually about 17), thickness 10-29, median 21, length of polar spines 35-79. Length of medullary shell 47-57, width 44-47. Description based on 100 specimens from cores V20-109, 107, and V21-148. Dimensions based on measurements made on 25 individuals from cores V20-109 and V21-148.

Druppatractus acquilonius is apparently restricted to the North Pacific. It is most abundant in cores just south of the Aleutians. It is rare to absent in V21-145 (34° N., 165° E.) and absent in cores from the equatorial Pacific and Antarctic.

Lamprocyclas heteroporos? Hays
Pl. 1, fig. 3

Shell rough, campanulate with indistinct collar stricture but distinct lumbar stricture, cephalis hemispherical with pores circular to subcircular unequal in size scattered over surface of cephalis, pore diameter approximately equal to bar width. Cephalis bears a stout, three-bladed, vertical, approximately apical spine as long to three times as long as cephalis. Many specimens have a shorter lateral oblique three-bladed spine. Thorax campanulate, rough, pores circular to subcircular often with raised hexagonal borders of similar size to those of cephalis arranged in irregular longitudinal rows increasing in size distally. Abdomen cylindrical, slightly inflated to slightly conical; pores large, irregular in size and shape separated by heavy bars with raised hexagonal frames, abdominal pores are one to ten times diameter of thoraxic pores. Aperature slightly constricted with 5-8 short terminal teeth extended outward and down from a thick peristomal ring.

Length of apical horn 15-70, of cephalis 21-35, of thorax 41-82, of abdomen 50-110. Width of cephalis 26-44, of abdomen 88-120. Diameter of pores, thorax 3-12, usually 6, abdomen 3-50, usually 24.

The individuals of this species from the North Pacific resemble in nearly all respects the Antarctic individuals. They differ only in being on average larger than the Antarctic forms and some have a lateral oblique spine rising from the cephalis. It is possible that the stratigraphic range in the North Pacific is longer than in the Antarctic.

In the North Pacific it is found from V20-109 in the North to V21-145 in the South. In the Antarctic it is found only in cores north of the Polar Front; therefore, it seems to have been a sub-polar species.

Stylatractus universus n. sp.
Pl. 1, figs. 1, 2

Skeleton consists of 1 cortical and 2 medullary shells, medullary shells spherical cortical shell prolate. Innermost shell thin-walled pores circular with hexagonal borders. Second shell thin-walled pores regular to irregular in size and shape. Cortical shell wall very thick. Pores circular to oval, 11-14 across equatorial diameter, surface varying from smooth to rough. Medullary shells connected to cortical shell by numerous stout radial beams, two lying along the major axis project through cortical shell as stout polar spines; other beams radiate out in all directions from bases attached to inner medullary shell. Some beams penetrate through cortical shell and form short primary spines. Shell bears two large nearly equal polar spines as long to half as long as major axis of cortical shell.

Diameter of innermost shell 15-20, of 2nd shell 40-50, of cortical shell (minor axis) 106-115, (major axis) 109-123, length of spines 40-120.

Stylatratus universus was wide-spread in its occurrence when it lived, having been found in equatorial Pacific and Indian Ocean sediments, Antarctic Ocean sediments in all sectors and the North Pacific. It apparently disappeared from all these areas about 400,000 years B.P. It is morphologically uniform throughout its geographic range. The only exception to this uniformity is the slightly more robust Antarctic forms, but even these individuals do not have significantly different dimensions.

ACKNOWLEDGMENTS

The writer wishes to acknowledge the able assistance of Mrs. Sally Turner whose careful measurements and counts of species abundances were essential to the completion of this work. The illustrations of species in Plate 1 are her work. The writer is grateful to Doctors Norman Newell, Roger Batten, Neil Opdyke, David Ericson, Theodore Moore, Jr., and William Berggren for reading the manuscript and making numerous helpful suggestions. Unpublished paleomagnetic data were freely contributed by Dr. Opdyke and Mr. Foster and are gratefully acknowledged.

Deep appreciation is extended to the able and hard working crews of our Research Vessels, *Vema* and *Conrad,* who gathered the cores used in this study, and to Dr. Maurice Ewing who directed the cruises. Roy R. Capo's able custodianship of our core collection greatly facilitated this effort.

The work was supported by grants GA 1193 and GA 4499 from the National Science Foundation, and grant N-00014-67-A-0108-0004 from the Office of Naval Research.

REFERENCES CITED

Bailey, J. W., 1856, Notice of microscopic forms found in the soundings of the Sea of Kamchatka: Am. Jour. Sci., Ser. 2, v. 22, p. 1-6.

Bandy, O. L., 1963, Miocene-Pliocene boundary in the Philippines as related to late Tertiary stratigraphy of deep-sea sediments: Science, v. 142, p. 1290-1292.

Bandy, O. L., and Wade, M. E., 1967, Miocene-Pliocene-Pleistocene boundaries in deep-water environments: Progress in Oceanography, v. 4, p. 51-66.

Banner, F. T., and Blow, W. H., 1965, Progress in the planktonic foraminiferal biostratigraphy of the Neogene: Nature, v. 208, p. 1164-1166.

——1967, The origin, evolution and taxonomy of the foraminiferal genus *Pulleniatina* Cushman, 1927: Micropaleontology, v. 13, p. 133-162.

Berger, W. H., and Heath, G. R., 1968, Vertical mixing in pelagic sediments: Jour. Marine Research, v. 26, p. 134-143.

Berggren, W. A., Phillips, J. D., Bertels, A., and Wall, D., 1967, Late Pliocene-Pleistocene stratigraphy in deep-sea cores from the south-central North Atlantic: Nature, v. 216, p. 253-255.

Black, D. J., 1967, Cosmic ray effects and faunal extinctions at geomagnetic field reversals: Earth and Planetary Sci. Letters, v. 3, p. 225-236.

Blow, W. H., 1968, Late middle Eocene to Recent planktonic foraminiferal biostratigraphy: Archives Sci. (in press).

Bolli, H. M., 1966, The planktonic Foraminifera in Well Bodjonegoro-I of Java: Ecologae Geol. Helvetiae, v. 59, p. 449-465.

Bullard, E. C., 1955, The stability of a homopolar dynamo: Cambridge Philos. Soc. Proc., v. 51, p. 744-760.

Carson, H. L., 1961, Relative fitness of genetically open and closed experimental populations of *Drosophila robusta*: Genetics, v. 46, p. 553-567.

Dodimeed, A. J., Farorite, F., and Hirano, T., 1963, Review of oceanography of the subarctic Pacific region: Internat. North Pacific Fisheries Comm. Bull., no. 13, 195 p.

Donahue, J. G., 1970, Pleistocene diatoms as climatic indicators in North Pacific sediments: Geol. Soc. America Mem. 126, p. 121-138.

Dymond, J., 1969, Age determinations of deep-sea sediments: A comparison of three methods: Earth and Planetary Sci. Letters, v. 6, p. 9-14.

Ericson, D. B., and Wollin, G., 1968, Pleistocene climates and chronology in deep-sea sediments: Science, v. 162, p. 1227-1234.

Ericson, D. B., Wollin, G., and Ewing, M., 1963, Pliocene-Pleistocene boundary in deep-sea sediments: Science, v. 139, p. 727-737.

Fager, E. W., and McGowan, J. A., 1963, Zooplankton species groups in the North Pacific: Science, v. 140, p. 453-460.

Friend, J. K., and Riedel, W. R., 1967, Cenozoic orosphaerid radiolarians from tropical Pacific sediments: Micropaleontology, v. 13, p. 217-232.

Glass, B., Ericson, D. B., Heezen, B. C., Opdyke, N. D., and Glass, J. A., 1967, Geomagnetic reversals and Pleistocene chronology: Nature, v. 216, p. 437-442.

Harrison, C. G. A., 1968, Evolutionary processes and reversals of the earth's magnetic field: Nature, v. 217, p. 46-47.

Hays, J. D., 1965, Radiolaria and late Tertiary and Quaternary history of Antarctic seas: Biology of the Antarctic Sea II: Antarctic Research, Ser. 5, p. 125-184.

Hays, J. D., and Berggren, W., 1971, Quaternary boundaries: Micropaleontology of marine bottom sediments: Cambridge Univ. Press (in press).

Hays, J. D., and Ninkovich, D., 1970, North Pacific deep-sea ash chronology and age of present Aleutian underthrusting: Geol. Soc. America Mem. 126, p. 263-290.

Hays, J. D., and Opdyke, N. D., 1967, Antarctic Radiolaria, magnetic reversals, and climatic change: Science, v. 158, p. 1001-1011.

Hays, J. D., Saito, T., Opdyke, N. D., and Burckle, L. H., 1969, Pliocene-Pleistocene sediments of the equatorial Pacific—their paleomagnetic, biostratigraphic and climatic record: Geol. Soc. America Bull., v. 80, p. 1481-1514.

Horn, D. R., Horn, B. M., and Delach, M. N., 1970, Sedimentary provinces of the North Pacific: Geol. Soc. America Mem. 126, p. 1-21.

Ku, T. L., and Broecker, W. S., 1966, Atlantic deep-sea stratigraphy: Extension of absolute chronology to 320,000 years: Science, v. 151, p. 448-450.

Nigrini, C., 1970, Radiolarian assemblages in the North Pacific and their application to a study of Quaternary sediments in core V20-130: Geol. Soc. America Mem. 126, p. 139-183.

Ninkovich, D., Opdyke, N. D., Heezen, B. C., and Foster, J. H., 1966, Paleomagnetic stratigraphy rates of deposition and tephrachronology in North Pacific deep-sea sediments: Earth and Planetary Sci. Letters, v. 1, p. 476-492.

Opdyke, N. D., and Foster, J. H., 1970, Paleomagnetism of cores from the North Pacific: Geol. Soc. America Mem. 126, p. 83-119.

Opdyke, N. D., Glass, B., Hays, J. D., and Foster, J. H., 1966, Paleomagnetic study of Antarctic deep-sea cores: Science, v. 154, p. 349-357.

Parker, F. L., 1967, Late Tertiary biostratigraphy (planktonic Foraminifera) of tropical Indo-Pacific deep-sea cores: Am. Paleont. Bull., v. 52, no. 235, p. 115-203.

Ried, J. L., 1962, On circulation, phosphate-phosphorous content, and zooplankton volumes in the upper part of the Pacific Ocean: Limnology and Oceanography, v. 7, p. 287-306.

Riedel, W. R., 1957, Radiolaria: A preliminary stratigraphy: Swedish Deep-Sea Expedition Rept., v. 6, p. 61-96.

——1959, Oligocene and lower Miocene Radiolaria in tropical Pacific sediments: Micropaleontology, v. 5, p. 285-302.

——1969, Neogene radiolarian zones: (in press).

Riedel, W. R., and Funnell, B. M., 1964, Teritary sediment cores and microfossils from the Pacific Ocean floor: Geol. Soc. London Quart. Jour., v. 120, p. 305-368.

Riedel, W. R., Parker, F. L., and Bramlette, M. N., 1963, "Pliocene-Pleistocene" boundary in deep-sea sediments: Science, v. 140, p. 1238-1240.

Stommel, H., 1957, A survey of ocean current theory: Deep-Sea Research, v. 4, p. 149-184.

Uffen, R. J., 1963, Influence of the earth's core on the origin and evolution of life: Nature, v. 198, p. 143.

Waddington, C. J., 1967, Paleomagnetic field reversals and cosmic radiation: Science, v. 158, p. 913-915.

Watkins, N. D., and Goodell, H. G., 1967, Geomagnetic polarity change and faunal extinction in the Southern Ocean: Science, v. 156, p. 1083-1087.

LAMONT-DOHERTY GEOLOGICAL OBSERVATORY CONTRIBUTION NO. 1487
MANUSCRIPT RECEIVED APRIL 10, 1969
REVISED MANUSCRIPT RECEIVED JUNE 30, 1969

THE GEOLOGICAL SOCIETY OF AMERICA, INC.
MEMOIR 126, 1970

Ice-Rafted Detritus in Northwest Pacific Deep-Sea Sediments

J. R. CONOLLY
AND
M. EWING

Lamont-Doherty Geological Observatory of Columbia University, Palisades, New York

ABSTRACT

The distribution of ice-rafted detritus in piston cores (8 to 17 m long) from siliceous oozes in the northwest Pacific indicates that up to six major zones of ice-rafted sediment were deposited during the Brunhes normal epoch. The easily recognizable detritus first appeared after the end of the Gauss magnetic epoch about 2.2 m.y. ago and was relatively rare until about 1.5 m.y. ago, but became particularly abundant during the last million years. The rafted material consists mainly of altered or zeolitized intermediate to basic volcanic rocks and associated sediments rich in volcanic detritus probably derived from the Kurile-Kamchatka-Aleutian arc. The distribution and abundance of the rafted detritus suggests that currents responsible for distribution of icebergs in the Pleistocene were similar to those that exist today.

CONTENTS

Figure

Plate

Table

INTRODUCTION

Piston cores taken during 1964 (*Vema* 20) and 1965 (*Vema* 21) cruises of Lamont Geological Observatory research vessel *Vema* in the northwest Pacific contain variable percentages of erratic detritus to which an ice-rafted origin can be attributed. These cores sample an extensive area in which the upper sediment layer, called "acoustically transparent" by Ewing and others (1968) is pelagic and varies in thickness from 300 to 800 m. No analyses were made of the silt and clay fractions because of the difficulty in recognizing rafted detritus in these size fractions.

The distribution of this detritus in a vertical and lateral sense throughout the upper 10 to 20 m of siliceous ooze that makes up the deep-sea floor in this part of the Pacific can probably be used to delineate the major cold climatic periods of the late Pleistocene. Most of the cores studied in this investigation occur in areas where the biogenous sedimentation rate was high, and hence, even some of the longest (17 m) of these cores fail to penetrate the zone where the last major magnetic reversal occurred about 700,000 years B.P. (Ninkovich and others, 1966; Opdyke and Foster, 1970, this volume).

The techniques used in this investigation are similar to those used by the authors in previous studies of ice-rafted detritus (Conolly and Ewing, 1965a, 1965b). These studies led the authors to believe that during colder periods of the Pleistocene, icebergs and sea-ice could be expected to travel from continentally glaciated areas in

Kamchatka, Siberia, and probably even parts of Alaska and western Canada, into the Pacific Ocean. As expected, an abundance of ice-rafted detritus was found in certain cores from the northwest Pacific. In this investigation the nature and distribution of this detritus is discussed.

TECHNIQUES

Samples taken every 10 to 30 cm from siliceous ooze cores from the northwest Pacific (Figs. 1 and 2) were weighed and sieved through a 62μ sieve; the greater than 62μ fraction was retained, weighed and examined under a binocular microscope. The abundance of different grain varieties was estimated by counting a total of up to 200 grains for each sample from certain cores (Figs. 3 and 4; Tables 1 and 2).

IDENTIFICATION AND COMPOSITION OF RAFTED GRAINS

The larger (greater than 0.5 mm in diameter) grains in the sand fractions are obviously rafted (Pls. 1 and 2) as there seems to be no other mechanism to transport these particles such great distances from land into the deep ocean. In those samples where there is a large number of these particles, it could be assumed that a major portion of the finer grained detritus of sand and silt size may also be rafted.

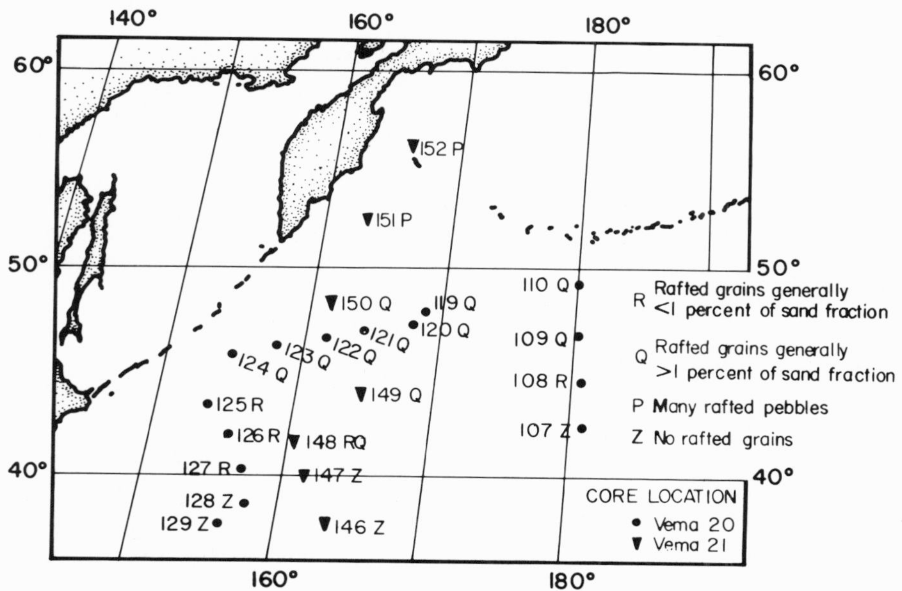

Figure 1. Location of deep-sea cores from the northwest Pacific Ocean sea floor.

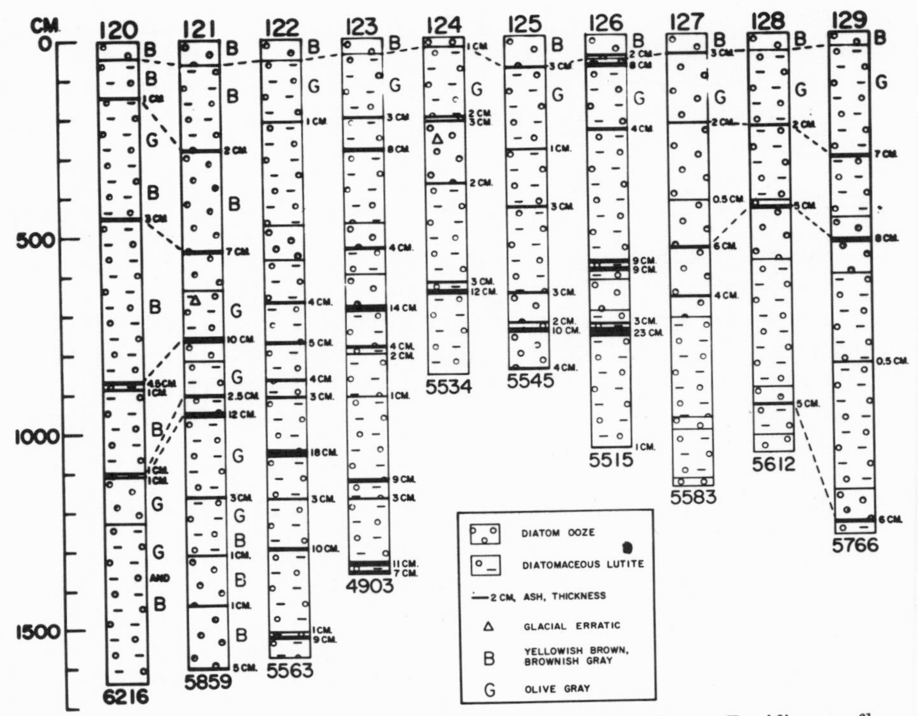

Figure 2. Lithology of *Vema* 20 cores from the northwest Pacific sea floor. None of these cores reaches sediment deposited before the first magnetic reversal 0.7 m.y. B.P.

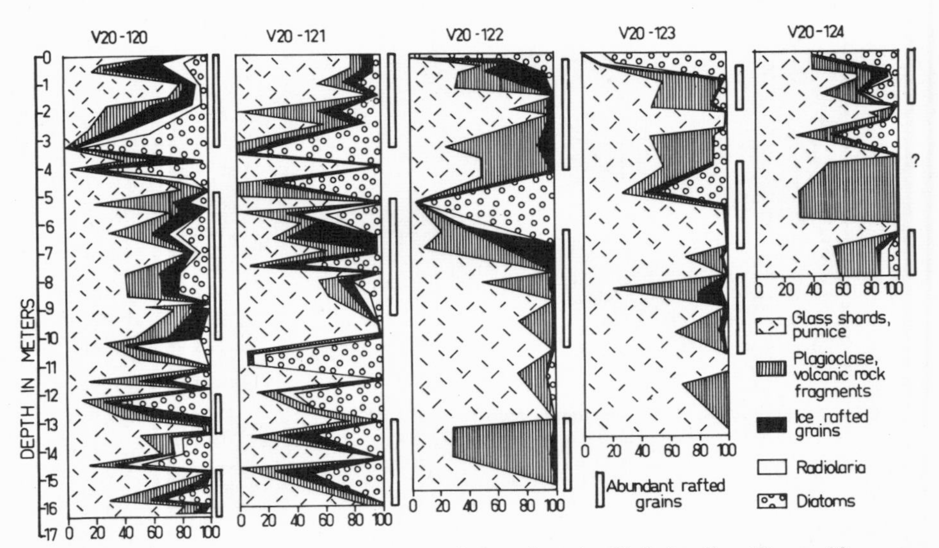

Figure 3. Composition of the sand fraction ($>62\mu$) in five *Vema* 20 cores from the northwest Pacific sea floor. Ice-rafted detritus occurs in three major zones.

Fresh volcanic detritus of andesitic to basaltic composition, presumably derived from the Aleutian Islands, occurs in abundance in the cores. Only a small proportion of the volcanic detritus in the sand fraction from the siliceous oozes in the northwest Pacific cores is fresh. Instead, most sand-sized volcanic detritus in the siliceous ooze cores (apart from glass and pumice) is altered (Pls. 1 and 2). The devitrification and common occurrence of low-grade metamorphic minerals, such as epidote, albite, prehnite and pumpellyite, in these fragments indicates they have suffered considerable secondary alteration.

The volcanic fragments in the sand fraction from the siliceous ooze cores consist mostly of altered fragments with minor percentages of fresh volcanic detritus. This indicates, first, that a minor percentage, namely, the fresh volcanic detritus, glass and pumice could be derived from contemporaneous volcanic eruptions, and second, that the remainder is derived from Quaternary or older volcanic rocks in the Kurile-Kamchatka-Aleutian land areas. Since these altered volcanic particles occur with other rocks, such as red siltstones and sandstones, graywacke, quartzite (Pls. 1 and 2) and rarely grains of granite and gneiss, it is reasonable to assume that the source rocks for the rafted grains consist mainly of altered andesites, basalts and associated volcanic sediments, with minor percentages of continental sedimentary, metamorphic and plutonic rocks.

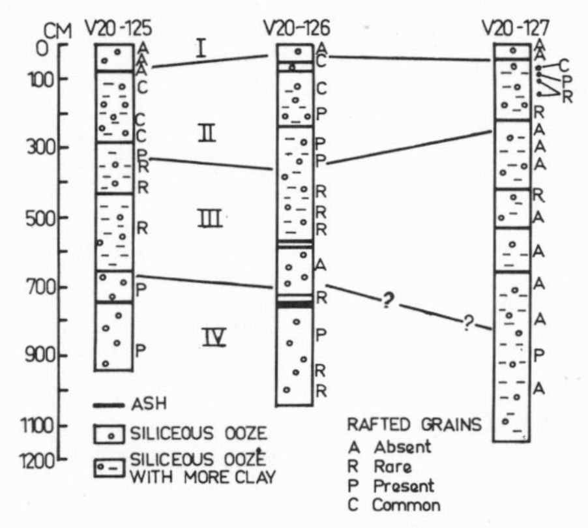

Figure 4. Relative amounts of ice-rafted grains found in the sand fraction (>62μ) of three cores from the northwest Pacific. At least two major zones containing rafted detritus occur (Zones II and IV) and are overlain by a siliceous ooze with no rafted grains (Zone I), presumably deposited during the Holocene and separated by a similar zone (Zone III) which probably corresponds to the penultimate warm period.

DISTRIBUTION OF ICE-RAFTED DETRITUS

Figure 3 shows the percentage composition of the sand fraction at different sampled levels in cores V-20-120 to V-20-124, inclusive. Many of the levels sampled were layers composed almost entirely of glass shards and pumice. Other levels consist mainly of a mixture of volcanic detritus (mostly plagioclase and volcanic rock fragments) with diatoms, Radiolaria, and rafted detritus. The rafted detritus includes mainly grains of rounded to angular black volcanic rocks, siltstones, tuffs, graywackes and minor proportions of plagioclase, quartz, gneiss, schist, quartzite and other miscellaneous sedimentary, igneous and metamorphic rocks.

Abundant rafted detritus tends to occur at particular levels in the cores. For instance, rafted detritus is abundant in a zone 100 to 300 cm thick immediately underlying the upper 5 to 50 cm of each core. At least two zones rich in rafted detritus occur beneath this uppermost zone. The two most westerly cores (V20-123 and V20-124) generally contain a higher percentage of volcanic detritus, as could be expected since these two cores lie closer to the possible volcanic source areas in the Kamchatka arc.

Figure 4 indicates the relative amounts of ice-rafted detritus in cores V20-125, 126 and 127 at different levels down the core. There is much less rafted detritus in these three cores than the cores situ-

TABLE 1. CORE V21-148

Sample depth (cm)	Percent sand	Percent in sand fraction					Approx. percent rafted sand in total sediment
		Ash	Minerals	Radiolaria	Diatoms	Rafted grains	
25	25	40	10	25	15	10	2.5 A
53	16.2	30	10	45	5	10	1.6
75	7.5	30	10	3	2	55	3.8 B
100	15.4	50	20	5	5	20	3.1
125	11.1	48	10	20	20	2	0.08
152	12.2	30	15	10	5	40	4.9 C
175	10.7	65	15	10	5	5	0.5
200	13.8	75	15	5	3	2	0.2
225	11.2	25	25	5	5	40	4.5 D
253	3.5	35	10	35	10	10	0.3
272	5.0	10	5	30	54	1	0.05
301	5.1	35	10	25	25	5	0.1 E
323	7.4	75	15	5	4	1	0.07
355	6.2	55	25	3	2	15	1.0
375	7.7	30	45	3	2	20	1.5 F
400	12.0	50	20	20	5	5	0.6
425	4.2	75	10	4	10	1	0.04
475	11.9	20	20	25	40	5	0.6 G
502	6.2	20	20	55	5	trace	trace
522	4.7	35	5	10	50	—	—
552	9.3	20	10	45	25	trace	trace
571	11.8	40	15	25	20	trace ?	trace ?
603	7.8	20	20	10	50	trace	trace
627	48	10	5	25	60	—	—, Mn
651	7.3	15	5	40	40	trace	trace, Mn
677	7.4	5	5	40	50	—	—, Mn
700	16.6	60	10	10	20	—	—, Mn
727	6.4	5	5	20	70	—	—, Mn
752	6.7	5	5	55	35	—	—, Mn
776	8.1	5	5	20	70	—	—, Mn
801	5.4	15	5	20	60	—	—, Mn

Mn—Manganese micronodules present
A, B, C, D: Rafted maxima

ated farther north (Fig. 1). For instance, the greatest amount of rafted detritus occurs in a zone 200 to 300 cm thick just below the top of each core, but only amounts to about 1 to 5 percent of a 3 to 5 percent sand fraction.

Another zone containing a concentration of rafted detritus occurs at depths of about 700 to 900 cm in all three cores. These results correlate with those in cores V20-120 to 125, suggesting that there were major periods of ice-rafting during the deposition of these cores. Cores V20-125, 126, and 127 penetrated only two major zones of glacial marine sediment.

TABLE 2. CORE V20-109

Sample depth (cm)	Percent sand	Percent in sand fraction						Approx. percent rafted sand in total sediment
		Ash	Minerals	Radio-laria	Diatoms	Rafted grains	Others*	
5	0.2	20	25	5	5	25	20*	0.05 A
30	0.6	40	30	5	5	10	10*	0.06
50	0.01	30	20	5	5	40	—	0.005
80	0.2	10	25	40	10	10	5*	0.02
110	0.3	30	10	10	10	40	trace	0.1 B
150	0.4	20	10	20	10	40		0.2
176	0.2	30	40	10	10	10		0.02
220	0.05	30	15	5	5	40	5 Mn	0.02 C
300	0.1	30	40	trace	trace	20	10 Mn	0.02
350	0.2	20	25	20	20	trace	15 Mn	trace
400	0.2	40	40	trace	trace	trace	20 Mn	trace
450	0.6	40	20	10	trace	30	10 Mn	0.2 D
500	0.5	10	15	30	10	30	5 Mn	0.2
600	0.2	5	5	50	trace	10	30 Mn	0.02
650	0.2	10	5	20	40	10	15 Mn	0.02
700	0.5	5	5	5	85	trace	trace	trace
750	0.1	trace	trace	40	50	trace	10 Mn	trace
800	0.1	10	10	40	20	5	15 Mn	0.005 E
850	0.3	5	5	40	5	20	25 Mn	0.06
900	0.1	10	10	50	5	trace	25 Mn	trace
950	0.3	20	10	10	55	—	5 Mn	—
1000	0.2	5	5	45	35	trace	10 Mn	trace
1050		trace sand only					Mn	
1100	0.8	10	5	45	30	—	10 Mn	—
1150	0.1	60	10	5	5	10	10 Mn	0.01 F
1200	0.5	10	10	50	10	—	20 Mn	—
1250	0.02	10	30	450	—	—	10 Mn	—
1300	0.5	5	trace	75	20		—	—
1350	0.8	5	trace	70	20	5	—	0.04 G
1400	0.7	—	—	95	5	—	—	—
1443	0.8	—	—	95	5	—	—	—

*Mostly arenaceous Foraminifera
Mn—Mostly manganese micronodules

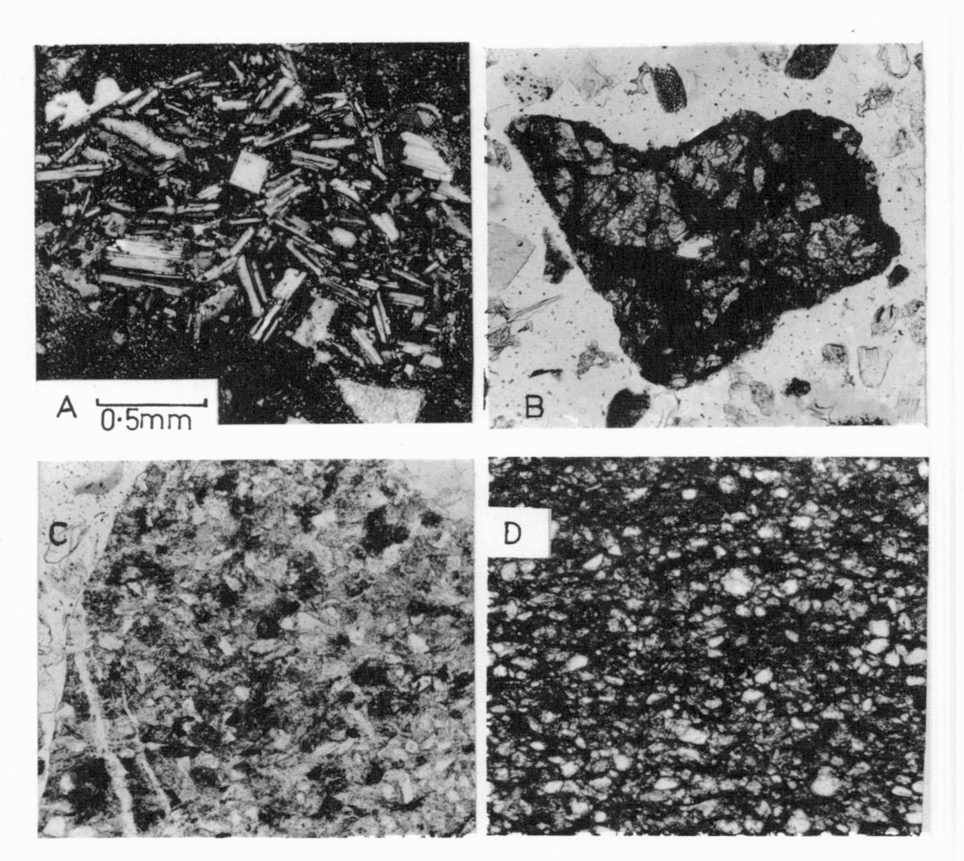

THIN SECTIONS OF THE >62μ FRACTION CONTAINING ICE-RAFTED GRAINS FROM CORES V20-120 AND V20-121

A. Rounded grain of porphyritic andesitic basalt from a sample at 850 cm in core V20-121. Viewed with crossed nicols. This volcanic grain is fairly fresh, with little or no alteration of the plagioclase phenocrysts and groundmass.

B. Subrounded grain of basic volcanic from a sample at 542 cm in core V20-120. Viewed in plane polarized light. This grain has been completely devitrified and weathered to a mixture consisting mainly of albite, quartz and clay minerals. Other small grains of glass shards, plagioclase, pyroxene, rock fragments and Radiolaria occur surrounding the volcanic grain on this strewn slide.

C. Portion of a rounded fine-grained graywacke from same sample as figures in A (above). Viewed in plane polarized light. The graywacke is moderately sorted and consists mainly of angular feldspar and volcanic rock fragments in a clay matrix. It was probably derived from andesitic-to-basaltic volcanic rocks. Fragments of this composition are common among rafted grains in the northwest Pacific and commonly show alteration most likely caused by weathering.

D. Portion of a large, rounded grain of a brown, fine-grained sandstone. Viewed in plane polarized light. The sandstone consists mostly of very fine and angular quartz and iron-stained or altered clay pellets in a clay matrix that is also colored by secondary iron oxide minerals. This grain was presumably derived from a red bed sequence of continental origin.

None of the V20 cores discussed above penetrate sediment deposited prior to the beginning of the Bruhnes magnetic epoch (Opdyke and Foster, 1970, this volume), which occurred 0.7 m.y. ago.

The upper and lower boundaries of these glacial marine zones are difficult to establish from the present data because there is great lateral variation in the amount of ice-rafted detritus. Glacial marine zones, rich in rafted grains in the sand fraction, also tend to contain a greater amount of silt and clay. This silt and clay was probably deposited by ice-rafting. The upper 20 to 50 cm of siliceous ooze was probably deposited during the recent world warming since about 17,000 years B.P. This zone is not only characterized by less rafted material, but also by a higher content of diatoms and Radiolaria.

The variation in intensity of the ice-rafting (assuming fairly constant biogenous sedimentation) between the three major ice-rafting zones indicates that the upper zone contains more rafted detritus. This suggests that the intensity of ice-rafting was greatest during the last glacial epoch.

CORES V21-148 AND V20-109

Cores V21-148 and V20-109, raised from near the southern limit of abundant rafted detritus (Fig. 1), penetrate sediment that was dated using paleomagnetic techniques (Opdyke and Foster, 1970, this volume), as being deposited through the Pleistocene (Pl. 2).

Analyses of the sand fraction from samples from these two cores (Tables 1 and 2) show the variation of the amounts of rafted detritus, fresh volcanic detritus (glass shards, pumice, fresh plagioclase), biogenic sand (arenaceous Foraminifera, diatoms and Radiolaria) and diagenetic particles (manganese micronodules). The rafted detritus is similar in composition to that already described from the *Vema* 20 cores. Even though the total amount of rafted sand in the core (expressed as a percentage of the total sediment) is not large, the distribution of this detritus in both cores is similar (Fig. 5) and leads to the following conclusions:

Maximum ice-rafting occurs back to a time approximately midway between the Jaramillo and Olduvai paleomagnetic events (on Tables 1 and 2), which is about 1.5 m.y. B.P., using the time-scale of Cox (1969).

Ice-rafted detritus is rare between this period and a time lying about midway between the Olduvai and the end of the Gauss (X on Figure 5), which is about 2.2 m.y. B.P., using the time-scale of Cox (1969).

Several periods with more intense ice-rafting occur within the Brunhes normal epoch. The density of the sampling from cores V21-

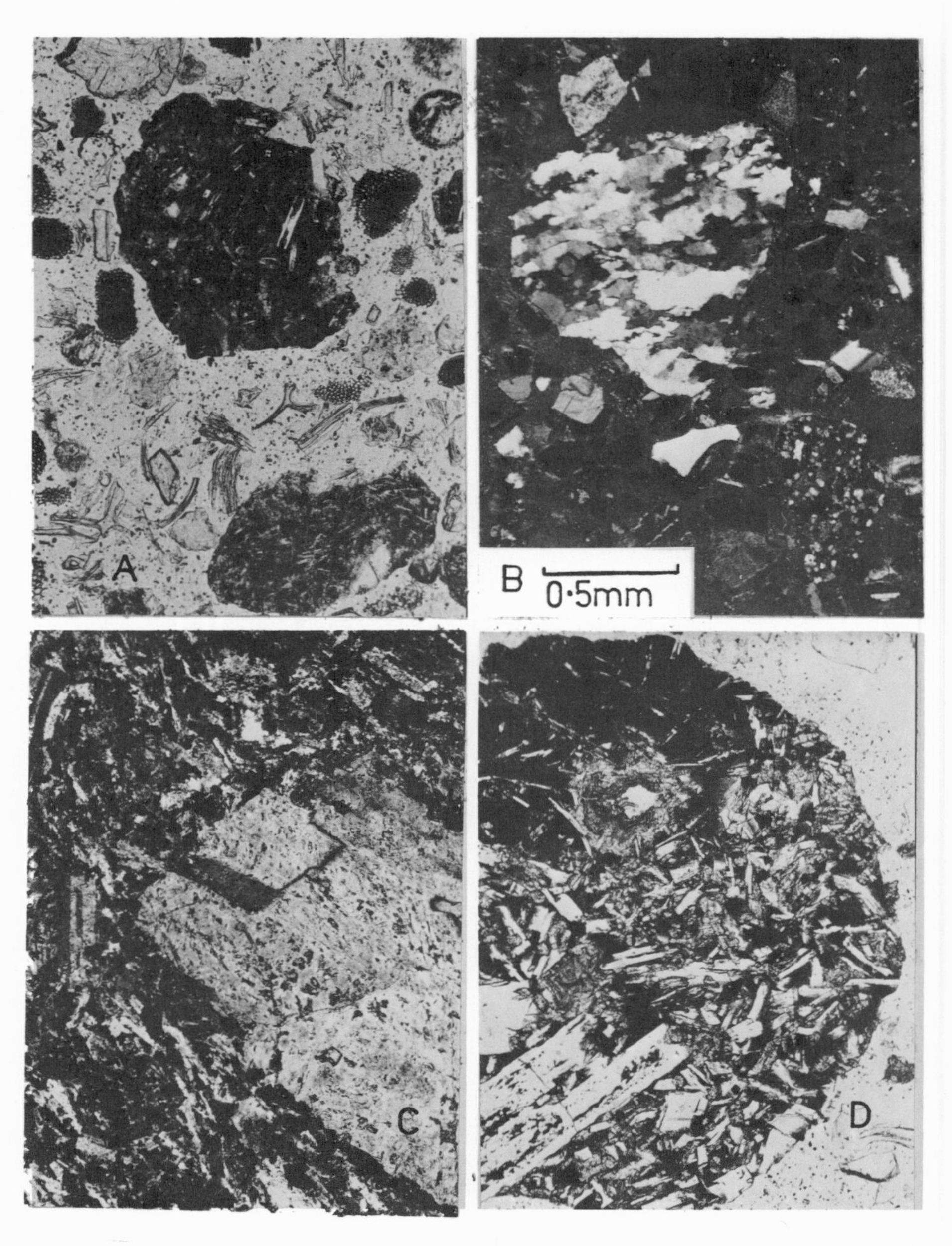

THIN SECTIONS OF THE $>62\mu$ FRACTION CONTAINING ICE-RAFTED GRAINS FROM SAMPLES FROM CORES V20-110, V20-120 AND V20-121

CONOLLY AND EWING, PLATE 2
Geological Society of America Memoir 126

THIN SECTIONS OF THE >62μ FRACTION CONTAINING ICE-RAFTED GRAINS FROM SAMPLES FROM CORES V20-110, V20-120, AND V20-121

A. General field of view with plane polarized light from a sample at 1075 cm in core V20-120, showing two large, rounded grains of basalt, surrounded by finer-grained fragments of glass, Radiolaria, plagioclase, pyroxene and rock fragments. The larger two volcanic fragments and the large clinopyroxene grain at the top of the photograph are obviously rafted, as is a large portion of the remainder of the sand fraction.

B. A large, rounded grain of metaquartzite from a sample at 60 cm in core V20-110. Viewed with crossed nicols. A smaller grain of finer-grained quartzite is situated beneath this grain and there are many sand-sized fragments of feldspar and other rock fragments in the remainder of the strewn slide.

C. Part of a large, rounded grain of devitrified porphyritic andesitic basalt, from a sample at 160 cm in core V20-120. Viewed with crossed nicols. All the plagioclase now consists of albite, and the groundmass is mostly altered to a mixture of albite and clay minerals.

D. Part of a large, rounded grain of porphyritic basalt from a sample at 650 cm in core V20-V121. Viewed with plane polarized light. This grain is typical of the rounded, black basaltic grains that occur commonly as rafted grains.

148 and V-20-109 is such that only six such maximum ice-rafting periods (A to F) can be delineated in core V21-148, and three (A to C) in core V20-109 (Fig. 5).

Cores raised from more tropical waters in the Pacific Ocean contain eight zones of high biogenic carbonate deposition during the Brunhes epoch; these zones have been correlated with colder waters and hence glacial periods (Hays, 1970, this volume).

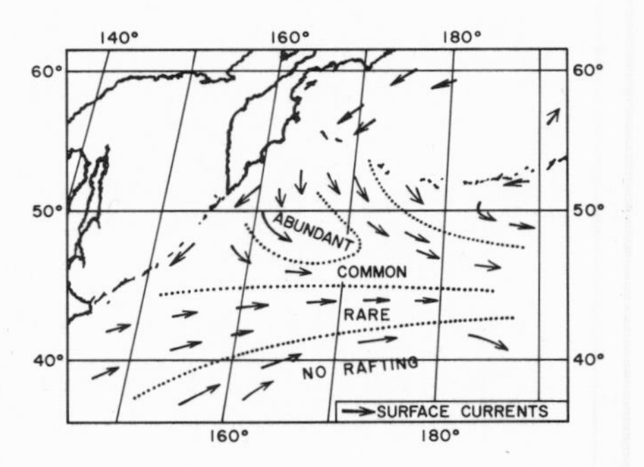

Figure 5. Cores V21-148 and V20-109 showing paleomagnetic stratigraphy and distribution of ice-rafted detritus.

Since up to six such periods have been identified in the northwest Pacific cores described herein, closer sampling and careful examination may eventually show that eight glacial marine zones also occur in these cores during the Brunhes epoch, as shown by Donahue (1970, this volume).

LATERAL DISTRIBUTION OF RAFTED DETRITUS

The greatest amount of rafted detritus occurs near Kamchatka where abundant pebbles and grains occur in cores V21-152 and V20-151 (Fig. 1). Such large (1 to 2 cm in diameter) pebbles occur rarely in most other cores. Rafted detritus commonly makes up 1 to 10 percent of a 2 to 10 percent sand fraction in cores V20-119, 120, 122, 123, 124 and 110, whereas rafted detritus commonly makes up less than 1 percent of the total sand fraction cores taken farther south.

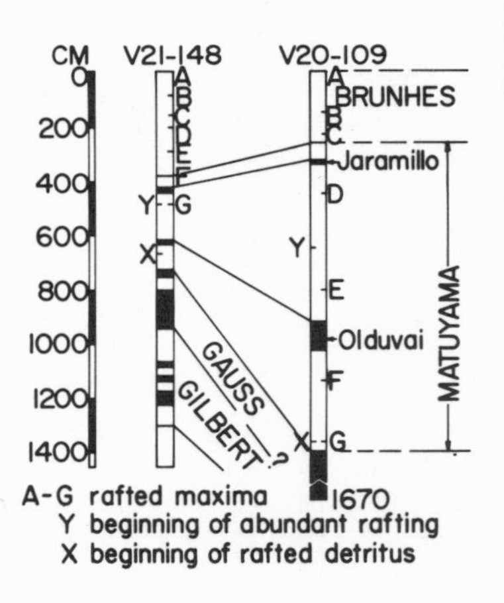

Figure 6. Distribution of ice-rafted detritus in the northwest Pacific Ocean. Cores containing the most abundant rafted debris occur near Kamchatka. The present surface current distribution is also shown indicating that past currents similar to these were probably responsible for distributing the rafted grains.

Cores V20-128, 129, 107 and V21-147 and 146 contain no rafted debris. The relative distribution of rafted debris is shown on Figures 1 and 6. Since there is a strong correlation between the abundance of rafted detritus and the present surface current distribution (Fig. 6), it is suggested that the surface currents of the Pleistocene had a similar distribution of those of today and were responsible for redistributing icebergs and sea-ice that carried rafted detritus from the land areas of the northwest Pacific.

ACKNOWLEDGMENTS

The writers thank the scientists and crew of the R.V. *Vema* during Cruises *Vema*-20 and *Vema*-21 in the North Pacific. The shipboard work has been supported by the U. S. Navy, Office of Naval Research, Contract N00014-67-A-0108-0004, and by grants from the National Science Foundation (GA 1193, GA 10635, GA 1299 and their predecessors). The study was initiated during tenure of a Ford Foundation Fellowship at Columbia University and completed during tenure of a Queen Elizabeth II Fellowship at the University of Sydney. Help with the sampling was received from Dr. J. Hays and scientists of the Core Laboratory at Lamont-Doherty Geological Observatory and is gratefully acknowledged.

REFERENCES CITED

Conolly, J. R., and Ewing, M., 1965a, Pleistocene glacial-marine zones in North Atlantic deep-sea sediments: Nature, v. 204, p. 135-138.

——1965b, Ice-rafted detritus as a climatic indicator in Antarctic deep-sea cores: Science, v. 150, p. 1822-1824.

Cox, A., 1969, Geomagnetic reversals: Science, v. 163, p. 237-245.

Donahue, J. G., 1970, Pleistocene diatoms as climatic indicators in North Pacific sediments: Geol. Soc. America Mem. 126, p. 121-138.

Ewing, J., Ewing, M., Aiken, T., and Ludwig, W. J., 1968, North Pacific sediment layers measured by seismic profiling: Am. Geophys. Union Mono. 12, p. 147-173.

Hays, J. D., 1970, The stratigraphy and evolutionary trends of Radiolaria in North Pacific deep-sea sediments: Geol. Soc. America Mem. 126, p. 185-218.

Ninkovich, D., Opdyke, N. D., Heezen, B. C., and Foster, J. H., 1966, Paleomagnetic stratigraphy, rates of deposition and tephrachronology in North Pacific deep-sea sediments: Earth and Planetary Sci. Letters, v. 1, p. 476-492.

Opdyke, N. D., and Foster, John H., 1970, Paleomagnetism of cores from the North Pacific: Geol. Soc. America Mem. 126, p. 83-120.

LAMONT-DOHERTY GEOLOGICAL OBSERVATORY CONTRIBUTION NO. 1488

PRESENT ADDRESS (CONOLLY): DEPARTMENT OF GEOLOGY, UNIVERSITY OF SOUTH CAROLINA, COLUMBIA, SOUTH CAROLINA

MANUSCRIPT RECEIVED APRIL 10, 1969
REVISED MANUSCRIPT RECEIVED JULY 18, 1969

THE GEOLOGICAL SOCIETY OF AMERICA, INC.
MEMOIR 126, 1970

Holocene Palynology of the Middle America Trench near Tehuantepec, Mexico

DANIEL HABIB
Department of Geology, Queens College of the City University of New York, Flushing, New York, and Lamont-Doherty Geological Observatory, Palisades, New York

DAVID THURBER
Department of Geology, Queens College of the City University of New York, Flushing, New York, and Lamont-Doherty Geological Observatory, Palisades, New York

DAVID ROSS
Woods Hole Oceanographic Institution, Woods Hole, Massachusetts

JACK DONAHUE
Department of Geology, Queens College of the City University of New York, Flushing, New York

ABSTRACT

Pollen profiles of two cores from deep on the seaward flank of the Middle America Trench are used to interpret the Holocene climatic history of adjacent southwestern Mexico. Four broad pollen zones are distinguished, which carbon-14 analysis dates to 8000-9000 yrs B.P.

The total pollen and spores per gram of sediment, together with the abundance and relative percentages of certain pollen types (pine, oak, alder, fir) and altered specimens, are interpreted to reflect variations in moisture and temperature during the Holocene in cordilleran Mexico.

Source vegetation and the proportionate roles of transportation to the sea by streams and winds are considered to be the principal factors in emplacing the pollen zones. Large numbers of specimens per gram sediment, diversity of types, higher percentages of altered grains, and abundance of oak, alder, and fir pollen are presented as evidence for greater available moisture and increased stream activity on the western slopes of the Sierra Madre del Sur. On the other hand, fewer specimens and the preponderance of relatively few types (mostly pine pollen and fern spores) are considered to represent diminished stream activity.

The oldest zone (Zone IV) represents a climatic phase which varied from cool-dry to cool-moist. Zone III is interpreted to reflect the period of maximum warmth and moisture. The next higher zone reflects the driest climatic phase represented in the cores. The uppermost zone (Zone I) marks the return to more moist conditions.

CONTENTS

INTRODUCTION

Studies of pollen grains and spores in the oceans date back to the middle 1950's, and many papers on their distribution in Holocene sediments have been published since that time. The present study has two goals. One is to examine the methods by which grains were transported to the marine basin and distributed by marine currents. The other is to interpret the vertical distribution of pollen and spores in two cores from the Middle America Trench in terms of the Holocene climatic history of adjacent Mexico.

PREVIOUS STUDIES

Since 1959, several important papers have been published in marine palynology. Among these is the study of the Orinoco drainage system in northern Venezuela by Muller (1959). He was the first to emphasize the importance of stream processes in delivering polliniferous sediments to the Gulf of Paría, and pointed out that marine processes further segregated the pollen according to their size and morphology. He also presented data indicating that the larger and more diverse assemblages occur nearest the coastline. As an example of the role played by marine currents, Muller showed that the pollen of red mangrove (*Rhizophora mangle*), because of its small size and smooth surface, is carried farther offshore than larger and more ornamented grains from the same habitat. Mangrove pollen was observed to increase offshore *relative to other types,* although the absolute number of grains decreased. Because of the concentration of grains near the areas of major stream discharge, and the occurrence of specimens apparently transported long distances across land and against the prevailing wind patterns, Muller concluded that for the Orinoco drainage system the effect of wind transport is small, and that streams and marine currents are the major media of transportation.

Traverse and Ginsburg (1966, 1967) investigated the distribution of pollen and spores in the marine carbonate banks of the Bahamas. They concluded from pine pollen (*Pinus*) distribution in the Great Bahama Bank, and the inverse relationship observed between its concentration in the surface sediments and water above, that these grains were most useful for determining sedimentation patterns, particularly water turbulence in the locus of deposition. As stated by Traverse and Ginsburg (1966), pine pollen is particularly sensitive to hydrodynamic processes by its possession of air sacs.

Cross and others (1966) reported on the distribution of pollen and spores in sediments of the southern Gulf of California. They showed a relationship between the grains in the Gulf and the plant communities of the surrounding region, but were careful to point out the significance of marine sedimentation processes in explaining their distribution in the marine sediments. They also reported the widespread distribution of pine in the sediments in the Gulf of California, as did Koreneva (1966) for the Sea of Okhotsk. Cross and others (1966) showed the close correspondence of pine pollen to the total amount of pollen and spores per gram of sediment. The amount of pine pollen was found to fluctuate directly with the overall abundance of pollen. Oak pollen (*Quercus*), on the other hand, was observed to have a more limited areal distribution, as did the pollen of fir (*Abies*) and red mangrove (*Rhizophora*). The scarcity of fir pollen was explained by the fact that these grains are not transported as far by wind. The occurrence of mangrove pollen corresponded to the distribution of mangroves along the coasts.

GEOLOGIC AND OCEANOGRAPHIC SETTING OF THE MIDDLE AMERICA TRENCH

The Middle America Trench (Fig. 1) extends along the west coast of Mexico and Central America from Islas Tres Marias, Mexico, to the Cocos Ridge southwest of Costa Rica. It is located from 30 to 165 km offshore and conforms rather closely to the outline of the west coast of Central America (Fisher, 1961). Recent seismic profiling studies indicate that it extends as far south as Panama, although it is filled with sediment (Ewing and others, 1968).

Seismic reflection and sedimentological studies in the northern part of the trench indicate that the sediments in the trench axis are fine-grained turbidites derived from nearshore environments (Ross, 1965). Terrigenous sediments prevail on the landward flank, and a mixture of biogenous and terrigenous material is typical of the seaward flank. The fine-grained turbidites of the axis are differentiated from sediments of either flank mainly by the mineral composition of the sand-sized (2 mm to 62 microns) fraction (Ross, 1965).

The general bathymetry (Fisher, 1961) and structure (Ross and Shor, 1965) in this area suggest less sediment accumulation in the axis than that in the northern part of the trench.

Cores V18-338 and V18-339 were collected by the research vessel *Vema,* during October, 1962. They were taken from deep within the trench on the seaward flank. Core 339, 1003 cm in length, was collected at lat. 15° 09.5′ N. and long. 95° 37′ W. at a depth of 5369 m, approximately 50 km offshore (Fig. 1). Core 338 was collected just to the west of 339. It is 758 cm in length, and was collected from 15° 08′ N. and 95° 43′ W. in 5253 m of water.

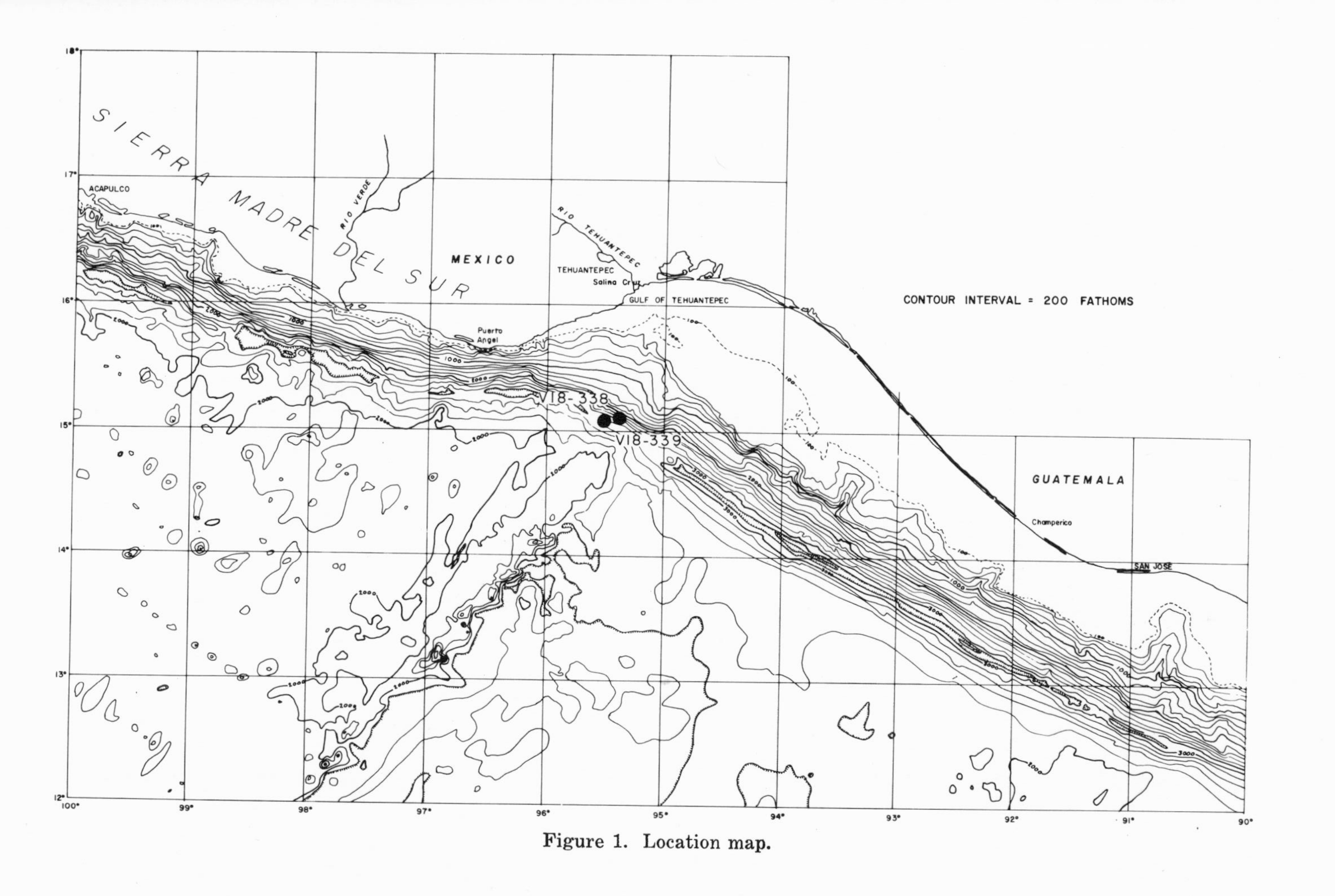

Figure 1. Location map.

In the area of investigation, the trench trends to the southwest, away from the coast of Mexico. A seismic reflection profile in this area (Ross and Shor, 1965) shows a broad synclinal continental shelf separating the mainland from the trench and containing as much as 2000 m of sediment. A narrow trough filled with as much as 500 m of slumped material occurs between 1100 and 1700 m depth on the landward flank of the trench. Seaward of this trough is a similar unfilled trough.

The coast of Mexico near Tehuantepec is formed by coastal plain clastic deposits underlain by granites (de Cserna, 1961). To the east is a thick sequence of Paleozoic metamorphics overlain locally by Mesozoic sediments. These rocks are separated from coastal plain sediments by intrusive volcanics.

ENVIRONMENT AND VEGETATION IN CENTRAL AMERICA

The west coast of Central America is formed by a series of mountain ranges. In the area of the Gulf of Tehuantepec, the principal range is the Sierra Madre del Sur, with a majority of the peaks at elevations from 2500 to 3000 m, although the highest peaks reach over 3500 m. West of the Gulf, near Acapulco, the highest peaks approach 3500 m, and decrease in elevation slightly southeastward across the Isthmus of Tehuantepec to 3140 m near Guatemala (Fig. 1). The range is interrupted in the immediate vicinity of the Gulf by markedly lower elevations across the Isthmus. Chief among the major streams draining the Sierra Madre del Sur are the Verde and Tehuantepec Rivers, emptying into the Pacific Ocean in and west of the Gulf of Tehuantepec.

Annual temperatures range from 69 to 86° F at the coast to 50 to 68° F at higher elevations. The average annual precipitation is 40 to 60 inches at the coast, 20 to 40 inches at slightly higher elevations, and to over 80 inches farther inland. The vegetation is tropical savannah near the coast, with highlands vegetation on the slopes. Prevailing winds are northeasterly across the region in January (Northeast Trades) and easterly in July (Tropical Convergence Zone). Tropical storms converge in the general area, approaching westerly (onshore) and easterly across the Isthmus from the Caribbean Sea. Storms also travel northward across the area toward the Gulf of California.

Leopold (1950) distinguished twelve major vegetation types in Mexico, five of which pertain to Pacific Mexico in the region of Tehuantepec (Fig. 2). These are: tropical savannah, tropical deciduous forest, pine-oak forest, cloud forest, and boreal forest.

Tropical Savannah

Of limited areal distribution, savannah vegetation occurs in the low, restricted coastal plain of the Gulf of Tehuantepec. The dominant plants are coarse tropical grasses with only scattered trees. Its distribution is primarily edaphic, and is not representative of the climatic climax vegetation.

Tropical Deciduous Forest

Leopold (1950) reported the main area occupied by the deciduous forest as the more humid foothills below the pine-oak uplands along the west coast of Mexico. His vegetation map (Fig. 2) shows its distribution along the coast to the immediate west and east of the Gulf (tropical savannah), and directly inland to the savannah along the Gulf. It is a broad linear belt extending north to just south of Baja California. The principal vegetation includes *Ipomoea, Bombax, Acacia,* and *Bursera.*

Pine-Oak Forest

The pine-oak forest is the major climatic climax vegetation in the region encompassing Tehuantepec. It occupies the greatest single zone of vegetation, forming the upland forest on the western slopes of the Sierra Madre del Sur. Many species of oak and pine are present. Leopold subdivided this zone into (1) pine forest, with dense

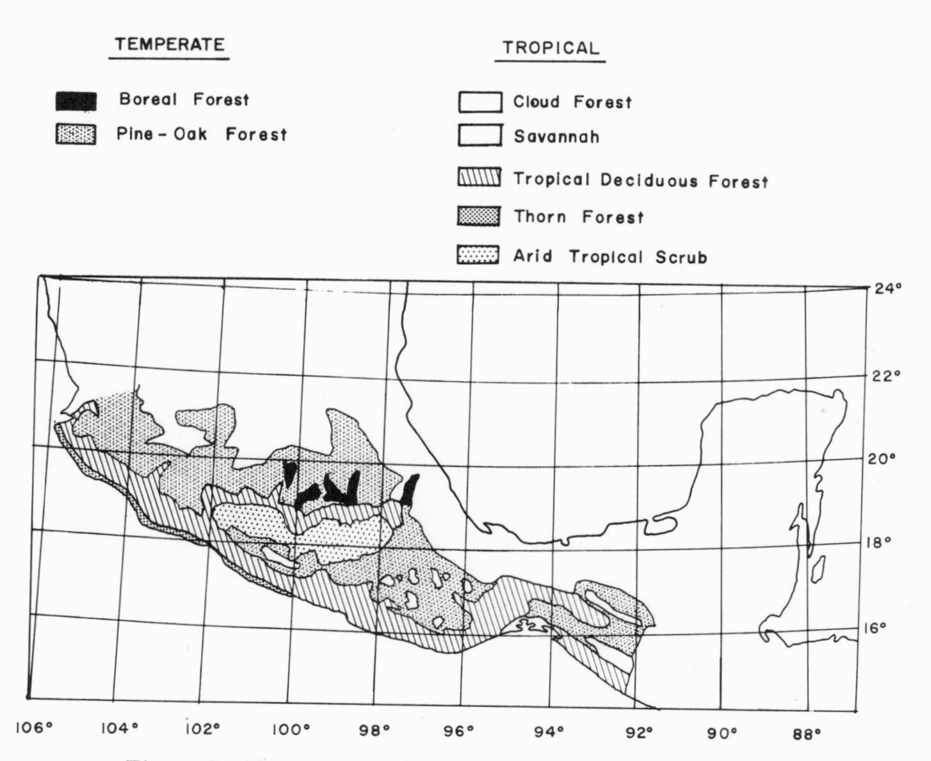

Figure 2. Vegetation zones of Mexico (*after* Leopold, 1950).

stands of pine, chiefly *Pinus montezumae*; (2) pine-oak woodland, with the principal species *P. montezumae* and *Quercus arizonica*; (3) pinon-juniper woodland; and (4) oak scrub at the lower elevations. Pine and pine-oak woodlands occupy the highest slopes in the zone.

Cloud Forest

In the area of this investigation, the cloud forest occupies the highest elevations. It is composed of temperate montane hardwoods along with pine and oak, generally at elevations above 1600 m which are swept by moisture-laden air currents. The dominant trees are temperate species of *Pinus, Quercus, Liquidambar, Fagus, Nyssa,* and *Tilia,* with tree ferns forming the understory. This forest type is rather restricted areally, forming isolated patches in the pine-oak belt.

Boreal Forest

According to Leopold (1950), the boreal zone in Mexico is restricted to the high volcanic peaks occurring between 18° N. and 20° N. lat., in the general area around Mexico City. In Figure 2, it appears as isolated northeast-trending linear belts in the pine-oak forest. This vegetation type is rather remote from the area of study, but is described because of its climatic significance. Several subdivisions were distinguished: (1) open pine and grasses, occurring at the highest elevations, approaching 4000 m, and composed mainly of pine, scattered juniper, and coarse grasses; (2) fir forest, at elevations to 3200 m in the belt of heaviest fog, and composed of almost pure stands of *Abies religiosa;* (3) pine-alder-fir, at slightly lower elevations, with *Pinus montezumae, Abies religiosa,* and *Alnus glabrata* (alder).

The vegetation zones of western Mexico continue in the main through Guatemala (Steyermark, 1950). Steyermark illustrates a continuous band of mangrove vegetation along the coast of Guatemala. Leopold (1950) did not discuss the mangrove vegetation in any detail, but did report its extent northward to Baja California.

PALYNOLOGICAL STUDIES IN CENTRAL AMERICA

Sears and Clisby (1955) discussed the forest vegetation of Mexico with respect to a Pleistocene palynological investigation of two cores near Mexico City. They listed fir, alder, oak, and pine as the four important genera for climatic interpretation. These genera largely form the present upland forest near Mexico City, ranging up in elevation from approximately 2300 m to the timberline. Pine is considered to be an indicator of dry climate, with a relatively wide range in elevation. It occurs at the higher elevations with fir, which is confined to the more moist valleys. At lower elevations, oak and alder are associated with pine, with the former again found mainly in more moist areas. Sears and Clisby (1955) concluded that oak and alder pollen associated in the sediments near Mexico City indicated warmth and moisture; fir pollen also indicated more moist conditions, but at cooler temperatures than did oak.

Using oak and alder pollen as moisture indicators, Sears and Clisby distinguished major episodes and interpreted their pollen profiles as representing a series of moist-dry oscillations within longer trends of changing temperature. The moist periods (high oak and alder) were warm, while the temperatures of dry periods (high pine) were less easy to determine. They concluded that the periods of major glacial advance were also moist and warm.

Martin (1964) investigated the vertical distribution of pollen in a bog-filled lake high in the Andes of Costa Rica. He distinguished four pollen and spore zones, composed of alternating assemblages of (1) abundant tree pollen, especially oak, indicating post-glacial and interglacial intervals, and (2) abundant non-arboreal pollen and lycopod spores, indicating Pleistocene glacial intervals. Martin (1964) concluded that his pollen profile provided fossil evidence for a major depression of the vegetation zones on the mountainous slopes of Central America, associated with cooling during the Wisconsin Stage.

Van der Hammen and Gonzalez (1960) investigated the pollen and spores in sediments of an ancient lake bed located in the Colombian Andes (Sabana de Bogotá). They interpreted their profiles as representing climatic variations during the Pleistocene. With the support of radiocarbon age determinations, they stated that the major climatic phases of the late Pleistocene and Holocene affected the tropics with the same magnitude and at almost exactly the same time as the areas in more northerly latitudes.

In the glaciated terrain of the Colombian Andes to the north of Sabana de Bogotá and at slightly higher elevations, Gonzalez and others (1966) studied and calibrated by carbon-14 the post-glacial pollen sequences derived from lake sediments. They supported the view that major climatic events in the high altitude tropics were essentially synchronous with those of North America and Europe.

METHODS

Lamont-Doherty cores V18-338 and V18-339 were sampled at 10 cm levels for the uppermost 100 cm, and at each 50 cm interval thereafter to the bottom of each core; 24 samples were taken from V18-338 and 29 from V18-339. The dry-weight of the samples varied from 9 to 16 gms.

The cores were macerated differently. Core V18-339 was studied first and, in order to guarantee a recoverable residue, the zinc-chloride flotation method was used.

A minimum of 250 grains was counted from each sample of V18-339. In those slides with a particularly large number of grains, a minimum of 500 specimens was counted.

Because of the method used in preparing the pollen and spores of core 339 for study, only the relative pollen frequencies could be determined. Crude visual estimates of the absolute abundance of grains were attempted, however, from the amount of the residue recovered by flotation, and the abundance of grains on the glass microslides. The most abundant assemblages were recovered from samples from the upper 100 cm of the core, and from the sampling interval 400 cm to 700 cm from the top. The smallest recoverable residues were obtained from the interval 150 to 350 cm.

Core 338 was treated to obtain the frequency distribution of pollen and spores per gram of dried sediments (absolute pollen frequency) as well as the relative frequency (total pollen and spores equal 100 percent).

The number of grains per gram of sediment in core 338 varied systematically from less than 300 at the 100 cm depth sample to over 17,500 in the 450 cm level, and to almost 30,000 in the 750 cm sample. For the most part, however, the number of grains varied from high values of 10,000 to 15,000 and low values of less than 1000.

LITHOLOGIC DESCRIPTIONS

The lithology of core 338 is very similar to that of 339, and to other cores collected from the seaward flank of the Middle America Trench (Ross, 1965).

The coarse fraction (greater than 62 microns) was separated from samples taken at the same levels as those used for pollen-and-spore analysis; every 10 cm through the uppermost 100 cm of each core, and every 50 cm for the remainder. The coarse fraction is usually less than 10 percent of the total sediment by weight, and averages 5 percent.

Figure 3 illustrates the relative frequency distribution of the coarse fraction. For both cores the predominant component is Radiolaria (Table 1). Quartz, feldspar, and other terrigenous grains are

TABLE 1. AVERAGE WEIGHT-PERCENTAGE COMPOSITION OF THE COARSE FRACTION

Core	Radiolaria	Fecal Pellets	Quartz and Feldspar	Other Terrigenous Grains	Mica
V18-338	33.4	18.7	24.1	24.9	6.7
V18-339	45.4	13.1	18.6	9.5	13.2

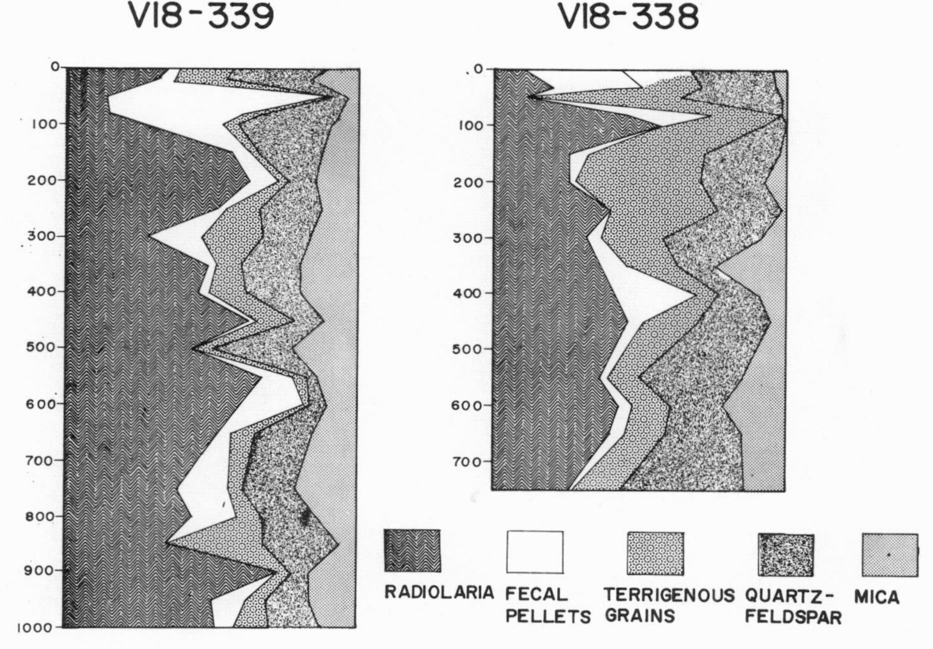

Figure 3. Distribution of course fraction in cores V18-339 and V18-338.

of secondary importance, and fecal pellets and mica comprise the remainder of the coarse fraction. Core 339 has slightly more biogenous material.

The general similarity of the cores suggests that they were deposited by particle settling, and that turbidite deposition was not significant.

DESCRIPTION OF PRINCIPAL POLLEN TYPES

Pinus
Pl. 1, figs. 1 and 2

Haploxylon bisaccate pine pollen, averaging 35-45 microns in maximum diameter, although a few grains as large as 80-90 microns were observed. The larger grains were distinguished from fir pollen by their lack of a structured proximal cap. Apparently, there are more than 30 species of pine in Mexico, and all but a few can be found in the pine-oak forest (Leopold, 1950).

Quercus
Pl. 1, figs. 3 and 4

Prolate tricolporate oak pollen, ranging from 20-30 microns in diameter. The pores are weakly developed. The grains commonly are rounded-tetragonal in equatorial view, and almost circular in polar view. The ornamentation is typically irregularly scabrate. According to Standley

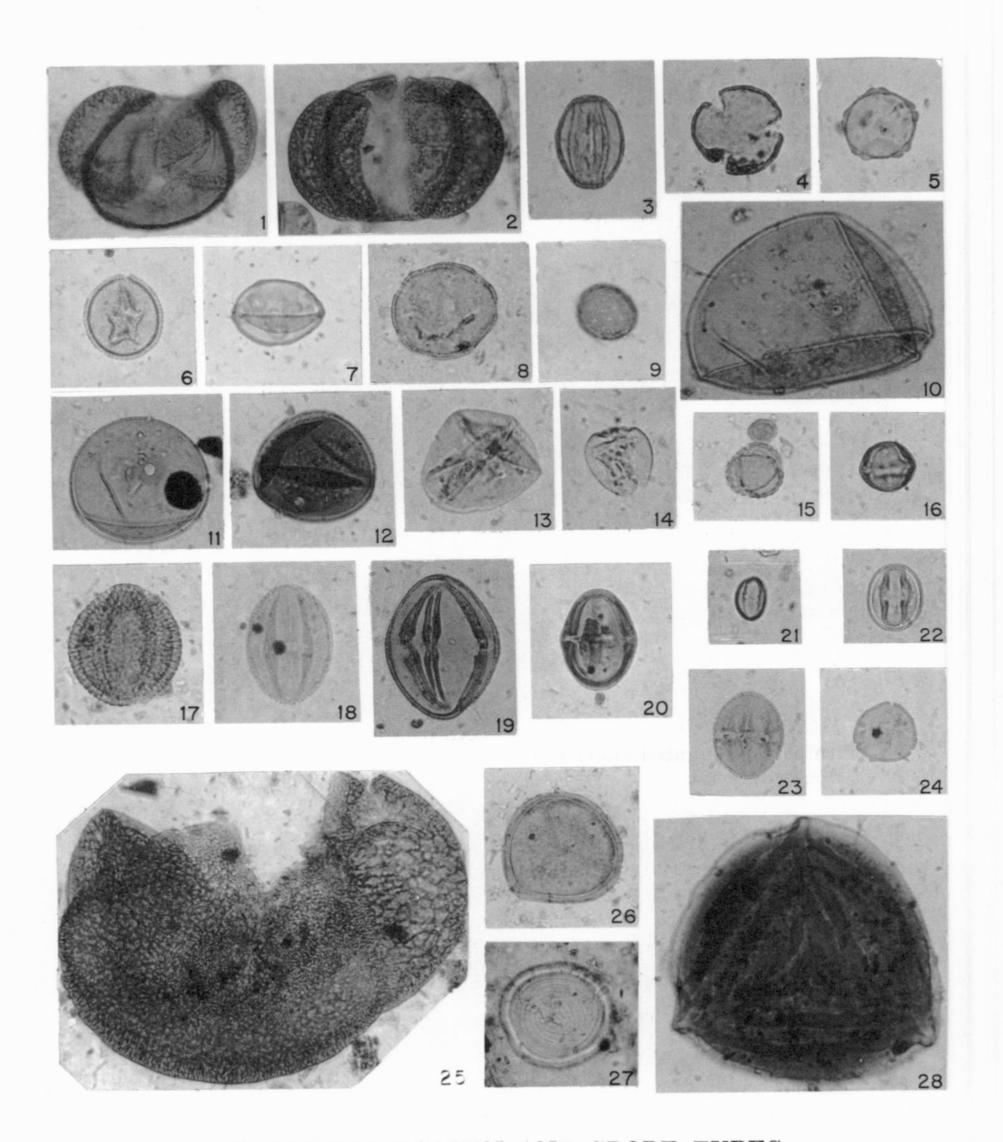

PRINCIPAL POLLEN AND SPORE TYPES

HABIB AND OTHERS, PLATE 1
Geological Society of America Memoir 126

PRINCIPAL POLLEN AND SPORE TYPES

All figures approximately ×500

Figure

1-2. *Pinus*. 1. Near-equatorial view. 2. Polar view.
3-4. *Quercus*. 3. Equatorial view. 4. Polar view.
5. *Alnus*. Five-porate grain, with well-developed vestibules and arci.
6. *Hedyosmum*. Tectate, single-aperturate (?) grain.
7. *Juglans*. Periporate grain; an annulus circumscribes each pore.
8. *Liquidambar*-type. Large specimen; the pores are relatively large and with well-developed "pits."
9. CHENOPODIACEAE. Small, subcircular, periporate grain.
10-13. GRAMINEAE. 10. Large grain, similar in form and size to the pollen produced by *Zea*. 11. Circular grain, with well-developed pore and annulus; exine smooth. 12. Specimen with same morphology as that of Figure 11; this specimen is dark brown in color, and apparently did not accept stain. 13. Folded specimen, with scabrate ornamentation.
14. CYPERACEAE. Specimen is scabrate; a single pore is discernible.
15. COMPOSITAE. *Ambrosia*-type.
16. *Rhizophora*. Small, circular grain with psilate ornamentation. Tricolporate, with pores elongated along the equator.
17-23. Tricolporate types. 17. *Gentiana*. Tegillate grain with columellae arranged in reticulate design. 18. Specimen similar to the pollen produced by *Anacardium* (ANACARDIACEAE). Exine striato-reticulate. 19. Large grain. Furrows are split and extended to very near the poles. 20. SAPOTACEAE. 21.*Weinmannia*-type. Exine is faintly microreticulate. 22. *Alchornea*-type. Exine thick and psilate. 23. Specimen with microreticulate exine and pores elongated along the equator. This grain is similar to the pollen produced by *Avicennia nitida* (black mangrove).
24. *Engelhardtia*-type (JUGLANDACEAE). Triporate pollen with simple apertures.
25. *Abies*. Bisaccate grain. The proximal cap is well-developed.
26. *Hymenophyllum*. Fern spore with trilete sutures bifurcating subequatorially.
27. Palynomorph assigned to the form genus *Concentricystis*.
28. Reworked Mesozoic palynomorph assigned to the form genus *Appendicisporites*. The specimen is dark brown, the same color as the specimen illustrated in Figure 12.

(1926), there are as many as 112 species of oak in Mexico. Leopold (1950) reports that all can be found in the pine-oak forest. Oak also occurs on high peaks where there is available moisture. Species of oak form the cloud forest in the Colombian Andes (van der Hammen and Gonzalez, 1960).

Alnus
Pl. 1, fig. 5

Stephanoporate pollen of alder, 4 to 6 pores connected by arci. Polygonal outline in polar view. Grains compressed in equatorial view appear tetragonal. Size range mostly between 20 and 30 microns. The exine is psilate. Sears and Clisby (1955) reported alder pollen is a moisture indicator. Alder occurs with pine and fir, and with traces of cloud forest elements in valleys of Mexico at elevations above 2500 m (Leopold, 1950). In the Colombian Andes it is found in the high cloud forest (van der Hammen and Gonzalez, 1960).

Abies
Pl. 1, fig. 25

Fir pollen; large bisaccate grains from 80 to more than 130 microns in diameter. The sacs are microreticulate. The central body is microreticulate to infragranulate, with a thick proximal cap. Fir is exemplary of the boreal forest in Mexico. It stands best in high cold and wet valleys.

Hedyosmum
Pl. 1, fig. 6

Tectate pollen grains, apparently single-aperturate or inaperturate, with a subcircular or oval outline. Size approximately 30 to 35 microns; van der Hammen and Gonzalez (1960) listed a species of *Hedyosmum* (*H. bonplandianum*) as a woody climber in the oak forest of Colombia. The genus evidently exists also in Mexico.

Grains corresponding to *Hedyosmum* were counted because of their relative abundance, ease of identification, and close association with oak. Pollen production is high in the species given by van der Hammen and Gonzalez (1960).

Juglandaceae
Pl. 1, figs. 7 and 24

Pollen grains attributed to the Family Juglandaceae correspond to two morphological types; *Juglans* pollen, psilate circular periporate grains, annulate, with pores confined to the equatorial and distal surfaces of the exine, and *Engelhardtia*-type pollen, small triporate, with simple pores and slightly convex inter-radial margins. *Juglans* (walnut) is representative of the temperate Appalachian flora and is most likely an occupant of the cloud forest in Mexico. *Engelhardtia* apparently has a wider distribution in the tropical flora.

Liquidambar
Pl. 1, fig. 8

Periporate circular grains, scabrate, with small "pitting" in the pores. Grains reach to 35 microns in diameter. *Liquidambar* is one of the many temperate hardwoods forming the cloud forest. It has been recorded south into Guatemala (Steyermark, 1950).

Small Prolate Tricolpate Pollen
Pl. 1, fig. 21

Prolate tricolpate and tricolporate pollen grains, 20 microns in diameter or less, were counted together. Several types were difficult to distinguish. Most of the grains correspond closely to the pollen of *Weinmannia*, described by van der Hammen and Gonzalez (1960). These are typically tricolporate grains with an inconspicuous microreticulate ornamentation. Others are morphologically similar to *Castanea* (chestnut) pollen of the Appalachians. The pollen grains probably represent the more temperate species of Mexico.

Gramineae and Cyperaceae
Pl. 1, figs. 10-13

Pollens of the non-arboreal grasses (Gramineae) and sedges (Cyperaceae) were counted together. The grass pollen is typically subcircular-circular in unfolded compressions, with a single pore and erect annulus. Three morphological varieties were recognized; a smaller (25-40 microns) circular psilate type, a scabrate type of approximately the same size, commonly with several secondary folds giving it an angular outline, and a larger type (65 microns and larger) which is psilate and may be folded to a tetragonal outline. This latter type is similar to the pollen produced by maize (*Zea*). Its lowest recorded occurrence is in the 550 cm sample in core 339.

The sedge pollen is subrounded or angular in outline, commonly with a single pore which may be difficult to discern in some specimens. The ornamentation is typically scabrate. Grasses and sedges have a wide distribution in Mexico and are controlled to a large degree by edaphic as well as a wide range of climatic factors. Grasses, for example, occur along the coast and also near the top of the high peaks.

Compositae
Pl. 1, fig. 15

Circular, echinate, tricolporate pollen grains. A non-arboreal family, the Compositae have a wide distribution in Mexico.

Rhizophora
Pl. 1, fig. 16

Pollen grains morphologically identical to red mangrove (*Rhizophora mangle*) are subcircular to elliptical in outline, with a relatively thick exine. The grains are tricolporate with very well-developed pores (transverse furrows) in the equatorial region. Ornamentation is psilate. Mangrove forms the coastal vegetation in tropical and subtropical latitudes. It indicates coastal swamps.

Pteridophyte Spores
Pl. 1, fig. 26

Monolete and trilete spores of various structure and ornamentation were observed, including representatives of the Sphagnales (Bryophyta),

Lycopodiales (Lycopodophyta) and families of the Filicales (for example, Matoniaceae, Polypodiaceae, Hymenophyllaceae, Gleicheniaceae, Cyatheaceae, Schizaeaceae). Several of the highly structured triangular forms apparently are restricted to tropical latitudes.

Large Tricolporate Grains
Pl. 1, figs. 17-20; 22-23

A number of relatively large (25-60 microns) tricolporate grains were not counted because of their low percentage even as a single group. They have various plant affinities.

Type 1. (Pl. 1, fig. 22) Psilate tricolporate pollen grains, approximately 25 microns in maximum diameter. The furrows and pores are very well-developed, with the pores commonly protruding from the exine, in polar view. These grains closely resemble the pollen produced by the tropical genus *Alchornea.*

Type 2. (Pl. 1, fig. 17) *Gentiana.* Tegillate tricolpate pollen grains with club-shaped columellae. The columellae form clusters of circles (reticulate pattern) on the surface of the exine. *Gentiana* occurs high on the slopes of the Andes in Central America.

Type 3. (Pl. 1, fig. 18) Tectate tricolporate pollen grains. The outline equatorial view is elliptical, with slightly tapering polar areas. The furrows are long, commonly extending from one polar region to the other; the pores tend to be circular. The exine is ornamented with a striato-reticulate pattern oriented parallel with the polar axis. Pollen of this morphology is produced by members of the temperate Family Anacardiaceae.

Type 4. (Pl. 1, fig. 19) Tectate, tricolporate pollen, with a psilate or scabrate exine. Outline in equatorial view oval, with markedly tapering polar areas. The furrows are distinct and extend to the polar areas; the pores are less well-defined, and are circular in form. These grains are morphologically similar to members of the Anacardiaceae, but lack the ornamentation common to the species of this family.

Type 5. (Pl. 1, fig. 20) Sapotaceae. Psilate tricolporate pollen grains with a thick exine. Outline in equatorial view is elliptical, with well-rounded polar areas. Pores well-developed; they are elliptical in form, and are oriented in the transverse direction. Members of the subtropical and tropical family Sapotaceae occur in the tropical deciduous forest.

Type 6. (Pl. 1, fig. 23) Tectate tricolporate pollen, with elliptical outline in equatorial view. The furrows are long, extending most of the distance to the poles; the pores are elongated transversely along the equator. The ornamentation is microreticulate. Pollen of this type resemble those dispersed by the subtropical and tropical genus *Avicennia* (black mangrove), which is closely associated with red mangrove in swamp vegetation.

Sedimentary Altered Grains
Pl. 1, figs. 12 and 28

Sedimentary altered pollen and spores were distinguished either by dark discoloration of the exine and/or total lack of stain acceptance. Many specimens, especially those assignable to pine, oak, alder, grass, and fern species, were found to be well-preserved in form but chocolate brown or dark yellow in color. These grains exhibited none or very little of the red color (Safranin O) which readily stained other grains.

Most of these grains occurring in the Middle America Trench most likely represent reworking by streams of bedded Pleistocene deposits. They may also represent contemporaneous grains which have endured chemical and organic alteration (Groot, 1966). The sedimentary altered pollen and spores form only a small percentage of the total assemblage, but appear to have meaningful fluctuations in the pollen diagrams (Figs. 4 and 5).

cf. *Concentricystis*
Pl. 1, fig. 27

Unknown subcircular to oval specimens of the form genus *Concentricystis* occur irregularly throughout both cores in very low numbers. It has been reported in modern sediments from the southern Gulf of California (Cross and others, 1966) and from the Mediterranean Sea opposite Israel (Rossignol, 1961). Palynomorphs of this form have also been recovered from the Upper Mesozoic of western Canada (Pocock, 1962).

Dinoflagellate cysts and cysts of animal affinity were also observed. Fungal spores occur throughout both cores.

RADIOCARBON DATES

The core, V18-338, was very rich in organic material from which CO_2 was extracted by burning after treatment with acid to remove the carbonate fraction. The radiocarbon concentration was determined as described by Broecker and others (1959) and the ages were calculated according to the conventions now in use (Deevey and others, 1968). The results are shown in Table 2.

The initial C^{14}/C ratio was assumed to be that of surface sea water, 5 percent below the standard atmosphere. The latter assumption is based on the observation that although much of the organic material is coarse and woody, the αC^{13} for L#1237 (Table 2) was 20.2 percent, corresponding to the average for warm temperature marine plankton (Sackett and others, 1965). If half of the organic material is terrestrial, the age would be 200 years older than that calculated here, and 400 years older if all of the material were terrestrial. Sackett and Thompson (1963) have pointed out that marine derived material commonly predominates even in near-shore sediments.

TABLE 2. RADIOCARBON DATES AND SEDIMENTATION RATES OF CORE V18-338

L#	Sample Level	Age (B.P.)	Sedimentation Rate
1237	90-100cm	1750 ± 150	0-95 cm: 54cm/1000 yr
1282	230-235cm	3300 ± 300	95-230 cm: 82cm/1000 yr
1284	450-455cm	4050 ± 550	230-450 cm: 293cm/1000 yr

Average sedimentation rate of interval 0-450 cm is 111 cm/1000 yr

DESCRIPTION OF POLLEN DIAGRAMS

Assemblage zones were established for cores 338 and 339 using the distribution of the most abundant types. Relative percentages form the basis for delineating zones, in order to conform to the criteria used by Pleistocene palynologists. In the case of core 338, numbers of grains per gram of sediment are considered equally important, if not more so, for describing and interpreting each of the zones. In accordance with the suggestions made by Faegri and Iversen (1964), only the major fluctuations are considered to be significant for distinguishing zones.

Contrary to the standard procedure of numbering palynological zones from the bottom of the section to the top (for example, Gonzalez and others, 1966; Faegri and Iversen, 1964), we have chosen to number them from the tops of the cores downward, because longer cores with lower sedimentation rates may become available, and more detailed pollen analysis may permit recognition of older climatic events.

Pollen and spore diagrams of cores 338 and 339 were constructed to illustrate the relative percentage distribution of the principal types (Figs. 4 and 5). Pine, oak, alder, and fir pollen are considered to be most diagnostic and together they comprise 80 percent and more of each sample. Sears and Clisby (1955) and Leopold (1950) have shown these genera to represent the climatically significant vegetation of Mexico. Other types included in the diagrams are pollen attributed to cloud forest vegetation, the nonarboreal plants such as the grasses and sedges, and the pteridophytes. The occurrence of mangrove pollen and sedimentary-altered grains is also indicated.

Pine is the single most abundant pollen in the cores, and is followed by oak. These two types appear to be inversely related in the

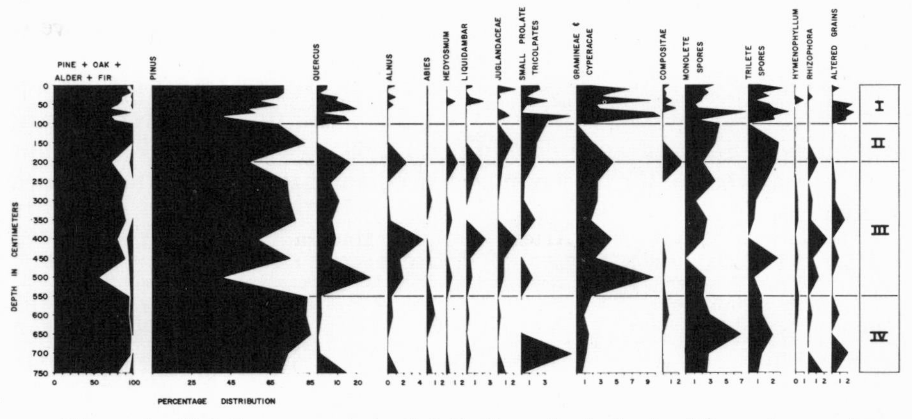

Figure 4. Relative percentage distribution of principal pollen and spore types in V18-338.

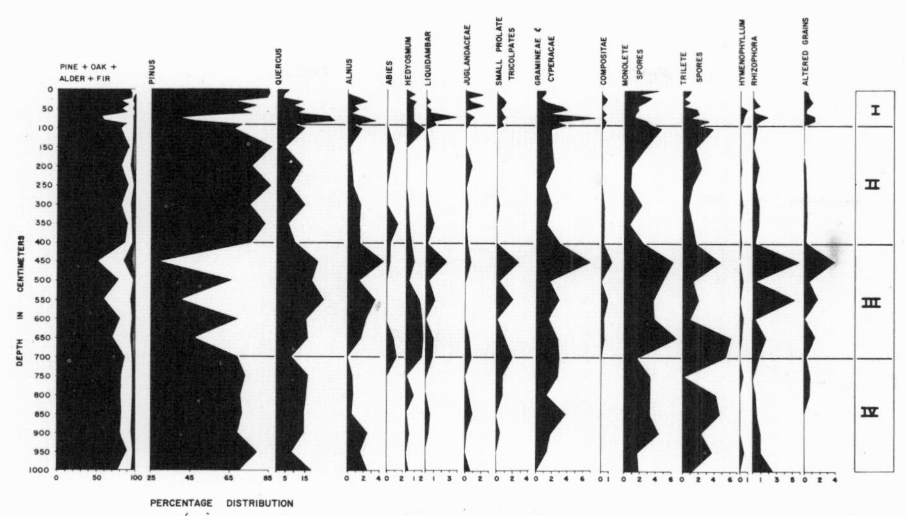

Figure 5. Relative percentage distribution of principal pollen and spore types in V18-339.

relative pollen frequencies, pine being highest when oak values are lowest, and lowest when oak values are highest. The distribution of alder is somewhat more complicated, but appears in the relative pollen frequencies to be more closely related to the trends of oak. Fir pollen is sparse, only rarely exceeding 1 percent in any sample. It tends to be more closely related to the distribution of pine, although in some samples it is found associated with oak.

Correlation coefficients were calculated for pine, oak, alder, and fir to obtain a comparison with the values determined by Sears and Clisby (1955; Tables 3 and 4).

The coefficients of the cores from Mexico City show a close similarity to the coefficients of the cores from the trench. In both terrestrial and marine sediments, oak and alder have a positive cor-

TABLE 3. CORRELATION BETWEEN POLLEN PERCENTAGES BY PAIRS OF UPLAND FOREST GENERA

Relation	Coefficient (r) Belles Artes	Madera	Significance
Pinus-Quercus	— .68	— .73	Inverse
Pinus-Alnus	— .76	— .60	Inverse
Pinus-Abies	— .27	— .11	Slight inverse
Abies-Quercus	— .09	.17	Slight direct
Quercus-Alnus	.32	.35	Direct
Abies-Alnus	.05	— .06	No correlation

(*After* Sears and Clisby, 1955)

V18-339

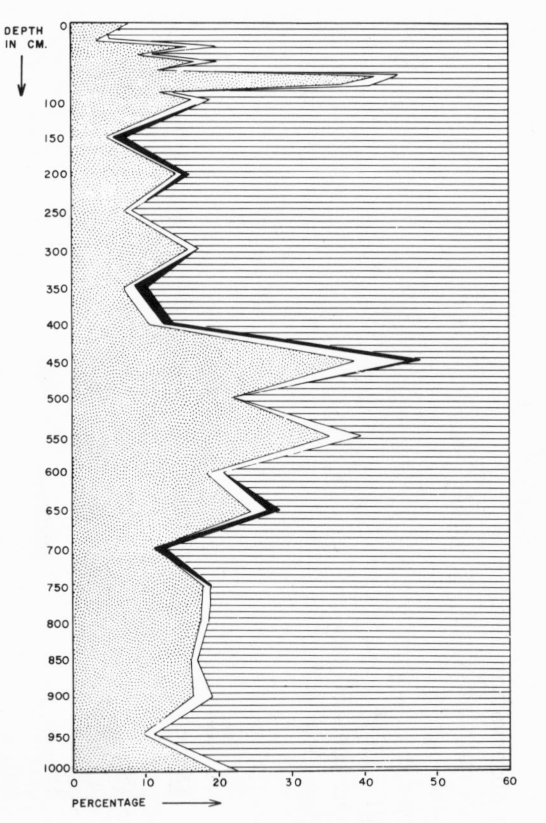

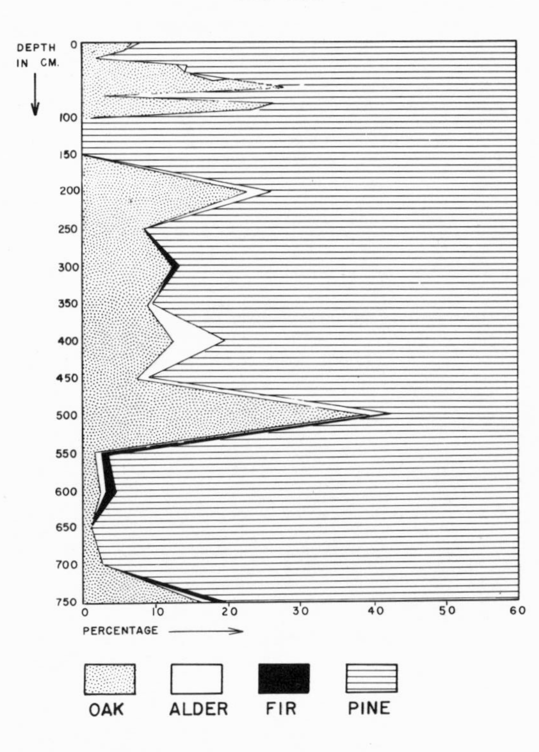

Figure 6. Relative percentage distribution of pine, oak, alder and fir.

TABLE 4. CORRELATION BETWEEN POLLEN PERCENTAGES BY PAIRS FROM MIDDLE AMERICA TRENCH SEDIMENTS

Relation	Coefficient (r)		Significance
	V-338	V-338	
Pinus-Quercus	—.74	—.84	Inverse
Pinus-Alnus	—.37	—.63	Inverse
Pinus-Abies	.15	.25	Slight direct
Abies-Quercus	—.17	—.23	Slight inverse
Quercus-Alnus	.24	.54	Direct
Abies-Alnus	—.15	—.33	Slight inverse

relation. Also, they agree in that the pairs pine-oak and pine-alder show a negative correlation. In the case of the three pairs, pine-fir, fir-oak, and fir-alder, our results differ from those of Sears and Clisby, but the differences are not great.

Figure 6 illustrates the relative percentage distribution of oak, alder, and fir, compared with pine, after the method used by Sears and Clisby (1955). Oak and alder, especially, appear to be inversely related to the distributioin of pine.

The number of grains per gram of pine, oak, alder, and fir in core 338 is illustrated in Figure 7. Pine pollen is at least five times more abundant than oak in most samples, and appears to vary directly with distribution of the total number of grains per gram, an observation similar to that made by Cross and others (1966) for the southern Gulf of California. Oak varies directly with pine in the upper half of the core, in contrast to the relative pollen frequencies for this portion, but is inverse to pine through much of the lower half. Alder appears to vary with oak, although in the sample taken from the 450 cm level, its higher value correlates with a high pine value. Fir pollen varies with pine in the lower two-thirds of the core.

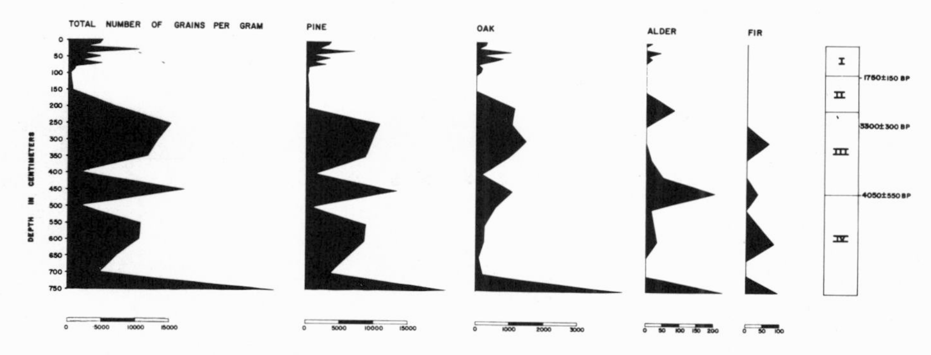

Figure 7. Absolute pollen frequency in core V18-338. Number of grains per gram calculated on a dry-weight basis.

CORE 338

Zone IV. (750-550 cm) The lowermost zone in core 338 contains small amounts of oak pollen and larger amounts of pine. Values for pine are lower in the lower part, but rise to the upper boundary, as do the total amount of pollen per unit weight (Fig. 7).

Zone III. (550-200 cm) Zone III is characterized by higher percentages of oak, alder, *Hedyosmum,* and non-arboreal (grasses and sedges) pollen. *Rhizophora* and altered grains are also higher. Pine is relatively low. Pteridophyte spores are represented by slightly lower percentages. The total amount of pollen is the highest of all the zones in core 338. The increase in relative percentages of oak and other types is not at the expense of pine frequencies, as is indicated in Figure 4, but occurs *in addition* to high pine values (Fig. 7). This zone also contains the most diverse assemblage, including many types of which only a few individuals were found.

TABLE 5. COMPARATIVE POLLEN STRATIGRAPHY OF NORTHERN EUROPE, NORTHEASTERN UNITED STATES, ANDEAN COLOMBIA AND CORE V18-338

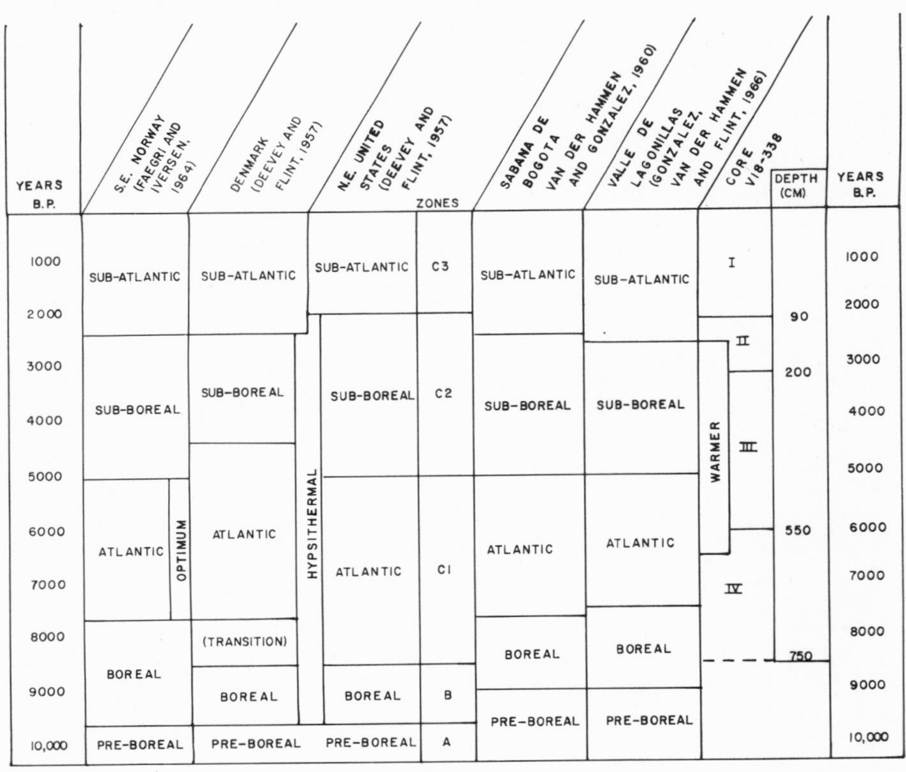

YEARS B.P.	S.E. NORWAY (FAEGRI AND IVERSEN, 1964)	DENMARK (DEEVEY AND FLINT, 1957)	N.E. UNITED STATES (DEEVEY AND FLINT, 1957)	ZONES	SABANA DE BOGOTA VAN DER HAMMEN AND GONZALEZ, 1960	VALLE DE LAGONILLAS (GONZALEZ, VAN DER HAMMEN AND FLINT, 1966)	CORE V18-338	DEPTH (CM)	YEARS B.P.
1000	SUB-ATLANTIC	SUB-ATLANTIC	SUB-ATLANTIC	C3	SUB-ATLANTIC	SUB-ATLANTIC	I		1000
2000								90	2000
3000							II	200	3000
4000	SUB-BOREAL	SUB-BOREAL	SUB-BOREAL	C2	SUB-BOREAL	SUB-BOREAL	WARMER III		4000
5000									5000
6000	ATLANTIC OPTIMUM	ATLANTIC HYPSITHERMAL	ATLANTIC	C1	ATLANTIC	ATLANTIC		550	6000
7000							IV		7000
8000	BOREAL	(TRANSITION)			BOREAL	BOREAL		750	8000
9000		BOREAL	BOREAL	B					9000
10,000	PRE-BOREAL	PRE-BOREAL	PRE-BOREAL	A	PRE-BOREAL	PRE-BOREAL			10,000

Zone III may also be subdivided into a lower and upper part. The lower part contains the lowest relative percentages of pine and highest of alder and grasses-sedges.

The radiocarbon date obtained from the sample at 450 cm is 4050 ± 550 B.P. (Table 2). The sample analyzed at 230 cm indicates an age of 3300 ± 300.

Zone II. (200-90 cm) This zone is characterized by high relative percentages of pine and very low percentages of oak, alder, *Hedyosmum,* and grasses-sedges, with higher values for small tricolporate grains and pteridophyte spores. Comparison with the numbers of grains/gram shows this zone to possess the smallest amount of pollen in the core. It contains mostly pine pollen and monolete and trilete spores, and few other types. Oak and alder occur at less than one grain per gram of sediment, less than but on the same order as values obtained from the lower part of Zone IV.

The sample analyzed at 90-100 cm yielded a radiocarbon date of 1750 ± 150 B.P.

Zone I. (90-0 cm) The uppermost zone possesses higher percentages of pine and oak, along with high percentages of grasses, sedges, and pteridophyte spores. Alder is higher at the top. *Rhizophora* and altered grains are higher than below as well. The grains/gram distributions show similar high values for pine, oak, and alder.

CORE 339

Zone IV. (1000-700 cm) The zone at the bottom of core 339 shows high percentages of pine pollen, and percentages of oak, alder, *Hedyosmum,* and grasses-sedges which are lower than in other zones. It can be subdivided into two parts, based on the higher percentages of alder in the lower part (Fig. 5).

Zone III. (700-400 cm) This zone is characterized by high percentages of oak, alder, and *Hedyosmum* pollen. Grasses-sedges and pteridophytes fluctuate from high values to low, but do not appear to vary significantly from their distributions in the zone below. *Rhizophora* pollen and altered grains are highest in Zone III. There is also the greatest number of pollen types, although many form less than 1 percent of the assemblage.

Zone II. (400-90 cm) This zone is similar to Zone IV in that oak, alder, and *Hedyosmum* occur at lower percentages. Other types are also of low frequency, although the pteridophytes rise to higher percentages in the upper part. This zone contains the least variation in types of pollen. Pine comprises as much as 85 percent of the assemblage through most of this zone.

Zone I. (90-0 cm) The uppermost zone is distinguished by high percentages of pine, oak, alder, *Hedyosmum,* and grasses-sedges. *Rhizophora* and altered grains are also higher than in the next lower zone.

CONCLUSIONS

Fluctuations of major pollen types in cores 338 and 339, held within the framework of radiocarbon dates, can be used for interpreting the climatic history of adjacent Mexico during the last 8000 to 9000 years.

Interpretation of the pollen assemblages is complicated by the variety of processes responsible for their emplacement in the Middle America Trench. The two principal methods of transportation are water and wind. Muller (1959) attributed concentrations of *Podocarpus* pollen and spores of the tree fern *Hemitelia* in surface sediments north and west of Trinidad, to the prevailing southeasterly winds in that area. Both these genera are anemophilous in pollen and spore dispersion. *Podocarpus* pollen is very similar in morphology to pine, and presumably is equally well-dispersed.

Koreneva (1964) discovered the most abundant types in surface sediments of the western Pacific Ocean to be bisaccate pollen of the conifers and trilete spores, removed hundreds of kilometers from the nearest source. She recovered 150 grains per gram of sediment 200 to 300 km off the coast of New Zealand, and fewer grains per gram farther offshore. These presumably were also transported at least some distance by winds. Fern spores are also the most abundant type in surface marine sediments adjacent the southwestern coast of Puerto Rico (Habib and others, 1968), where there is a lack of permanent stream drainage. Only 50 to 250 grains per gram were recovered, and their deposition could be attributed to the prevailing northeasterlies across Puerto Rico. Bisaccate pollen-producing conifers are not common in Puerto Rico.

Muller (1959, p. 5) determined the total pollen distribution in the Gulf of Paría north of the Orinoco drainage system and Trinidad. The most abundant grains (6000 to 10,000 per gram sediment) were recorded adjacent the delta, and were largely deposited by distributaries. Fewer grains (500 to 2000) were deposited by the winds north of Trinidad, the bulk of which is *Podocarpus* and *Hemitelia* (Muller, 1959, p. 16).

The pollen and spore zones in V18-338 and V18-339 are interpreted to be the result of climatically controlled shifts in the Holocene vegetation of western Mexico. The interpretation of paleoclimates in Mexico is further enhanced by the fact that the bulk of the pollen and spores are assignable to an upland flora of climatic significance (Leopold, 1950; Sears and Clisby, 1955).

Large numbers of grains, diverse morphological types, and increased numbers of altered grains are found during times of high sedimentation rates. We suggest they reflect greater atmospheric moisture and increased stream activity during the time these zones were formed. Smaller amounts of pollen, together with the predominance of a few types (notably pine pollen and fern spores), are the result of lesser stream activity. Fluctuations of terrigenous sediments (quartz, feldspar, and other terrigenous grains) in the coarse fraction of core 338 (Fig. 3) suggest that there was less detrital influx at the 90-150 cm and 400-450 cm intervals than elsewhere. Similar intervals can be found in core 339 as well.

Thus, from variations with depth in the total number of grains per gram of sediment in core 338, we suggest that the streams draining the western slopes of the Sierra Madre del Sur have varied in discharge and in ability to deliver pollen and other sediments to the sea in this area.

Zones I through IV represent Holocene climatic events in southwestern Mexico dating back approximately 8000 to 9000 years B.P. (Fig. 4). On the basis of the radiocarbon date of 1750 ± 150 B.P., Zone I is interpreted to be contemporary with the sub-Atlantic climatic phase of northwestern Europe and C2 phase of northeastern United States. The large number of specimens in core 338 (from 4000 to 13,000 per gram), high pine and oak numbers, and variety of types, indicate a moist climate. The decrease in total grains in the upper 50 cm of the core suggests a drying trend through approximately the last 1000 years.

Zone II is interpreted as corresponding in time to the sub-Boreal climatic phase of northern Europe. A radiocarbon date at the 230 cm level indicates it to be as old as 3300 ± 300 B.P. The climate is interpreted to have been relatively dry, primarily because it contains the lowest total amount of pollen found in the core, and because most of it is pine pollen, small tricolporate grains, and fern spores. According to the composition of this assemblage, stream activity is interpreted to have been low.

Zone III contains the largest amount of pollen grains and highest percentages of oak and pine (Fig. 7). This zone probably represents a warm and moist climate during its period of formation, because it contains the highest percentages of oak, an abundance of alder, *Hedyosmum,* grasses-sedges, and altered grains, and relatively diverse types. Radiocarbon analysis of the sample at the 450 cm level indicates an age of 4050 ± 550 years B.P. This zone corresponds in time to the Atlantic phase of Europe, and Cl phase of the northeastern United States.

The lowest zone in the cores, Zone IV, is interpreted to have formed during cooler and less moist climates than Zone III. The cooler climate is indicated by lower relative percentages of oak, despite the abundance of pollen in the sediments. Because of the high pollen content in these sediments, the climate is considered to have been sufficiently moist in the upper part of the zone for streams to have participated in the deposition of the assemblage.

By extrapolation of carbon-14 dates higher in core 338, at least the upper part of Zone IV corresponds to the climatic phase during which the Boreal zone was formed in higher northern latitudes.

The relative frequencies of pollen in cores 338 and 339 (Figs. 4 through 6), are very similar in the "dry" Zones II and IV. Both zones show very low percentages of oak and alder, relative to pine. They thus appear to represent a similar source for pollen and depositional environment for the two cores. The grains-per-gram distribution is different between Zones II and IV in core 338, however, suggesting different conditions were present when these zones formed. Zone II contains very low total pollen, mostly pine. In contrast, a high abundance of pine pollen is found in Zone IV, especially the upper part, again largely restricted to this genus. Cross and others (1966) suggest that large numbers of pine pollen in bottom sediments from the Gulf of California are water transported, since the pine forests on the mainland are down-wind from the Gulf at the time the pollen is shed. They stated that a significant amount of pine pollen (5000 grains per gram and more) may have been contributed by streams, including the Colorado River, in addition to that brought in by winds.

This conclusion may be useful in distinguishing between Zones II and IV in the cores. As stated previously, pine pollen (and fern spores) may have been transported to the site of deposition in Zone II in a dry climate by the winds. During the formation of Zone IV, however, streams may have played a more important role in depositing pine, and for this reason the climate is interpreted to have been more moist than during the phase of Zone II. On the other hand, the phase of Zone IV is considered to have been cooler and less moist than the next succeeding zone (Zone III), because of the lower relative percentages of oak and alder.

Oak pollen is usually considered to reflect periods of warmth, although larger amounts reflect increased moisture as well, especially when associated with alder and fir pollen. The interval of large numbers of oak in core 338 is from 600 cm through 200 cm, a period of time which appears to have extended from approximately 6500 to 3000 years B.P. Although of shorter duration, this period falls within the hypsithermal interval given as 9000 to 2500 B.P. for northeastern United States and northern Europe (Deevey and Flint, 1957). This period of oak abundance appears to coincide also with the period of warmth suggested by van der Hammen and Gonzalez (1960) and Gonzalez and others (1966) for the alpine regions of Colombia.

The interval of large numbers of oak, alder, and fir in core 338 extends from the upper part of Zone IV through Zone III and the lower part of Zone II, and corresponds in time to the Boreal (B), Atlantic (C1), and sub-Boreal (C2) zones, respectively, of the northern United States and northwestern Europe. The rise of oak in the upper part of Zone IV indicates increasing warmth and moisture in adjacent Mexico during this time. Its consistently high abundance in the lower part of Zone II, despite diminishing amounts of pollen and spores upward in the core, reflects a warm but dry period during this phase. This interpretation corresponds closely to the xerothermic period hypothesized by Deevey (1949) for the C2 zone of the United States.

The relative pollen frequencies do not indicate subdivisions in Zone IV even though they were recognized by Gonzalez and others (1966) south of Mexico in cordilleran Colombia. The grains-per-gram distribution in core 338 may be interpreted to represent a cool and dry climate, drier than the upper part of Zone IV. The 750 cm level, however, contains the largest number of pollen in core 338, and the relative percentages show higher oak than do other parts of Zone IV. No interpretation is offered, however, because it is represented by a single sample and is at the bottom of the core.

Because of the limited number of cores, the conclusions drawn in this study must be regarded as preliminary, and must await the substantiating or modifying evidence brought about by the examination of additional independently dated cores from the area of Tehuantepec. Examination of longer and older cores outside and seaward of the Middle American Trench (for example, V18-337, V18-340) has yielded a series of oscillating moist-dry zones which extend below the Wisconsin stage. The occurrence of diatoms and radiolarians, as well as dinoflagellate cysts, thus makes possible the relationship of climatic events in terrestrial and oceanic environments (*see* the papers in this volume on Pleistocene diatom stratigraphy by Jessie G. Donahue and radiolarian stratigraphy by Catherine Nigrini).

ACKNOWLEDGMENTS

Appreciation is extended to Thomas van der Hammen (Hugo deVries Laboratorium, Amsterdam), James D. Hays, Lloyd Burckle (Lamont-Doherty Geological Observatory), Andrew McIntyre, Walter S. Newman (Queens College, City University of New York), Calvin J. Heusser (New York University), and David Wall (Woods Hole Oceanographic Institution) for reviewing the manuscript.

Thanks are also extended to Matsuo Tsukada (University of Washington, Seattle) and Calvin J. Heusser for providing us with pollen reference material and for discussing with us the flora and palynology of Latin America.

Support for the research and publication of this paper was provided from grants awarded by the National Science Foundation (NSF-GA-558, GP-1346, and GA-1193) and Office of Naval Research (ONR-N00014-67-A-0108-004) to Lamont-Doherty Geological Observatory.

REFERENCES CITED

Broecker, W. S., Tucek, C. S., and Olson, E. A., 1959, Radiocarbon analysis of oceanic CO_2: Internat. Jour. Appl. Radiation and Isotopes, v. 7, p. 1-18.

Cross, A. T., Thompson, G. G., and Zaitzeff, J. B., 1966, Source and distribution of palynomorphs in bottom sediments, southern part of Gulf of California: Marine Geology, v. 4, no. 6, p. 467-524.

Cserna, Z. de, 1961, Tectonic map of Mexico: Geol. Soc. America.

Deevey, E. S., 1949, Biogeography of the Pleistocene, Pt. I, Europe and North America: Geol. Soc. America Bull., v. 60, p. 1315-1416.

Deevey, E. S., and Flint, R. F., 1957, Postglacial hypsithermal interval: Science, v. 125, p. 182-184.

Deevey, E. S., 1949, Biogeography of the Pleistocene, Pt. I, Europe and statement: Radiocarbon, v. 10, no. 1, p. 0.

Ewing, J., Ewing, M., Aitken, T., and Ludwig, W. J., 1968, North Pacific sediment layers measured by seismic profiling: Am. Geophys. Union Mono. 12, p. 147-173.

Faegri, K., and Iversen, J., 1964, Text-book of modern pollen analysis: New York, Hafner Pub. Co., p. 237.

Fisher, R. L., 1961, Middle America Trench: topography and structure: Geol. Soc. America Bull., v. 72, p. 703-720.

Gonzalez, E., Hammen, Th. van der, and Flint, R. F., 1966, Late Quaternary glacial and vegetational sequence in Valle de Lagunillas, Sierra Nevada del Cocuy, Colombia: Leidse Geol. Meded., v. 32, p. 157-182.

Groot, J. J., 1966, Some observations in the estuary of the Delaware River: Marine Geology, v. 4, no. 6, p. 409-416.

Habib, D., Donahue, J., and Krinsley, D. H., 1968, Modern pollen distribution off southwestern Puerto-Rico—a preliminary report (abs.): Preprinted Program, 5th Caribbean Geol. Conf., 1968, St. Thomas, Virgin Islands.

Hammen, Th. van der, and Gonzalez, E., 1960, Upper Pleistocene and Holocene climate and vegetation of the "Sabana de Bogotá" (Colombia, South America): Leidse Geol. Meded., v. 25, p. 261-315.

Koreneva, E. V., 1964, Distribution and preservation of pollen in sediments of western part of the Pacific Ocean: Acad. Sci. Trans., Geol. Inst., Moscow, v. 109, p. 1-88 (in Russian; translated by H.

Solnick, Geol. Bull., Queens College Press, Flushing, N.Y., no. 2, 1968).

—— 1966, Marine palynological researches in the U.S.S.R.: Marine Geology, v. 4, no. 6, p. 565-574.

Leopold, A. S., 1950, Vegetation zones of Mexico: Ecology, v. 31, p. 507-518.

Martin, P. S., 1964, Paleoclimatology and a tropical pollen profile: 6th Internat. Congr. Quaternary Rept., Warsaw, 1961, p. 319-323.

Muller, J., 1959, Palynology of Recent Orinoco delta and shelf sediments: Micropaleontology, v. 5, p. 1-32.

Pocock, S. A. J., 1962, Microfloral analysis and age determination of strata at the Jurassic-Cretaceous boundary in the western Canada Plains: Paleontographica, v. III, p. 1-95.

Ross, D. A., 1965, Ph.D. thesis, The University of California, Scripps Institution of Oceanography.

Ross, D. A., and Shor, G. Jr., 1965, Reflection profiles across the Middle America Trench: Jour. Geophys. Research, v. 70, no. 22, p. 5551-5572.

Rossignol, M., 1961, Analyse pollinique de sediments marine Quaternaires en Israel; sediments Recents: Pollen et Spores, v. 3, p. 303-324.

Sackett, W. M., and Thompson, R. R., 1963, Isotope organic carbon composition of Recent continental derived clastic sediments of the eastern Gulf coast, Gulf of Mexico: Am. Assoc. Petroleum Geologists Bull., v. 47, p. 525.

Sackett, W. M., Eckelmann, W. R., Bender, M. L., and Bé, A. W. H., 1965, Temperature dependence of carbon isotope composition in marine plankton and sediments: Science, v. 148, p. 235-237.

Sears, P. B., and Clisby, K. H., 1955, Palynology in southern North America. Part IV: Pleistocene climate in Mexico: Geol. Soc. America Bull., v. 66, p. 521-530.

Standley, P. C., 1926, Trees and shrubs of Mexico: U.S. Natl. Herb. Contr., v. 23, p. 1-1721.

Steyermark, J. A., 1950, Flora of Guatemala: Ecology, v. 31, p. 368-372.

Traverse, A., and Ginsburg, R. N., 1966, Palynology of the surface sediments of the Great Bahama Bank, as related to water movement and sedimentation: Marine Geology, v. 4, no. 6, p. 417-460.

—— 1967, Pollen and associated microfossils in the marine surface sediments of the Great Bahama Bank: Rev. Palaeobot., Palynol., v. 3, p. 243-254.

LAMONT-DOHERTY GEOLOGICAL OBSERVATORY CONTRIBUTION NO. 1489
WOODS HOLE OCEANOGRAPHIC INSTITUTION CONTRIBUTION NO. 2357
MANUSCRIPT RECEIVED APRIL 10, 1969
REVISED MANUSCRIPT RECEIVED AUGUST 25, 1969

THE GEOLOGICAL SOCIETY OF AMERICA, INC.
MEMOIR 126, 1970

North Pacific Deep-Sea Ash Chronology and Age of Present Aleutian Underthrusting

JAMES D. HAYS
AND
DRAGOSLAV NINKOVICH

Lamont-Doherty Geological Observatory of Columbia University, Palisades, New York

ABSTRACT

Volcanic ash layers in North Pacific deep-sea sediments are distributed in a 1000-km wide belt bordering the volcanic arcs of the North and Northwest Pacific. Paleomagnetic and radiolarian stratigraphy combined with study of the physical properties of the ash provide tools for correlating and dating the ash layers.

The belt of ash can be divided into two areas based on the chronology and composition of the ash layers; the first is in the Northwest Pacific with Japan, Kamchatka and the Kurils as its probable source, and the second is south of the active part of the Aleutian Islands (east of 175° E.), from which it was probably derived. In the northwest area the rates of sedimentation are high, in most cores exceeding 5 cm/1000 yrs. Some cores contain as many as 10 ash layers in sediments deposited in less than 300,000 years. The oldest ash encountered in this area is 1.3 m.y. old but it occurs near the outer limit of ash dispersal, so older ashes are to be expected.

Ashes in cores south of the Aleutians indicate that at least 20 large andesitic eruptions have occurred since 1.8 m.y. B.P. Sediments between 1.8 and at least 2.8 m.y. contain no ashes, suggesting that 1.8 m.y. (lower Pleistocene) was the beginning of the present cycle.

Previous work has shown that andesitic volcanism in island arcs is related to the depth of underthrusting of oceanic lithosphere beneath the arc. When the oceanic plate is underthrust to a depth of between 100 and 200 km the conditions are met for the formation of andesitic magma. By using measured rates and directions of sea-floor spreading it is possible to calculate that the underthrusting must have begun 10 to 14 m.y. B.P. in order to reach a depth of 100 to 200 km by 1.8 m.y. B.P. This age is in agreement with an age for the formation of the Aleutian Trench based on the end of turbidite deposition on the relict Aleutian abyssal plain.

The initiation of underthrusting 10 to 14 m.y. B.P. could be explained either by a pause in sea-floor spreading prior to this, or a change in direction of the ridge axis, which caused a reorientation of the stress system of the Pacific margin and initiated a new phase of underthrusting.

CONTENTS

INTRODUCTION

The presence of volcanic ash layers in North Pacific deep-sea sediments has been known for some time. These ashes are the product of violent volcanic eruptions along the margin of the North Pacific. Menard (1953) correlated a single ash in a core south of Alaska with the 1912 eruption at Katmai. Later, Nayudu (1964) identified the same layer in several cores from this area and recorded the presence of two older ashes. Royce (1967) and Nelson and others (1968) identified volcanic glass shards in turbidites off the northwestern United States. They correlated these with ash deposits formed on land by the 6600 year B.P. eruption of Mount Mazama (Crater Lake) on the basis of their index of refraction. Shima and others (1967) used fission track dating methods to date a single ash layer in a deep-sea core east of Japan. Ninkovich and others (1966) used paleomagnetic stratigraphy to date ash layers in the central North Pacific, some of which were as old as 1.2 m.y. Index of refraction and isodynamic magnetic properties were also used to correlate the ashes from core to core. Horn and others (1969, 1970, this volume), from an examination of published data and over 200 deep-sea cores in the Lamont-Doherty collection, mapped the distribution of North Pacific ash.

In this paper we study the distribution of volcanic ash layers of various compositions in the North Pacific. Through the stratigraphy established by the paleomagnetic record and radiolarian zones, we date ash layers and correlate them from core to core. The correlations based on age and position in sequence are supplemented by study of the physical properties of the ash.

Andesitic volcanism in North Pacific island arcs and arc-like structures is related to underthrusting of the sea-floor beneath the arcs. The beginning of volcanic activity in these arcs, as evidenced by the earliest ash layers in deep-sea cores adjacent to the arcs, is critical to theories of episodic sea-floor spreading. The initiation of volcanism here is an important clue to unravelling the Neogene history of the North Pacific.

DISTRIBUTION, CORRELATION AND DATING OF NORTH PACIFIC ASH LAYERS

Distribution

Horn and others (1969, 1970, this volume) have analyzed all available North Pacific deep-sea cores from the Lamont-Doherty Geological Observatory collection north of 20° N. They found that the layers of volcanic ash have been distributed in a zone about 1000 km wide, bordering the active volcanic arcs of Japan, the Kurils, Kamchatka, the Aleutians and Alaska. No layers of ash, however, have been found in central Pacific deep-sea sediments surrounding the active volcanoes of Hawaii, in spite of the fact that some cores penetrate sediments deposited at least 3 m.y. ago (Opdyke and Foster, 1970, this volume).

The North Pacific zone of volcanic ash layers defined by Horn and others (1970, this volume) is divided into two areas based on the composition of the ash and its probable source (Fig. 1). One area, in the Northwest Pacific, has layers of only colorless ash, apparently originating from the volcanoes of Japan, the Kurils and Kamchatka. The second area, south of the eastern Aleutians and Alaska, has layers of both brown and colorless ash, but predominantly brown ash (Fig. 2).

Correlation and Dating Ash Layers

Radiolarian Stratigraphy. Five radiolarian species occurring in North Pacific sediments became extinct in the upper Pliocene and Pleistocene. They permit the division of these sediments into four stratigraphic zones (Hays, 1970, this volume).

The upper limits of three of these species (*Eucyrtidium matuyamai, Lamprocyclas heteroporos,* and *Eucyrtidium elongatum peregrinum*) fall near magnetic reversals, so although they serve as an independent check on the magnetic stratigraphy, they do not provide additional levels for correlation between cores. *Druppatractus acquilonius* and *Stylatractus universus,* on the other hand, have upper limits in the central portion of the Brunhes series, thus providing levels of correlation that can supplement the magnetic stratigraphy.

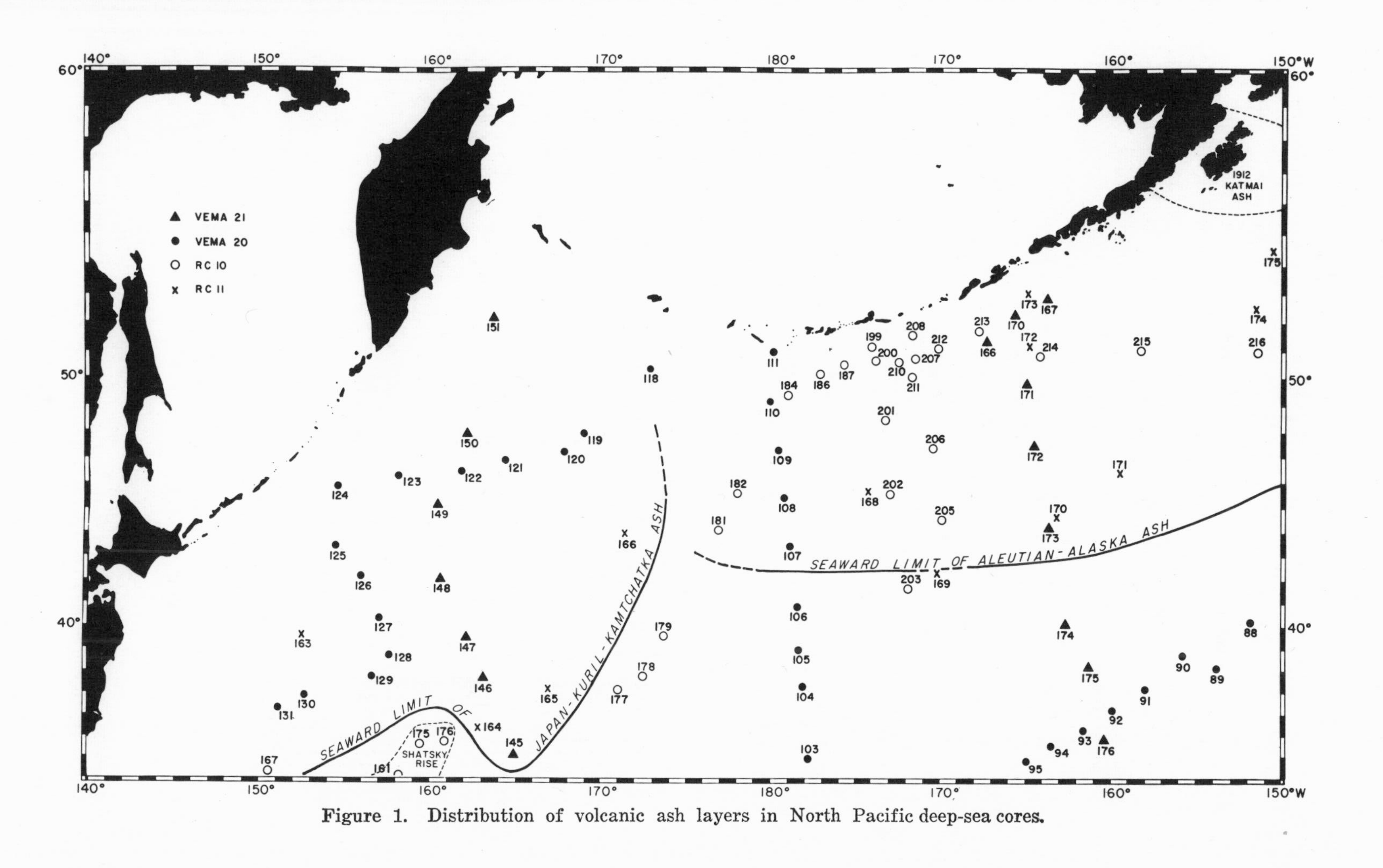

Figure 1. Distribution of volcanic ash layers in North Pacific deep-sea cores.

In this study we have determined the upper limits of both *D. acquilonius* and *S. universus* in 16 cores that penetrate into the Matuyama reversed series. We can estimate the ages of the upper limit of both species by assuming constant rates of accumulation through the Brunhes series after subtracting the thickness of the ash layers within this interval (Fig. 3). The upper limits of both species cluster about separate ages, 310,000 years B.P. for *D. acquilonius,* and 400,000 years B.P. for *S. universus.* The spread of points for the upper limit of *D. acquilonius* is less than for *S. universus,* which is probably due to the fact that because *D. acquilonius* is considerably more abundant than *S. universus,* it is easier to determine its upper limit. In some cores the age of the upper limit of these species is significantly different than the average age (for example,

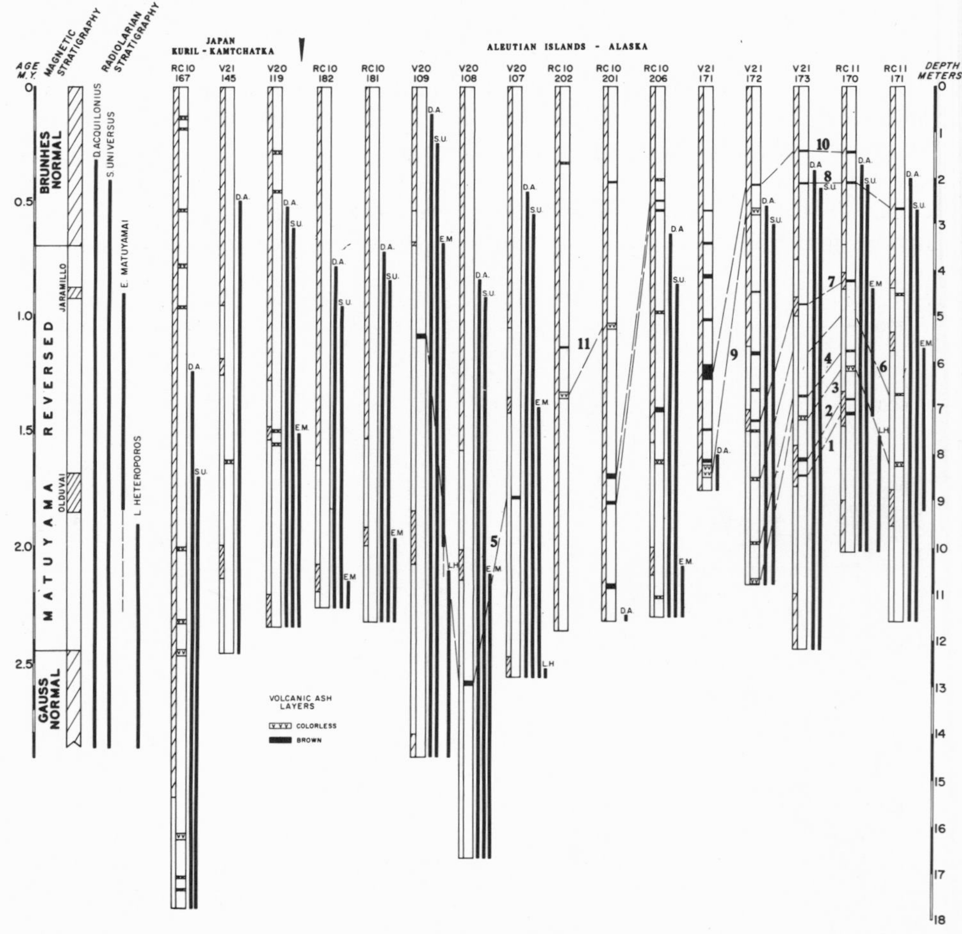

Figure 2. Paleomagnetic and radiolarian stratigraphy of deep-sea cores. Numbers 1 to 11 represent correlative ash layers.

V20-105, V20-109 and V21-148). These departures are probably best explained by hiatuses between the extinction of the species and the Brunhes/Matuyama boundary in cores V20-105 and V21-148, and between the species extinction and the top of the core in V20-109. This tighter stratigraphic control permits greater resolution of the location of hiatuses or accumulation rate changes within the Brunhes series.

Layers of volcanic ash from 16 North Pacific cores have been plotted within the stratigraphic framework of the paleomagnetic and radiolarian data (Fig. 2). The paleomagnetic stratigraphy (Fig. 2) is derived from the work of Ninkovich and others (1966) and Opdyke and Foster (1970, this volume).

Northwest Pacific Ash. In the basin bounded by Japan and the Kuril Islands on the west, the Emperor Seamounts on the east, and the Shatsky Rise on the south (Fig. 1), only four cores penetrated the 0.7 m.y. reversal. All others from this basin fail to reach the 0.31 m.y. *D. acquilonius* boundary. Therefore, judging by these 9- to 16-m long cores, the rate of deposition throughout most of this basin is greater than 5 cm/1000 yrs. All cores taken within about 1000 km of Japan, the Kurils and Kamchatka (with the exception of core V21-148), contain layers of volcanic ash (Fig. 1). The ash layers generally thin eastward and are white, gray or grayish-orange,

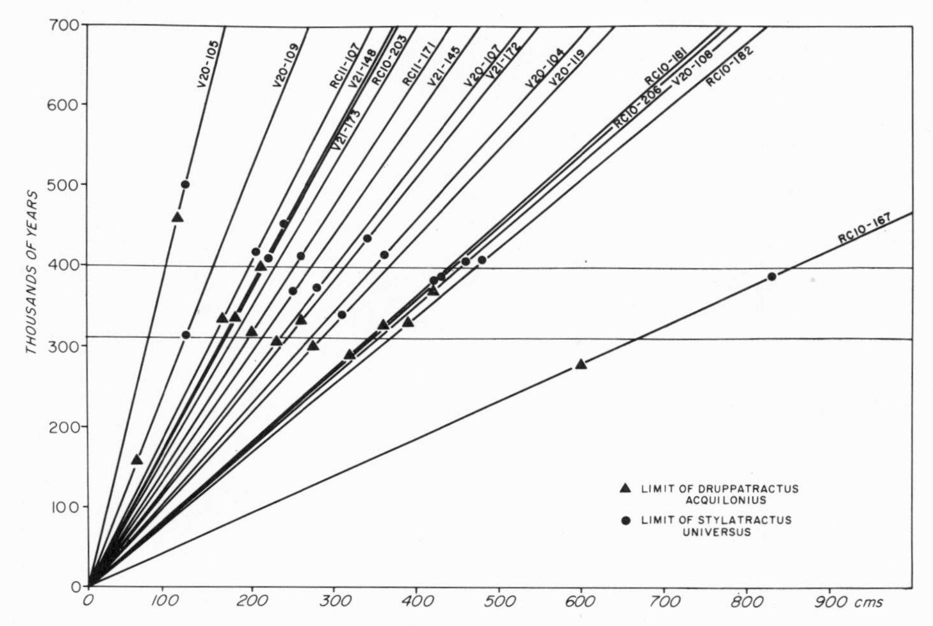

Figure 3. Time versus depth plot of the upper limits of *D. acquilonius* and *S. universus*.

TABLE 1. CORE RC10-167[1]

Magnetic Strati-graphy	Volcanic Ash Layers		Rates of Deposition cm/1000 yrs	Radiolarian Data			
				D. acquilonius		*S. universus*	
Epoch	Depth in cm	Age in m.y.		Depth in cm	Age in m.y.	Depth in cm	Age in m.y.
Brunhes normal 0-1535 cm	62-70	0.03					
	89-92	0.04					
	266-271	0.12					
	385-391	0.17					
	480-485	0.22	2.1	620	.280		
	1002-1007	0.46				850	.390
	1152-1162	0.53					
	1218-1230	0.55					
Matuyama reversed 1535-1758 cm	1615-1627	0.74					
	1720-1721	0.78					
	1735-1737	0.79					

[1]Tables 1 through 16 are ages of volcanic ash layers and upper limits of radiolarian species *D. acquilonius* and *S. universus* inferred from paleomagnetic data.

classified in this study as "colorless." The index of refraction of volcanic glass shards ranges from about 1.50 to 1.52.

Cores RC10-167, V21-145 and V20-119 penetrate the 0.7 m.y. geomagnetic reversal (Fig. 2), allowing estimates of the ages of ash layers in these cores (Tables 1, 2 and 3). Core V21-145 shows that the volcanic activity that produced the ash layers in the sediments of the Northwest Pacific started at least 1.3 m.y. ago and has lasted

TABLE 2. CORE V21-145

Magnetic Stratigraphy		Volcanic Ash Layers		Rates of Deposition cm/1000 yrs	Radiolarian Data	
					S. universus	
Epoch	Event	Depth in cm	Age in m.y.		Depth in cm	Age in m.y.
Brunhes normal 0-475 cm				0.68	250	.370
Matuyama reversed 475-1225 cm	Jaramillo 590-625 cm	812-818	1.32	0.50		
	Olduvai 995-1065					

TABLE 3. CORE V20-119

Magnetic Stratigraphy		Volcanic Ash Layer		Rates of Depo-sition	Radiolarian Data			
					D. acquilonius		*S. universus*	
Epoch	Event	Depth in cm	Age in m.y.	cm/1000 yrs	Depth in cm	Age in m.y.	Depth in cm	Age in m.y.
Brunhes normal 0-638 cm		138-140	0.15	0.91	275	.300	310	.335
		222-224	0.24					
Matu-yama reversed 638-1170 cm	Jaramillo 740-765 cm	749-751	0.88	0.62				
		777-780	0.92					
	Olduvai 1100-1170 cm							

nearly continuously throughout the upper Matuyama and Brunhes epochs (V20-119, RC10-167).

Core V21-123, taken immediately to the east of the Kuril Islands, is 1360 cm long and contains 10 ash layers. The core failed to penetrate to the range of *D. acquilonius,* indicating that in the Kurils alone there were at least 10 large eruptions during the last 0.3 m.y.

One would expect the Shatsky Rise to be included in this area of ash falls. However, no core on the Rise contains ash (Fig. 1). This is probably best explained by numerous hiatuses in the cores taken on the crest of the rise (Ewing and others, 1966; Opdyke and Foster, 1970, this volume).

The seaward limit of northwest Pacific colorless ash falls, which range in age from at least 1.3 m.y. to the present, is marked by a line lying to the east of cores V21-145, RC11-166 and V20-119 (Fig. 1).

Cores RC10-177 to 182 taken to the east of this line contain no ash layers. Of these, cores 177, 178 and 179 do not reach the 0.31 m.y. *D. acquilonius* boundary; however, cores 181 and 182 reach sediments deposited 1 m.y. ago (Fig. 2; Tables 4 and 5). Farther to the east, cores V20-103 to 109 penetrate through the Matuyama series (Opdyke and Foster, 1970, this volume), and of these, 107, 108 and 109 contain one layer of brown ash at 1.2 m.y. B.P. (Fig. 2) that does not correlate with any ash layers to the west.

The Northwest Pacific ash apparently originated from volcanoes on Japan, the Kurils and Kamchatka, suggesting westerly and northwesterly prevailing winds throughout the Pleistocene.

Central North Pacific Ash. To the east of the Emperor Seamounts and south of the Aleutian Islands, 13 cores have been studied (Figs. 2 and 6; Tables 6 to 16). Of these, three penetrate the Gauss

TABLE 4. CORE RC10-182

Magnetic Stratigraphy		Rates of Deposition cm/1000 yrs	Radiolarian Data			
			D. acquilonius		*S. universus*	
Epoch	Event		Depth in cm	Age in m.y.	Depth in cm	Age in m.y.
Brunhes normal 0-820 cm		1.17	390	.335	480	.410
Matuyama reversed 820-1130 cm	Jaramillo 1035-1095 cm					

TABLE 5. CORE RC10-181

Magnetic Stratigraphy		Rates of Deposition cm/1000 yrs	Radiolarian Data			
			D. acquilonius		*S. universus*	
Epoch	Event		Depth in cm	Age in m.y.	Depth in cm	Age in m.y.
Brunhes normal 0-765 cm		1.1	360	.330	420	.380
Matuyama reversed 765-1160 cm	Jaramillo 960-995 cm					

TABLE 6. CORE V20-109

Magnetic Stratigraphy		Volcanic Ash Layers		Rates of Deposition cm/1000 yrs	Radiolarian Data			
					D. acquilonius		*S. universus*	
Epoch	Event	Depth in cm	Age in m.y.		Depth in cm	Age in m.y.	Depth in cm	Age in m.y.
Brunhes normal 0-270 cm				0.35	60	.160	120	.315
Matuyama reversed 270-1400 cm	Jaramillo 337-346 cm	540-545*	1.2(5)	0.70				
	Olduvai 924-1037 cm							
Gauss normal 1400-1671 cm								

*Brown ash

TABLE 7. CORE V20-108

Magnetic Stratigraphy		Volcanic Ash Layers		Rates of Deposition cm/1000 yrs	Radiolarian Data			
					D. acquilonius		*S. universus*	
Epoch	Event	Depth in cm	Age in m.y.		Depth in cm	Age in m.y.	Depth in cm	Age in m.y.
Brunhes normal 0-792 cm					420	.370	460	.410
				1.13				
Matu-yama reversed 792-1670 cm	Jaramillo 1005-1070 cm							
		1290-1295*	1.1 (5)					

*Brown ash

TABLE 8. CORE V20-107

Magnetic Stratigraphy		Volcanic Ash Layers		Rates of Deposition cm/1000 yrs	Radiolarian Data			
					D. acquilonius		*S. universus*	
Epoch	Event	Depth in cm	Age in m.y.		Depth in cm	Age in m.y.	Depth in cm	Age in m.y.
Brunhes normal 0-525 cm					230	.305	280	.375
				0.75				
Matu-yama reversed 525-1282 cm	Jaramillo 678-712 cm							
		890-895*	1.2 (5)					
	Olduvai 1235-1282 cm							

*Brown ash

TABLE 9. CORE RC10-202

Magnetic Stratigraphy	Volcanic Ash Layers		Rates of Deposition cm/1000 yrs
Epoch	Depth in cm	Age in m.y.	
Brunhes normal 0-1180	162-164 566-569 665-680	0.04 0.12* 0.14(11)	4.7

*Brown ash

TABLE 10. CORE RC10-201

Magnetic Stratigraphy	Volcanic Ash Layers		Rates of Deposition cm/1000 yr	Radiolarian Datum	
				D. acquilonius	
Epoch	Depth in cm	Age in m.y.		Depth in cm	Age in m.y.
Brunhes normal 0-1160 cm	206-208*	0.06	3.5		
	515-528	0.14			
	842-850*	0.24			
	902-908*	0.25			
	1078-1080*	0.30		1150	

*Brown ash

TABLE 11. CORE RC10-206

Magnetic Stratigraphy		Volcanic Ash Layers		Rates of Deposition cm/1000 yrs	Radiolarian Data			
					D. acquilonius		*S. universus*	
Epoch	Event	Depth in cm	Age in m.y.		Depth in cm	Age in m.y.	Depth in cm	Age in m.y.
Brunhes normal 0-775 cm		200-202	0.18	1.1				
		248-250*	0.22		320	.290		
		268-271*	0.24				430	.390
		490-492	0.44					
		698-712*	0.62					
Matu-yama reversed 775-1150 cm	Jaramillo 100-1060 cm	817-823	0.73					
		1107-1110	0.98					

*Brown ash

series (V20-109, V21-173, RC11-170), seven penetrate the Matuyama series (V20-107, V20-108, V21-172, RC10-181, RC10-182, RC10-208, RC11-171; Opdyke and Foster, 1970, this volume), two reach the range of *Druppatractus acquilonius* (V21-171, RC10-201), and one contains only sediments younger than 0.31 m.y. (RC10-202).

Most of the cores taken between the eastern half of the Aleutian Islands and about 43° N. contain both brown and colorless ash layers. The westernmost of these cores containing ash are V20-107, 108 and 109, which contain one layer at 1.2 m.y. The same layer may extend farther west but has not been reached by cores RC10-

TABLE 12. CORE V21-171

Magnetic Stratigraphy	Volcanic Ash Layers		Rates of Deposition cm/1000 yrs	Radiolarian Data	
				D. acquilonius	
Epoch	Depth in cm	Age in m.y.		Depth in cm	Age in m.y.
	245-255*	0.10			
	270-272*	0.11			
	340-342*	0.14			
Brunhes	409-415*	0.17			
normal	506-509*	0.21	2.3		
0-880 cm	605-618*	0.25 (10)			
	746-749*	0.31			
	811-816*	0.33		800	
	822-850	0.34 (9)			

*Brown ash

TABLE 13. CORE V21-172

Magnetic Stratigraphy		Volcanic Ash Layers		Rates of Deposition cm/1000 yrs	Radiolarian Data			
					D. acquilonius		*S. universus*	
Epoch	Event	Depth in cm	Age in m.y.		Depth in cm	Age in m.y.	Depth in cm	Age in m.y.
Brunhes normal 0-565 cm		212-213*	0.27 (10)					
		265-280	0.34 (9)		260	.335		
		577-583*	0.72	0.78			360	.440
		660-662	0.82					
	Jaramillo 705-749 cm	727-729*	0.91 (7)					
		749-752	0.94					
Matu-yama reversed 565-1082 cm		850-855	1.14 (6)	0.58				
		990-991	1.40 (4)					
		1067-1080	1.53 (3)					

*Brown ash

181 and 182 (Fig. 2). Core V21-118, 960 cm long, taken immediately to the south of the western part of the Aleutian Islands (Fig. 1), contains no ash. Thus, the area of the ash to the south and southeast of the Aleutian Islands begins at 180° and extends to the Gulf of Alaska. The number of ash layers and the age of the oldest ash layer increase eastward. The oldest ash layers in the area occur in sediment deposited during the Olduvai event and are therefore of lower Pleistocene age.

TABLE 14. CORE V21-173

Magnetic Stratigraphy		Volcanic Ash Layers		Rates of Deposition cm/1000 yrs	Radiolarian Data			
					D. acquilonius		*S. universus*	
Epoch	Event	Depth in cm	Age in m.y.		Depth in cm	Age in m.y.	Depth in cm	Age in m.y.
Brunhes normal 0-375 cm		138-140*	0.26 (10)	0.53	180	.335	220	.410
		208-210	0.39 (8)					
Matuyama reversed 375-1100 cm	Jaramillo 460-500 cm	474-475*	0.89 (7)	0.39				
		673-675	1.38 (4)					
		718-726	1.51 (3)					
	Olduvai 780-870 cm	807-814*	1.72 (2)					
		844-846*	1.80 (1)					
Gauss normal 1100-1218 cm								

*Brown ash

TABLE 15. CORE RC11-170

Magnetic Stratigraphy		Volcanic Ash Layers		Rates of Deposition cm/1000 yrs	Radiolarian Data			
					D. acquilonius		*S. universus*	
Epoch	Event	Depth in cm	Age in m.y.		Depth in cm	Age in m.y.	Depth in cm	Age in m.y.
Brunhes normal 0-350 cm		140-142*	0.28 (10)	0.49	170	.340	205	.410
		206-209*	0.41 (8)					
Matu-yama reversed 350-900 cm	Jaramillo 405-445 cm	420-421*	0.89 (7)	0.30				
		574-575	1.40 (4)					
		609-618	1.51 (3)					
	Olduvai 665-740 cm	680-684*	1.72 (2)					
		710-715*	1.81 (1)					
Gauss normal 900-1012 cm								

*Brown ash

TABLE 16. CORE RC11-171

Magnetic Stratigraphy		Volcanic Ash Layers		Rates of Deposition cm/1000 yrs	Radiolarian Data			
Epoch	Event	Depth in cm	Age in m.y.		*D. acquilonius*		*S. universus*	
Brunhes normal 0-440 cm				0.63	200	.320	260	.410
		266-268*	0.42(8)					
Matuyama reversed 440-1160 cm	Jaramillo 535-575 cm	452-454	0.72	0.43				
		671-673	1.18(6)					
		817-826	1.53(3)					
	Olduvai 875-955 cm							

*Brown ash

Evidence for correlation of eastern Pacific volcanic ash layers on the basis of their stratigraphical position and inferred absolute age is given in Figure 6.

The correlations based on magnetic and radiolarian stratigraphy have been further investigated on the basis of ash color, magnetic isodynamic properties of ash, and refractive index of glass shards. These techniques of correlation have been previously described by Ninkovich and others (1966). The correlative ash layers are numbered 1 to 11 from oldest to youngest (Figs. 2 and 6; Tables 6 to 16).

There are two distinct types of ash layers in the cores: brown ash (closed circles) and colorless ash, which includes white, gray and grayish-orange layers (open circles). Some of the layers can be correlated on the basis of their characteristic color in wet cores. Among the acidic ashes, the most distinctive is a thick white layer (No. 9, Figs. 2 and 6), which occurs in both cores V21-171 and 172 just below the upper limit of *D. acquilonius.* Layer No. 3 can be correlated on the basis of its characteristic pinkish color in cores V21-172, V21-173, RC11-170 and RC11-171. Variations in color can also be detected between brown ashes, for example, layers No. 8 and 10 in Figure 4. The upper layer (No. 10) occurs above the upper limit of *D. acquilonius,* and the lower layer (No. 8) near the upper limit of *S. universus* (Figs. 2 and 6). A similar variation in color can be observed in two layers deposited during the Olduvai event in cores V21-173 and RC11-170 (layers 1 and 2).

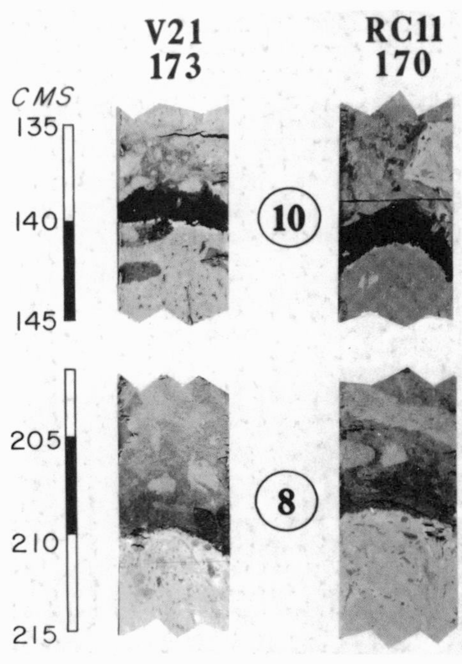

Figure 4. Photograph of ash layers (8 and 10) in two cores.

Further analysis of the ash within the size range .177 to .088 mm was made with a Frantz isolynamic magnetic separator (side tilt=15°, forward tilt=25°). Colorless ash layers have the greatest homogeneity of isodynamic magnetic properties. Eighty percent of the glass shards from individual colorless layers can be separated within an interval of about 0.2 amperes. Shards from discrete layers have the same color and identical indices of refraction, ranging from 1.50 to 1.53 (Fig. 5) and indicating a composition of 60 to 70 percent SiO^2 (Thorarinsson and Tomasson, 1967). The colorless ashes, therefore, can be most reliably correlated on the basis of their physical properties. Five colorless ash layers in the eastern North Pacific can be correlated (3, 4, 6, 9 and 11, Fig. 6).

Correlation of brown ash layers on the basis of their physical properties is more difficult than colorless ash layers because of the great variation in composition of individual layers. Although the over-all color of the layer is brown, it contains glass shards ranging in color from dark brown to colorless. Consequently, isodynamic magnetic properties and index of refraction of brown ashes are not reliable for purposes of correlation. However, about 90 percent of the volcanic glass shards from all brown ash layers can be separated from colorless ash layers on the basis of their magnetic isodynamic properties, with the boundary falling at about 0.7 amperes (Fig. 5). The bulk of the brown ash layers is composed of dark yellowish-brown shards with an index of refraction of 1.55 to 1.56, or about 55 to 60 percent SiO_2 (Thorarinsson and Tomasson, 1967).

On the basis of paleomagnetic, radiolarian stratigraphy, and the physical properties of the ash, 11 ash layers can be correlated (Figs. 2 and 6). Besides these 11, three layers of brown ash in the lower part of core RC10-201 probably correlate with layers of similar age from core V21-171. A layer about 0.73 m.y. old found in cores RC11-171, V21-172 and RC10-206 (Fig. 6) is correlative on the basis of paleomagnetic stratigraphy but has a different composition in all three cores. This suggests three simultaneous eruptions prior to the

0.7 m.y. reversal, each recorded in only one of the cores. The other layers cannot be correlated.

Thus, at least 20 volcanic eruptions have occurred in the Aleutian arc since the oldest one was recorded in the cores (which occurred during the Olduvai event; layer No. 1, Fig. 6). The interval between eruptions during this time was no more than 200,000 years. No volcanic ash layers older than 1.8 m.y. occur in three cores (V20-109, V21-173 and RC11-170) that penetrate sediments of the upper Gauss series, and which have an age of about 2.8 m.y., thereby establishing a pause in volcanic activity of at least 1 m.y. in the Upper Pliocene.

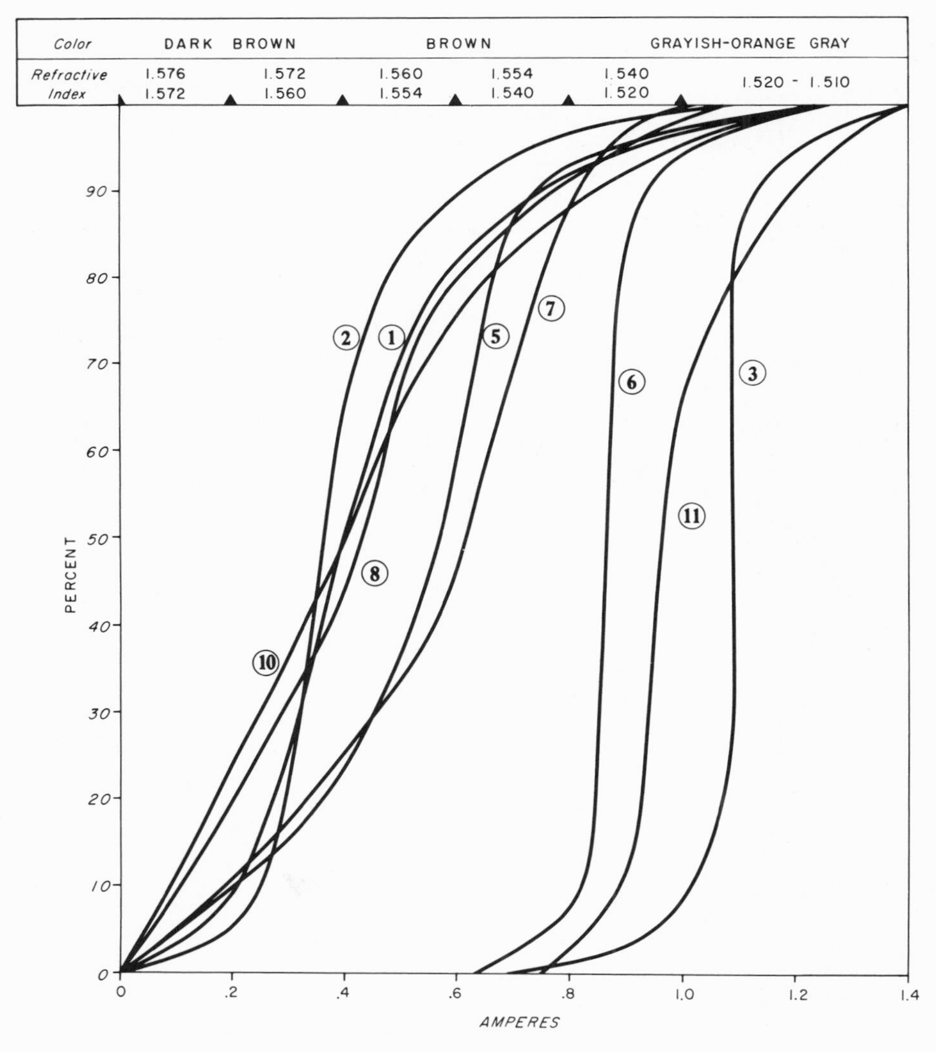

Figure 5. Cumulative separation curves for volcanic glass fractions separated with Frantz isodynamic magnetic separator. Curves represent averages of all correlative ash layers.

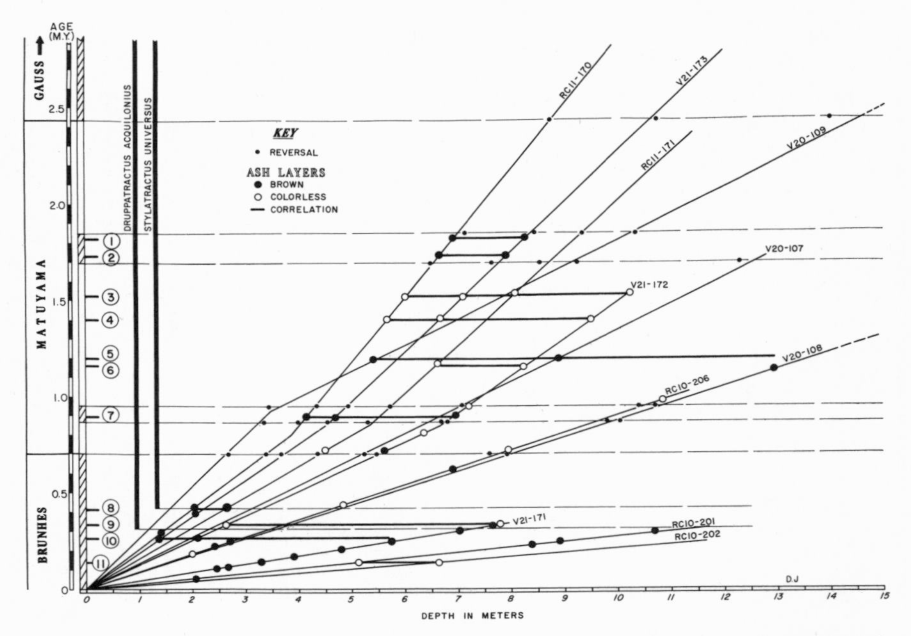

Figure 6. Chronology and correlation of ash layers from cores south of the Aleutian Islands, inferred from paleomagnetic and radiolarian stratigraphy. Numbers 1 to 11 represent ash layers correlated on the basis of physical properties of the ash, as well as position in sequence.

RELATIONSHIP BETWEEN ANDESITIC VOLCANIC ACTIVITY IN THE ALEUTIANS AND SEA-FLOOR SPREADING

A number of authors have suggested, from earthquake mechanism studies and from other evidence (Plafker, 1965; Stauder and Bollinger, 1966; McKenzie and Parker, 1967; Pitman and Hayes, 1968; Stauder, 1968; Isacks and others, 1968; Tobin and Sykes, 1968), that at present the North Pacific crustal plate is being underthrust at the Aleutian Trench.

Gutenberg and Richter (1954), Bäth (1958), and Isacks and others (1968) have noted the coincidence between the line of active volcanoes that stretches westward from the Alaska peninsula to Buldir Island (176° E.) and the zone in the Aleutians in which earthquakes have a focal depth of between 100 and 200 km. In the nonvolcanic portion of the arc to the west of Buldir Island (Fig. 7), the depth of earthquake foci is less than 100 km (Gutenberg and Richter, 1954; Barazangi and Dorman, 1969).

Dickinson and Hatherton (1967) have stated, after an examination of all published data on island arcs, that andesitic volcanoes never occur within the zone of shallow seismicity between the trench and the island arc. Typically, the first line of active volcanoes marks the arc-side boundary of the epicenters above the zone of shallow seismicity (0 to 80 km). In the series of arcs stretching between Tonga and the Macquarie ridge, the active andesitic and rhyolitic volcanoes range from the North Island of New Zealand northward, where the earthquakes also lie at depths greater than 100 km. Oliver and Isacks (1967) and Isacks and others (1968) related the depth of the epicenters to the depth of underthrust crust.

On North Island, in the belt where the earthquakes lie at depths greater than 100 km, late Cenozoic volcanism began with the eruptions of Taupo-Rotorua ignimbrites (Grindley and Harrington, 1961; Thompson and others, 1966). Measurements of K/A (Stipp, in prep.) and remnant magnetization (Cox, 1969) have shown that the Taupo-Rotorua ignimbrites range in age from 0.8 to 0.2 m.y. B.P., that is, from upper Matuyama through Brunhes geomagnetic epochs. Ninkovich (1968) has dated by paleomagnetic methods five rhyolitic ash layers in deep-sea cores east of New Zealand. The ash layers have similar ages to the Taupo-Rotorua ignimbrites. No ashes older than 0.8 m.y. are found in the cores although some penetrated sediments as old as 3 m.y. Late Pleistocene and Recent andesitic eruptions in the same belt on North Island have spread ashes having a maximum thickness of 15 cm along the eastern coast of North Island (Healy and others, 1964). These same ashes have been identified in sea bottom samples and deep-sea cores from the Chatham rise (Reed and Hornibrook, 1952; Norris, 1964). This indicates that ash layers in deep-sea cores adjacent to active island arcs accurately reflect the volcanic activity of the arc and can be used to date the initiation of volcanism.

If we assume that only when the underthrust oceanic plate reaches a depth of between 100 and 200 km are the conditions met for the production of andesitic magma that results in the formation of ash producing volcanoes, then the age of the earliest ash layer in deep-sea sediments adjacent to the arc should mark the time when the underthrust plate reached this depth.

Coats (1962) has suggested that the andesitic composition of Aleutian volcanoes, which rest on oceanic basaltic crust, may be explained by contamination to a depth of 100 km below the island arc, with eugeosynclinal sediment and sea water carried down from the trench along the major thrust.

In the present study we are concerned with the length of time necessary for the underthrust crust and sediments to reach a depth of between 100 and 200 km beneath the volcanic arc. By assuming an age of 1.8 m.y. for the beginning of andesitic volcanic activity (we infer this age on the basis of the oldest ash layer in cores V20-109, V21-173 and RC11-170), and knowing the rate of spreading of the North Pacific floor, and the length of the underthrust crust between the trench, and a depth of between 100 and 200 km below the volcanic arc, we can provide additional evidence that can be used to infer the age of the beginning of the recent episode of sea-floor underthrusting.

The present motion of the North Pacific plate is to the northwest (Vine, 1966; Pitman and Hayes, 1968; Hayes and Pitman, 1970, this volume), originating in spreading of the ridge passing from the Southwest Pacific to the Gulf of Alaska. The center of rotation of the North Pacific plate is south of Greenland (53° N., 47° W.; Morgan, 1968; Le Pichon, 1968). The equator of rotation passes between the East Pacific Rise at about 15° S. and Japan; thus, the spreading of the floor is nearly perpendicular to the Japan Trench. In the Aleutians the spreading is nearly perpendicular to the Alaska peninsula, but the angle between the direction of spreading and the axis of the trench decreases westward; in the westernmost portion of the Aleutians the spreading is nearly parallel to the axis of the trench (Fig. 7).

Rates of spreading have been measured over the Juan de Fuca-Gorda ridge and in the mouth of the Gulf of California. The rate of underthrusting under the Aleutians will depend on whether the ridge systems are fixed or migrating relative to North America. The rate of underthrusting will be equal to the half-rate of spreading if the ridge axis is fixed relative to North America. If, however, one limb of the ridge is fixed to the American continent, then the rate of underthrusting under the Aleutians will be equal to twice the half-rate of spreading. Pitman and Hayes (1968) have suggested that the eastern limb of the Juan de Fuca-Gorda ridge system is fixed to the North American continent, and therefore the ridge crest is migrating to the northwest at a rate of 2.9 cm/yr, relative to the North American continent. The near absence of intermediate depth earthquakes under western North America (Isacks and others, 1968) suggests that at the present time little or no underthrusting of the northwestern United States is occurring. However, the line of Quaternary andesitic and rhyolitic volcanoes (Wilcox, 1965) stretching from southern British Columbia to central California resembles an arc-like structure and suggests that underthrusting has occurred within the late Cenozoic. If such underthrusting has occurred, one might expect a trench off the northwest United States opposite the Juan de Fuca-Gorda ridge system.

Ewing and others (1968) show a buried trench (their profile 13) just west of the straits of Juan de Fuca. However, north of the Juan de Fuca-Gorda ridge system, north of Vancouver Island, there is no evidence of a trench (their profile 12). Curry (1966) has presented profiles across the Continental margin south of Cape Mendocino and has uncovered no evidence of a buried trench. Vine

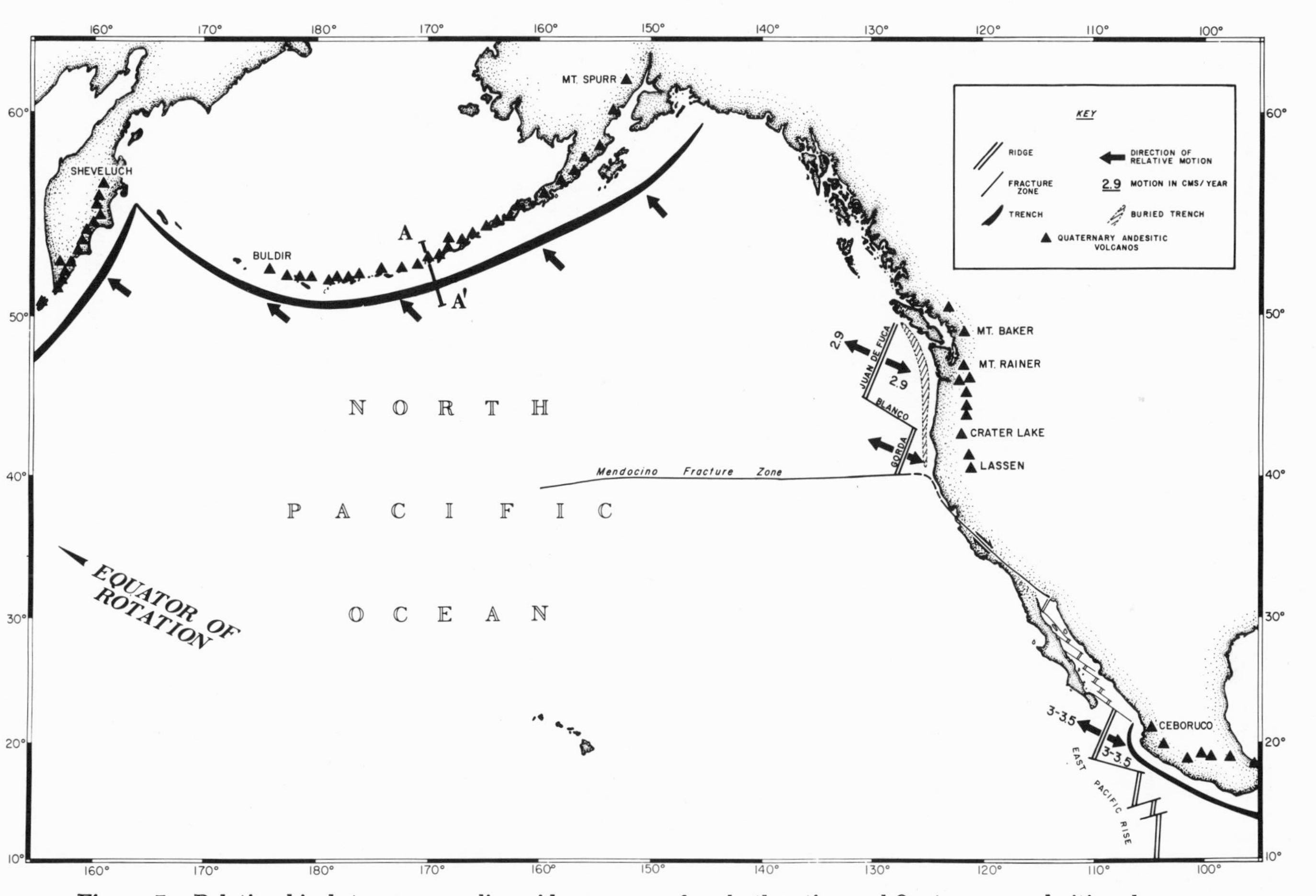

Figure 7. Relationship between spreading ridges, zones of underthrusting and Quaternary andesitic volcanoes.

(1966) suggested that the present trend of the Juan de Fuca-Gorda ridge was established about 10 m.y. ago and is different from a previous east-west direction of spreading. If east-west spreading prior to 10 m.y. ago (Hayes and Pitman, 1970, this volume) developed trenches along the western margin of North America, we would expect these trenches to extend beyond the region opposite the present Juan de Fuca-Gorda ridge system. Since the only evidence of a trench is opposite the present Juan de Fuca-Gorda ridge system, we can conclude that this buried trench was formed by underthrusting from the Juan de Fuca ridge system within the last 10 m.y.

Therefore, this evidence suggests that during the late Cenozoic there was underthrusting of the North Pacific plate under western North America, producing both the present buried trench off the straits of Juan de Fuca and the line of volcanoes to the east. If we assume that the line of volcanoes was formed in a similar way to volcanoes of island arcs, then the lithosphere to the east of the Juan de Fuca ridge underthrust the North American continent to a depth of between 100 and 200 km. On the opposite side of the ridge system under the Alaskan peninsula, earthquake foci reach depths of about 200 km today (Barazangi and Dorman, 1969). This suggests that underthrusting has been symmetrical on both sides of the ridge, and that for at least as long as it took to underthrust the North American continent in Alaska and the western United States to a depth of 100 to 200 km, the ridge axis was fixed relative to North America. Farther south, the spreading from the East Pacific Rise now is underthrusting both Central and South America and the island arcs of the North and Northwest Pacific (Fig. 7). We, therefore, can safely use a half-rate of spreading as the rate of underthrusting. The half-rate of spreading at the Juan de Fuca-Gorda ridge (Vine, 1966) and at the mouth of the Gulf of California (Larson and others, 1968; Moore and Buffington, 1968), are 2.9 cm/yr and 3 to 3.5 cm/yr, respectively. The underthrusting under the Aleutian Islands originates from the East Pacific Rise south of the Gulf of California and thus must have a rate of about 3.5 cm/yr.

In the eastern Aleutian Islands, from which the 1.8 m.y. ash layer originated, the direction of spreading inferred from the center of rotation of the North Pacific plate makes an angle of about 45° with the trend of the Aleutian arc.

Using a spreading rate of 3.5 cm/yr and an angle between the direction of spreading and the axis of the Aleutian Trench of 45°, we obtain a rate of underthrusting for the eastern Aleutians of 2.5 cm/yr perpendicular to the trench axis. An average distance between the Aleutian Islands and the trench is about 150 km; therefore, the length of the thrust between the trench and a depth of 100 to 200 km beneath the island arc would be about 200 to 250 km

(Fig. 8). We may then estimate the time to underthrust a slab of oceanic crust of this length to a depth of 100 to 200 km beneath the island arc, which would be between 8 and 10 m.y. Since andesitic volcanism appears to have been initiated about 2 m.y. ago, the most recent phase of underthrusting began between 10 and 12 m.y. ago, which would be upper Miocene.

This age for the initiation of underthrusting in the Aleutians is similar to the age suggested by Ewing and Ewing (1967) and Ewing and others (1968) for the beginning of the present episode of sea-floor spreading. They have concluded from the distribution of sediment thicknesses on the mid-ocean ridge flanks that the spreading of the ridge system may be episodic. They suggest that the present spreading episode started 10 to 12 m.y. ago and was preceded by an interval of 30 to 40 m.y. when little or no spreading occurred. We cannot tell from our data if there was a pause in spreading and it is possible that a change in orientation of the ridge crest, as suggested by Vine (1966), would be sufficient to reorient the stress system at the Pacific margin, thereby initiating a new phase of underthrusting.

It is probable that the Aleutian Trench was formed early in the present episode of underthrusting, and since its formation, has been an effective barrier to the distribution of turbidites from the Alaska peninsula to the south. Modern turbidites have been recognized in many cores in the Lamont-Doherty Geological Observatory collection

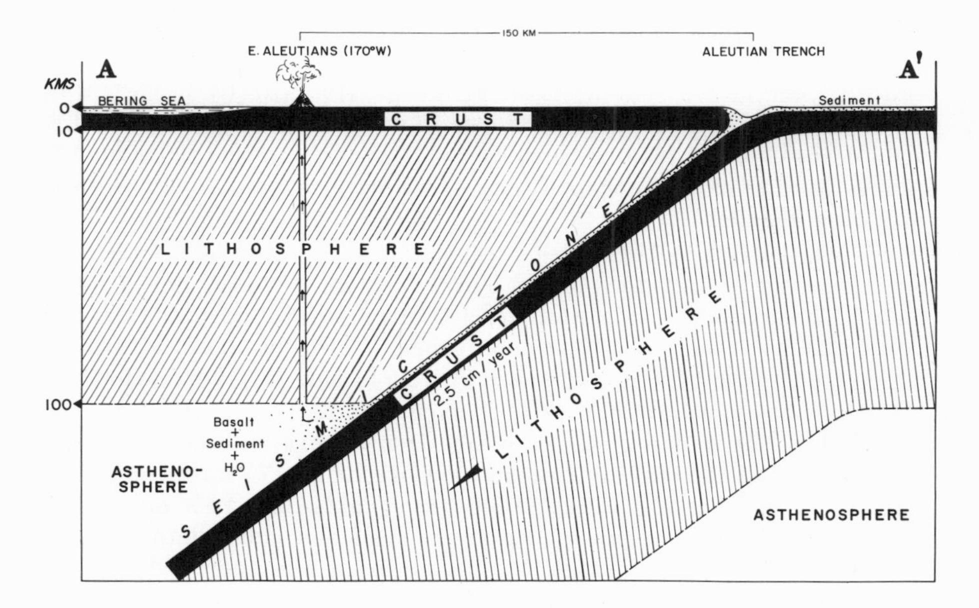

from the Bering Sea and the Aleutian Trench (cores RC10-187, 199, 200, 207, 208, 210, 212, 213, RC11-173, V21-167 and 170, Fig. 1; *also* Horn and others, 1970, this volume). In the trench they are distributed as far west as 175° W. No layer of turbidites has been found in deep-sea cores taken from the south of the trench (Horn and others, 1970, this volume). Hamilton (1967) and Ewing and others (1968), however, recognized a thick sequence of seismic reflectors beneath 40 to 100 m of seismically transparent sediment in the Aleutian relict abyssal plane. They interpreted this as a sequence of turbidites underlying 40 to 100 m of pelagic sediments. These earlier turbidites, like the modern ones from the trench, extend as far west as about 175° W. Hamilton (1967) estimated, on the basis of available data on rates of deposition in the North Pacific, that the beginning of deposition of the pelagic sediments overlying the relict turbidites occurred sometime between middle Cretaceous and Miocene, but most probably Eocene. He suggested that this event can be considered for dating the formation of the Aleutian Trench, which stopped the distribution of the turbidites to the south.

Grim (1969), using rates of sedimentation published by Ninkovich and others (1966) on cores along 180° (V20-107, 108 and 109, Figs. 1 and 2), inferred an age of 15 m.y. for the end of turbidite deposition on the Aleutian abyssal plain, and suggested that this was the age of formation of the Aleutian Trench.

Opdyke and Foster (1970, this volume) have estimated the age of the end of turbidite deposition based on rates of sedimentation of a number of paleomagnetically dated cores from the area. They concluded, using the minimum rate of sedimentation in the area and the maximum sediment thickness in Hamilton's profile, that the age of the end of turbidite deposition could be as old as 35 m.y. (Oligocene). They also suggest it could be as young as 10 m.y. We believe an age close to their younger estimate is most likely for the following reasons.

Two Lamont-Doherty deep-sea cores were taken along Hamilton's seismic reflection profile across the relict Aleutian abyssal plain. These are V21-172 and RC11-171 (Fig. 1). Rates of deposition in these cores have been established on the basis of paleomagnetic stratigraphy by Opdyke and Foster (1970, this volume). Core V21-172 has an average rate of deposition of about 0.68 cm/1000 yrs (Table 13) and was taken where the thickness of pelagic sediment overlying the turbidites is 96 m. Core RC11-171 has an average rate of deposition of 0.53 cm/1000 yrs (Table 16) and was taken where the thickness of pelagic sediment is 63 m. These data indicate the distribution of turbidites to the south of the trench ended 12 to 14 m.y. ago, suggesting that this was the time of formation of the trench.

The Aleutian ridge, on the other hand, may be much older. Since neither the modern turbidites in the trench (Horn and others, 1970, this volume), nor the relict turbidites extend west of 175° W. (Hamilton, 1967; Ewing and others, 1968), the source of both is probably the Alaskan peninsula. The Aleutian ridge, therefore, must have been a barrier to the southward extension of Bering Sea turbidites since early Tertiary time. The occurrence of shallow water Eocene and Oligocene Foraminifera on Amchitka Island (Todd, 1953) is additional evidence that the Aleutian ridge was a positive topographic feature at that time.

Therefore, evidence from a number of sources, including (1) age of andesitic volcanism in Aleutians, (2) age of formation of the Aleutian Trench, (3) age of reorientation of Juan de Fuca-Gorda ridge, and (4) depth of earthquake foci in the Aleutians and measured rates of sea-floor spreading in the North Pacific, converge on an age of 10 to 14 m.y. for the beginning of the present episode of Aleutian underthrusting.

ACKNOWLEDGMENTS

The writers are grateful to Doctors Lynn Sykes and Robert Page who read the manuscript and made numerous helpful suggestions, and to Dr. Walter Pitman who gave much valuable assistance. Unpublished paleomagnetic data were freely contributed by Dr. Neil Opdyke and John Foster and are greatly acknowledged.

Appreciation is extended to the able and hard working crews of our research vessels, who gathered the cores used in this study, and to Dr. Maurice Ewing, who directed the cruises of the *Vema* and *Conrad.* Roy R. Capo's able custodianship of our core collection greatly facilitated this effort.

The work was supported by grants from the National Science Foundation, GA 1193 and GA 4499, and the Office of Naval Research N-00014-67-A-0108-0004.

REFERENCES CITED

Barazangi, M., and Dorman, J., 1969, World seismicity maps compiled from ESSA, Coast and Geodetic Survey, epicenter data, 1961-1967: Seismol. Soc. America Bull., v. 59, p. 369-380.

Bath, M., 1958, Seismic exploration of the earth's crust: Recent developments: Geol. Fören. Stockholm Förh., v. 80, no. 3, p. 291-308.

Cots, R. R., 1962, Magma type and crustal structure in the Aleutian arc, *in* MacDonald, G. A., and Kuno, H., *Editors,* The crust of the Pacific basins: Am. Geophys. Union, Geophys. Mono. 6, p. 92-109.

Cox, A., 1970, A paleomagnetic study of secular variation in New Zealand: Earth and Planetary Sci. Letters (in press).

Curry, J. R., 1966, Geologic structure on the continental margin, from subbottom profiles, northern and central California, *in* Bailey, E. H., *Editor,* Geology of northern California: U. S. Geol. Survey Bull. 190, p. 337-342.

Dickinson, W. R., and Hatherton, T., 1967, Andesitic volcanism and seismicity around the Pacific: Science, v. 157, p. 801-803.

Ewing, J., and Ewing, M., 1967, Sediment distribution on the mid-ocean ridges with respect to spreading of the sea floor: Science, v. 156, p. 1590-1592.

Ewing, J., Ewing, M., Aitken, T., and Ludwig, W. J., 1968, North Pacific sediment layer measured by seismic profiling, *in* Knopoff, L., Drake, C., and Hart, P., *Editors,* The crust and upper mantle of the Pacific area: Am. Geophys Union, Geophys. Mono. 12, p. 147-173.

Ewing, M., Saito, T., Ewing, J., and Burckle, L., 1966, Lower Cretaceous sediments from the northwest Pacific: Science, v. 152, p. 751-755.

Grim, P. J., 1969, Seamap deep-sea channel: Tech. Rept., U. S. Commerce Dept., ESSA, 27 p.

Grindley, G. W., and Harrington, H. J., 1961, Late Tertiary and Quaternary volcanicity and structure in New Zealand (abs.): 9th Pacific Sci. Cong. Proc., 1957, v. 12, p. 198-203.

Gutenberg, B., and Richter, E. F., 1954, Seismicity of the earth and associated phenomena: 2d ed., Princeton, N.J., 310 p.

Hamilton, E. L., 1967, Marine geology of abyssal plains in the Gulf of Alaska: Jour. Geophys. Research, v. 72, p. 4189-4213.

Hayes, D. E., and Pitman, W. C., III, 1970, Magnetic lineations in the North Pacific: Geol. Soc. America Mem. 126, p. 291-314.

Hays, J. D., 1970, The stratigraphy and evolutionary trends of Radiolaria in North Pacific deep-sea sediments: Geol. Soc. America Mem. 126, p. 185-218.

Healy, J., Vucetich, C. B., and Pullar, W. A., 1964, Stratigraphy and chronology of late Quaternary volcanic ash in Taupo, Rotorua and Gisborn districts: New Zealand Geol. Survey Bull., v. 73, 88 p.

Horn, D. R., Delach, M. N., and Horn, B. M., 1969, Distribution of volcanic ash layers and turbidites in the North Pacific: Geol. Soc. America Bull., v. 80, p. 1715-1724.

Horn, D. R., Horn, B. M., and Delach, M. N., 1970, Sedimentary provinces of the North Pacific: Geol. Soc. America Mem. 126, p. 1-21.

Isacks, B., Oliver, J., and Sykes, L., 1968, Seismology and the new global tectonics: Jour. Geophys. Research, v. 73, p. 5855-5899.

Larson, R. L., Menard, H. W., and Smith, S. M., 1968, Gulf of California: A result of ocean-floor spreading and transform faulting: Science, v. 161, p. 781-783.

Le Pichon, X., 1968, Sea-floor spreading and continental drift: Jour. Geophys. Research, v. 73, p. 3661-3697.

McKenzie, D., and Parker, R. L., 1967, The North Pacific; an example of tectonics on a sphere: Nature, v. 216, p. 1276-1280.

Menard, H. W., 1953, Pleistocene and Recent sediments from the floor of the northeastern Pacific Ocean: Geol. Soc. America Bull., v. 64, p. 1279-1293.

Moore, D. G., and Buffington, E. C., 1968, Transform faulting and growth of the Gulf of California since the late Pliocene: Science, v. 161, p. 1238-1241.

Morgan, W. J., 1968, Rises, trenches, great faults, and crustal blocks: Jour. Geophys. Research, v. 73, p. 1959-1982.

Nayudu, Y. R., 1964, Volcanic ash deposits in the Gulf of Alaska and problems of correlation of deep-sea ash deposits: Marine Geology, v. 1, p. 194-212.

Nelson, C. H., Kulm, L. D., Carlson, P. R., and Duncan, J. R., 1968, Mazama ash in the northeastern Pacific: Science, v. 161, p. 47-49.

Ninkovich, D., 1968, Pleistocene volcanic eruptions in New Zealand recorded in deep-sea sediments: Earth and Planetary Sci. Letters, v. 4, p. 89-102.

Ninkovich, D., Opdyke, N., Heezen, B. C., and Foster, J. H., 1966, Paleomagnetic stratigraphy, rates of deposition and tephrachronology in North Pacific deep-sea sediments: Earth and Planetary Sci. Letters, v. 1, p. 476-492.

Norris, R. M., 1964, Sediments of Chatham Rise: New Zealand Dept. Sci. and Indus. Research Bull., v. 159, 39 p.

Oliver, J., and Isacks, B., 1967, Deep earthquake zone, anomalous structures in the upper mantle, and the lithosphere: Jour. Geophys. Research, v. 72, p. 4259-4275.

Opdyke, N., and Foster, J., 1970, The paleomagnetism of cores from the North Pacific: Geol. Soc. America Mem. 126, p. 83-119.

Pitman, W. C., III, and Hayes, D., 1968, Sea-floor spreading in the Gulf of Alaska: Jour. Geophys. Research, v. 72, p. 6571-6580.

Plafker, G. P., 1965, Tectonic deformation associated with the 1964 Alaska earthquake: Science, v. 148, p. 1675-1687.

Reed, J. J., and Hornibrook, N. de B., 1952, Sediments from the Chatham Rise: New Zealand Jour. Sci., Tech. Bull. 34 (3), 173 p.

Royce, C. F., Jr., 1967, Mazama ash from the continental slope off Washington: Northwest Sci., v. 41, p. 103-109.

Shima, M., Okada, A., and Amano, S., 1967, Age dating of volcanic glass collected from the mid-Pacific Ocean floor (Studies of deep-sea core V20-130, Part 5): Tokyo Nat. Sci. and Mus. Bull., v. 10, p. 467-470.

Stauder, W. S. J., 1968, Mechanism of the Rat Island earthquake sequence of February 4, 1965, with relation to island arcs and sea-floor spreading: Jour. Geophys. Research, v. 73, p. 3847-3858.

Stauder, W., S. J., and Bollinger, G. A., 1966, The focal mechanism of the Alaska earthquake of March 28, 1964, and of its aftershock sequence: Jour. Geophys. Research, v. 71, p. 5283-5296.

Thompson, B. N., Kermod, L. O., and Ewart, A., *Editors,* 1966, New Zealand volcanology, central volcanic region: New Zealand Dept. Sci. and Indus. Research Inf. Ser., v. 50, 212 p.

Thorarinsson, S., and Tomasson, J., 1967, The eruption of Hekla, 1947-1948, Part I: Soc. Scientiarum Islandica, Reykjavik, 183 p.

Tobin, D., and Sykes, L. R., 1968, Seismicity and tectonics of the northeast Pacific Ocean: Jour. Geophys. Research, v. 73, p. 3821-3845.

Todd, R., 1953, Foraminifera from the lower Tertiary of Amchitka Island, Aleutian Islands: Cushman Found. Foram. Research Contr., v. 4, pt. 1, p. 1-7.

Vine, F. J., 1966, Spreading of the ocean floor: New evidence: Science, v. 154, p. 1405-1415.

Wilcox, R. E., 1965, Volcanic-ash chronology, *in* Wright, H. E., Jr., and Frey, D. G., *Editors,* The Quaternary of the United States. A review volume for the 7th Congress of the Internat. Assoc. for Quaternary Research, Princeton, N.J., p. 807-816.

LAMONT-DOHERTY GEOLOGICAL OBSERVATORY CONTRIBUTION NO. 1492
MANUSCRIPT RECEIVED APRIL 10, 1969
REVISED MANUSCRIPT RECEIVED JUNE 30, 1969

THE GEOLOGICAL SOCIETY OF AMERICA, INC.
MEMOIR 126, 1970

Magnetic Lineations in the North Pacific

DENNIS E. HAYES
and
WALTER C. PITMAN III

Lamont-Doherty Geological Observatory
of Columbia University
Palisades, New York

ABSTRACT

The work of several investigators on the magnetic lineations in the northeast Pacific has been re-examined and a new lineation map of the area is presented. These lineations are presumed to have been generated by a process of sea-floor spreading. By adopting a time scale for the lineations we have presented a map of the inferred ages of the basement rocks and a maximum age for the overlying sediments. These ages are expressed in terms of the familiar geologic epochs. It has been postulated that the tectonic evolution of the northeast Pacific was largely dominated by a "Y"-shaped system of migrating ridge axes. The trenches that formed in response to this system of ridges, and the the ridges themselves, were characterized by periods of activity and inactivity. The barrier nature of these features should have had considerable influence on the distribution of terrigenous sediments in the northeast Pacific.

A new group of magnetic lineations has been discovered which trends northwest and is partially bounded by the Shatsky Rise, the Emperor Seamounts and the Hawaiian Arch. The lineations are thought to have been formed by a process of sea-floor spreading prior to the Late Cretaceous. Lineations trending east-northeast in the area just east of Japan are also thought to represent an old era of spreading. These two groups of lineations may comprise the two limbs of an ancient magnetic bight now centered in the northwest Pacific near the Shatsky Rise.

Quiet magnetic zones separate the major areas of magnetic lineations, giving rise to a distribution of four zones with distinctive magnetic signatures. These four zones may relate to similar magnetic zones in the North Atlantic.

CONTENTS

INTRODUCTION

In this paper the assumption of the validity of the sea-floor spreading hypothesis (Hess, 1960, 1962, 1968; Dietz, 1961, 1962, 1968) is implicit. Various evidence supporting this hypothesis has been reviewed, for example, by Pitman and others (1968) and Isacks and others (1968). Of particular importance to this paper is the magnetic anomaly pattern that is generated by the sea-floor spreading process. Hess and Dietz proposed that the oceanic crust spreads away from the mid-oceanic ridge axis in a bilaterally symmetric fashion. As new crust is formed at the axis it in turn is transported away. They regard the process as continuous. Vine and Matthews (1963) presumed that this new crustal material was basalt (which characteristically has high values of thermal remanent magnetization). As this basaltic material is formed at the ridge axis and cools through its Curie point it acquires a remanent magnetization in the direction of the earth's ambient magnetic field. They proposed that if the earth's magnetic field periodically reversed polarity (Cox and others, 1963) then the material formed at the axis of the ridge would appear as alternate strips of normally and reversely magnetized rock parallel to the ridge axis. Presuming that the spreading process is bilaterally symmetric with respect to the ridge axis, these strips of basalt and their associated magnetic anomalies should appear as a bilaterally symmetric pattern.

A pattern of bilaterally symmetric magnetic anomalies has been found in the North and South Pacific (Pitman and others, 1968); South Atlantic (Dickson and others, 1968); South Indian Ocean (Le Pichon and Heirtzler, 1968), and more recently, in the North Atlantic (Talwani and others, in prep.). The pattern extends from the ridge axis to the flanking basins. Because the time interval between magnetic polarity reversals is variable, the over-all pattern is unique. The pattern is similar in all the above-mentioned areas (Heirtzler and others, 1968), implying a simultaneity of spreading in all these regions.

Of particular importance has been the dating of the pattern. It is obvious from the sea-floor spreading hypothesis that the anomaly at the axis of a ridge should be the youngest and that the anomalies should be progressively older toward the flanks. Pitman and Heirtzler (1966), Vine (1966), and Heirtzler and others (1968), were able to show that a small portion of this anomaly pattern extending from the axis a short distance toward the flanks could be correlated directly with the detailed history of magnetic polarity reversals of the past 3.35 m.y. (Doell and others, 1966). By assuming a constant spreading rate and by extrapolation from this dated portion of the anomaly pattern, Vine (1966) and Heirtzler and others (1968) proposed ages for the entire pattern of anomalies. Using their time scale, Heirtzler and others (1968) dated the oceanic crust for those regions in which the magnetic pattern had been found. The time scales of Vine and of Heirtzler and others are quite similar; both assume continuous spreading from the present back to the Late Cretaceous.

Results from the deep-sea drilling project of JOIDES (Maxwell, 1969) provide confirming evidence for a constant average spreading rate for the South Atlantic and for the general validity of the Heirtzler and others (1968) time scale. The JOIDES results do not eliminate the possibility that in other regions spreading has been discontinuous or varied smoothly with time (Ewing and Ewing, 1967; Heirtzler and others 1968; Talwani and others, in prep.).

In this particular study we are primarily concerned with discussing the rough geographic distribution of basement isochrons in the North Pacific and in speculating on the evolution of the observed magnetic and morphologic fabric there. We allow that errors of a few percent are probable in the adopted time scale, but do not believe such errors significantly alter the results presented here. Because the time scale of Heirtzler and others (1968) is based on an extensive worldwide collection of data and is in good agreement with JOIDES results, it will be used here. By the use of this time scale and the Phanerozoic time scale (Anonymous, 1964), we propose geologic ages for much of the oceanic basement in the North Pacific and also propose a tectonic history of the region.

The paper will be divided into two sections according to region. The first region to be discussed is the northeast Pacific, bounded on the north by the Aleutian Trench, the east by continental North America, the west by the Emperor Seamount Chain and the southwest by the Hawaiian Islands. The data density in most of this region is sufficient to determine the detailed fabric of magnetic anomalies. The lineation pattern can be correlated with the worldwide magnetic pattern of Heirtzler and others (1968); therefore, an age for the basement rock in this region may be proposed. The unique configuration of the pattern in this area suggests an unusual tectonic history. The second part of the paper deals with the northwest Pacific. Here the data density is insufficient to describe the details of the magnetic lineations. However, there are in this region two groups of magnetic anomaly lineations. One group has been reported by Uyeda and others (1967) and Hayes and Heirtzler (1968). The other group is previously unreported. Neither of these sets of anomalies appear to be correlatable with the pattern of Heirtzler and others (1968). The age of these anomalies and the tectonic history of this region thus remains highly speculative.

THE MAGNETIC LINEATION PATTERN OF THE NORTHEAST PACIFIC

Much of the work in establishing a magnetic pattern in the northeast Pacific (that region east of the Emperor Seamount Chain and the Hawaiian Arch) and subsequently in correlating it with the patterns of Heirtzler and others (1968) has been published as studies of individual regions. We now bring results of these studies together, and with additional data present an overall picture of the magnetic lineations of the region, as shown in Plate 1. The numbers show the correlations of individual lineations with the worldwide pattern (Pitman and others, 1968; Heirtzler and others, 1968).

Portions of the pattern between the west coast of North America and east of 165° W., and from approximately 30° N. to the Aleutian Arc were found by Mason (1958), Raff and Mason (1961), Mason and Raff (1961), Raff (1966), Peter (1966) and Elvers and others (1967a, 1967b). These anomalies were later shown to be a part of the worldwide pattern (Vine, 1966; Pitman and others, 1968). On the basis of a few reconnaissance tracks, Pitman and Hayes (1968) identified and traced the magnetic anomaly pattern into the Gulf of Alaska. Peter and others (in press) and Menard and Atwater (1968), using additional data, derived a pattern for the Gulf of Alaska that varies somewhat from that of Pitman and Hayes. The interpretation used here approximates that of these other workers, with small modifications.

The east-west lineations between 165° W. and approximately 170° E. have been derived by Grim and Erickson (1969) and Hayes and Heirtzler (1968). These interpretations differ only in the detail concerning the presence or absence of minor fracture zones at 177° W. and 171° W. The interpretation of Grim and Erickson (1969), which is based on closely spaced data lines, is adopted here.

The most distinctive magnetic feature of the northeast Pacific is the sharp bend in the anomaly lineations at approximately 160° W. This feature, reported by Elvers and others (1967a, 1967b), has been named the Great Magnetic Bight. Raff (1968), Pitman and Hayes (1968) and Peter and others (in press) have offered explanations of this feature that are compatible with the spreading hypothesis. The explanation of Pitman and Hayes (1968) is fundamental to the tectonic interpretation that is to follow and will be reviewed and discussed in considerable detail.

By using the time scale of Heirtzler and others (1968) and the Phanerozoic time scale (Anonymous, 1964), we can express the age of the basement in the northeast Pacific in terms of geologic epochs, as shown in Plate 2.

The lightly stippled zone (Pl. 1) to the east of the Hawaiian Arch and west and south of anomaly 32, is an area in which most of the magnetic anomalies are of conspicuously smaller amplitude than in adjacent areas (Raff, 1966). It is difficult to correlate anomalies throughout this zone, and lineations are not apparent except for those clearly associated with the northeast-southwest trending fracture zones (Malahoff and Woollard, 1968). Figure 1 shows a typical magnetic anomaly profile through the area (Plate 1 shows the location of the track). Note in profile U-V the distinct change in the character of the anomalies in going from the magnetically quiet zone into the lineation pattern to the east of anomaly 32.

It must be pointed out that this area lies almost entirely to the west of the western edge of the East Pacific Rise as defined from

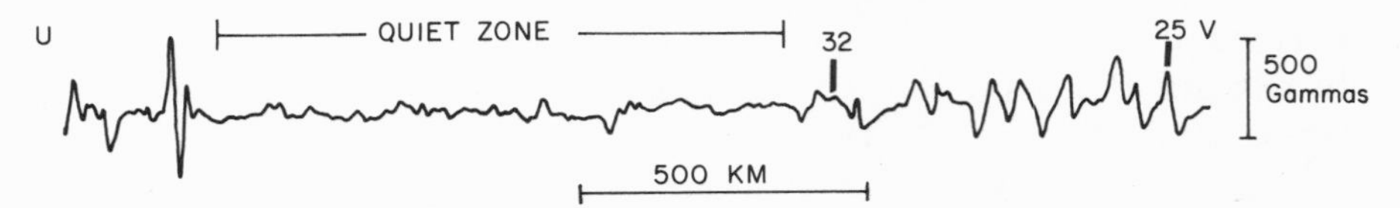

Figure 1. Total magnetic intensity anomaly profile along track U-V, as shown in Plate 1. Note the contrast in the amplitudes of the observed anomalies. The quiet zone may represent a long period of normal magnetic polarity during the Middle Cretaceous.

bathymetric evidence (Menard, 1966). Menard's boundary lies nearly along the northwest-southeast anomaly 32 lineations. However, as Menard pointed out, the Mendocino and Murray fracture zones continue through this area (Menard, 1966; *see also* Naugler and Erickson, 1968). This clearly suggests that the oceanic crust here has been generated by sea-floor spreading and was probably generated by the same ridge system as the area to the east. Except for the change in the magnetic amplitudes, there is no geophysical evidence that would suggest a structural or time discontinuity at the boundary between anomaly 32 and this quiet region. There is no apparent discontinuity in sediment thickness (Ewing and others, 1968) or in the value of terrestrial heat flow (Langseth and Von Herzen, in press).

Helsley and Steiner (1968), on the basis of paleomagnetic data, stated that although magnetic polarity reversals were present in the Lower Cretaceous and the uppermost Upper Cretaceous, the period from middle Cenomanian to middle Santonian (approximately 80 m.y.B.P. to 100 m.y.B.P.; Phanerozoic time scale, Anonymous, 1964) was one of dominantly normal polarity. They further suggested that this might be related to the quiet region beyond anomaly 32. We, therefore, assume that the spreading in this region, although older, has been continuous with the region to the east and north. Using that portion of the known pattern just south of the Mendocino Fracture Zone and extrapolating linearly to the west, one obtains an approximate geologic age of 135 m.y.B.P. (Jurassic-Cretaceous) for the western boundary of this region (*see* Pl. 2).

THE TECTONIC EVOLUTION OF THE NORTHEAST PACIFIC

The sharp bend in the magnetic lineation in the northeast Pacific, the Great Magnetic Bight (Elvers and others, 1967a, and Elvers and others, 1967b), and the lack of symmetry in the pattern are a problem of particular interest with regard to sea-floor spreading. Herron and Heirtzler (1967) suggested a "Y"-shaped configuration of ridge axes to explain the observed anomaly pattern at the intersection of the Galapagos Ridge and the East Pacific Rise. Raff (1968) proposed the same system to explain the "magnetic bight" of the northeast Pacific. Pitman and Hayes (1968) presented a similar model to explain the magnetic lineations and the tectonic evolution of the Great Magnetic Bight in the Gulf of Alaska.

It is this latter scheme that will be discussed here with some modifications and additions. Three properties are suggested regarding the sea-floor spreading process: (1) that it is possible for an actively spreading ridge (locus of opening between rigid crustal plates) to migrate toward an active trench; (2) that when a locus of opening (ridge) migrates very near or into a trench, the spreading process is stifled; (3) that even though the ridge is migrating it will still generate new crust in a bilaterally symmetric fashion about its axis.

The migration of a ridge into a trench and the cessation of spreading at that time and place may be difficult to accept. The single-sided configuration of magnetic lineations in the northeast Pacific demands that the ridges and the marginal trenches of North America moved, in a relative sense, toward one another. In this case, the crustal plate between the migrating ridge and the trench becomes smaller for two reasons: first, material plunges beneath the trench and, second, the locus of spreading migrates toward the trench. When the attrition of this plate has progressed to the point where little or none of the plate remains—the process stops (*see also* McKenzie and Morgan, 1969).

It is proposed here that in the Late Cretaceous (just prior to anomaly 32) the northeast Pacific consisted of four crustal plates (Pl. 3A) separated from one another by a system of trenches (hachured lines), ridge axes (dashed lines), and fracture zones (solid lines). Plate I (the central Pacific) and Plate III (continental North America plus the Bering Sea, the Aleutians and the two trenches) are assumed fixed with respect to each other. Plate II is moving northward into a fossil Aleutian Trench. Plate IV is moving eastward into a now inactive and sediment-filled trench along the western border of the United States and Canada (Shor, 1962; Hayes and Ewing, in press). It is presumed that Plate II was bordered on its western edge by a ridge-trench type transform fault (Wilson, 1965); perhaps the Emperor Seamounts (Vine and Hess, in press). Plate IV must likewise have been bordered on its southern edge by a ridge-trench transform fault (perhaps the Clipperton).

The assumption that there has been no relative motion between Plates I and III is perhaps erroneous, but is made for the sake of simplifying the discussion and illustration. We, of course, realize that simultaneous sea-floor spreading in the North Atlantic could have caused Plate III with its marginal trenches to drift toward the system of Pacific ridges shown here. Talwani and others (in prep.) have demonstrated that there was spreading in the North Atlantic at the rate of approximately 1.8 cm/yr (3.6 cm/yr is the total rate of opening) from 80 m.y.B.P. to ~ 40 m.y.B.P.

While the exact relative motion of the North American plate (III) with respect to the central Pacific plate (I) is an important problem, it is at present indeterminate. The only point critical to our discussion is that in terms of relative motion, the Pacific ridge axes have moved toward and into the marginal trenches of Plate III. Also, since only relative motion is implied in Plate 3, it is possible to add any other component of motion to the entire system. Vine (1968) has suggested that the crust that now forms the magnetic bight may have moved north by 20° of latitude, or more, since its formation. The precise location of the proposed fossil trenches is unknown. There is some evidence for a fossil trench along the west coast of North America (Shor, 1962; Hayes and Ewing, in press). To the north we presume a fossil trench in the approximate location of the Aleutian Trench. Menard and Dietz (1951) suggested that Kodiak Island and the Kenai Peninsula might have been the site of an older trench (*see also* Burk, 1965).

If the active ridge between Plates I and II is generating crustal material in a symmetric fashion, then this ridge will migrate toward the trench between Plates II and III. Likewise, the ridge between Plates I and IV will migrate toward the trench between Plates IV and III. The relative motion of Plates I, II, and IV is manifested by a "Y" configuration of ridges. It is important to note that anomaly 32 is the oldest identifiable anomaly of the magnetic bight. However, if as has been suggested, the quiet zone bounding anomaly 32 is merely a contiguous but older portion of crust that has been generated by the spreading, then the place of inception of this "Y" configuration of ridges may have been farther to the southwest. Thus, the onset of this "Y" pattern of spreading may have been considerably earlier than the Late Cretaceous.

With these various contingencies in mind, the proposed organization of crustal plates for the Late Cretaceous is as shown in Plate 3A. The paleogeography of the continental margins is unknown. The present-day continental margins are shown for reference only. To the north, an active trench blocks the passage of terrigenous sediments from the north to the deep basin in the south. To the east, along the west coast of North America, a trench likewise blocks the flow of turbidites and other terrigenous sediment to the basin areas. This generalized configuration remained until the late Paleocene. New crust was being formed at the migrating ridge axis, as Plates II and IV were moving toward and into the trenches. In early Paleocene the east-west ridge axis had migrated close to the northern bounding trench (Pl. 3B). By late Paleocene or perhaps early Eocene this segment of ridge reached the trench and this trench-ridge system

which were very convincing (Fig. 2). The first correlation required that the unknown pattern represent anomalies 7-12 (~ 26 to 37 m.y.B.P., Oligocene), the second correlation suggested the pattern represents anomalies 20-24 (~ 50 to 60 m.y.B.P., late Eocene-mid-Paleocene).

Figure 2 shows three observed magnetic anomaly profiles from the region. These profiles have been projected to simulate a normal crossing of the lineations. The computed anomaly profile at the top shows the possible correlation with anomalies 20-24. The anomaly profile at the bottom is the possible correlation with anomalies 7-12.

If the correlation with anomalies 20-24 is correct, then the pattern should be younger to the southeast and no simple continuation with the Aleutian anomalies is possible. However, immediately to the southeast of the lineations lies the Shatsky Rise (Pl. 1), and sediments have been dated by Ewing and others (1966) as at least as old as Early Cretaceous and possibly Late Jurassic. Furthermore, if the anomalies were generated by sea-floor spreading, there should be a ridge axis to the southeast and a bilaterally symmetric magnetic pattern. There is no evidence for either of these features.

If the correlations with anomalies 7-12 is correct, then the oceanic crust should be progressively younger toward the northeast. This relative chronology of younger crust lying toward the marginal trench is the same as seen in the northeast Pacific, and thus might be explained by a similar scheme of migrating ridges. However, if the correlation with anomalies 7-12 is correct, one would expect to see anomalies younger than anomaly 7 in one of the following configurations: adjacent to the present Kuril Trench, up to an intervening fossil trench, or up to an intervening fossil ridge, beyond which one would see the mirror image of the anomalies. It can be seen from

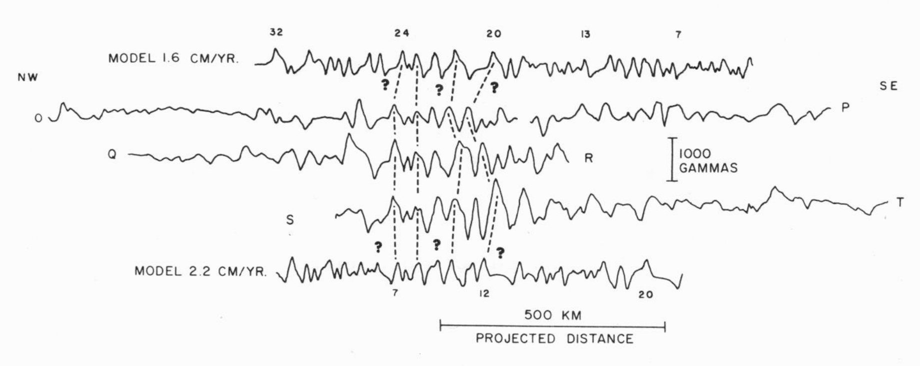

Figure 2. Projected profiles of total magnetic intensity anomalies from the northwest Pacific (*see* Pl. 1 for profile locations). Magnetic anomalies computed for the models of Pitman and others (1968) are shown at the top and bottom for comparison (*See* text for discussion).

Figure 2 that toward the northwest the amplitude of the anomalies diminishes rapidly to the point that we could call this area a magnetically quiet region. There is no evidence of younger anomalies in this region or of a fossil ridge axis with a symmetric pattern correlatable with any of the younger anomalies. Likewise, the seismic reflection work of Ewing and others (1968) gives no indication of an intervening fossil trench northwest of the pattern.

It is our opinion then that the lineations northeast of the Japan Trench were formed at a time before anomaly 32 and thus presumably before the Late Cretaceous. Several independent studies have suggested that the northwest Pacific represents a relatively old portion of the Pacific Ocean. The studies of Ewing and others (1968) strongly suggest the top of an opaque sediment horizon present in this area is approximately of Late Cretaceous age. In addition, the sediment comprising the opaque zone thins and disappears toward the northwest as the Kuril Trench is approached from the basin. This may imply relatively young basement near the Japan and Kuril trenches and relatively old basement away from them. This chronology of basement rocks would conform with that deduced in the basin south of the Aleutian Trench. The detailed analysis of worldwide heat flow led Langseth and Von Herzen (in press) to speculate that the Pacific basin east of Japan was quite old, as compared with other basins, because the heat flow is uniformly 0.3 to 0.4 μ cal/cm^2/sec lower than world or Pacific basin averages.

The magnetic pattern in the area is interrupted by the Shatsky Rise near 168° E. and 38° N. The observed magnetic anomalies west of the Emperor Seamount Chain are relatively small north of 45° N., except near the intersection of the seamount chain with the Aleutian-Kamchatka trenches. Similarly, east of the Emperor Seamount Chain, the observed magnetic anomalies south of 45° N. are small (*see* Pl. 1). This results in broad, magnetically quiet zones that are conspicuous in a regional sense, and which may prove to be interrelated.

Just south of the Shatsky Rise and near the Emperor Seamount Chain, the magnetic pattern is extremely complicated. Figure 4 shows a fan of tracks that converge at about 34° N. and 165° E. Hayes and Heirtzler (1968, Fig. 1) showed that similar wavelength anomalies are present along all azimuths in this region. They speculated that in this region the trend of anomalies probably changes by nearly 90°, giving rise to a magnetic bight.

NEW DATA IN THE CENTRAL NORTH PACIFIC

A previously unreported group of magnetic lineations has been found and mapped from reconnaissance ship surveys and Project Magnet flights. These anomalies trend northwest-southeast and lie in the area partially bounded by the Hawaiian Arch, the Emperor Seamount Chain, the Shatsky Rise, and the Marcus-Necker Ridge. Their westward extent is not known at present. These anomaly lineations are offset along one or more northeast-southwest-trending fracture zones. The anomalies are identified in an area of more than 2,000,000 sq km and individual lineations persist for distances of about 1000 km.

The trend of the new magnetic lineations, their known areal extent, their relationship to the dominant physiographic features of the area, and the magnetic lineations to the north are all schematically illustrated in Figure 3. Individual lineations are not shown in Figure 3; only the trends of the two major lineation patterns. The extent and location of the major fracture zones as shown are modified from published works on the basis of much of the magnetic and bathymetric data of this study. The existence of all of the indicated fracture zones and their relationship with previously known features is not firmly established, and this aspect of Figure 3 should be regarded as speculative.

The individual data control lines available to us for this portion of the western North Pacific are illustrated in Figure 4. This control is of a reconnaissance nature and the details of the inferred lineations will change as more information becomes available. It is remarkable and important that individual anomalies can be correlated with confidence across control lines often separated by a hundred or more miles.

It seems most probable that these magnetic lineations were formed by the same process and in a manner similar to other well-known and widely mapped magnetic lineations associated with mid-oceanic ridges. These lineations (*see* Fig. 5), consisting of about 20 characteristic peaks and troughs, have not been identified with respect to the known worldwide lineation pattern, as described by Pitman and others (1968). If the reported lineation pattern of Pitman and others (1968) represents substantially the complete history of spreading back to Late Cretaceous time, then the lineations described here must reflect a much older sequence of magnetic anomalies and pattern of sea-floor spreading. It is an interesting and significant fact that these lineations trend at nearly right angles to the lineations found northeast of Japan (*see* Pl. 1, and Fig. 3) which, as mentioned earlier, are also thought to be older than anomaly 32 ($\sim$ 77 m.y.B.P.). Both of the observed lineation patterns are interrupted or disturbed in the vicinity of the Shatsky Rise. Though definite anomaly correlations between these two patterns have not yet been made, it is possible that these two groups of anomalies represent the two limbs of an ancient magnetic bight (*see* Fig. 6) similar to that presently observed in the northeast Pacific (Elvers and others, 1967b).

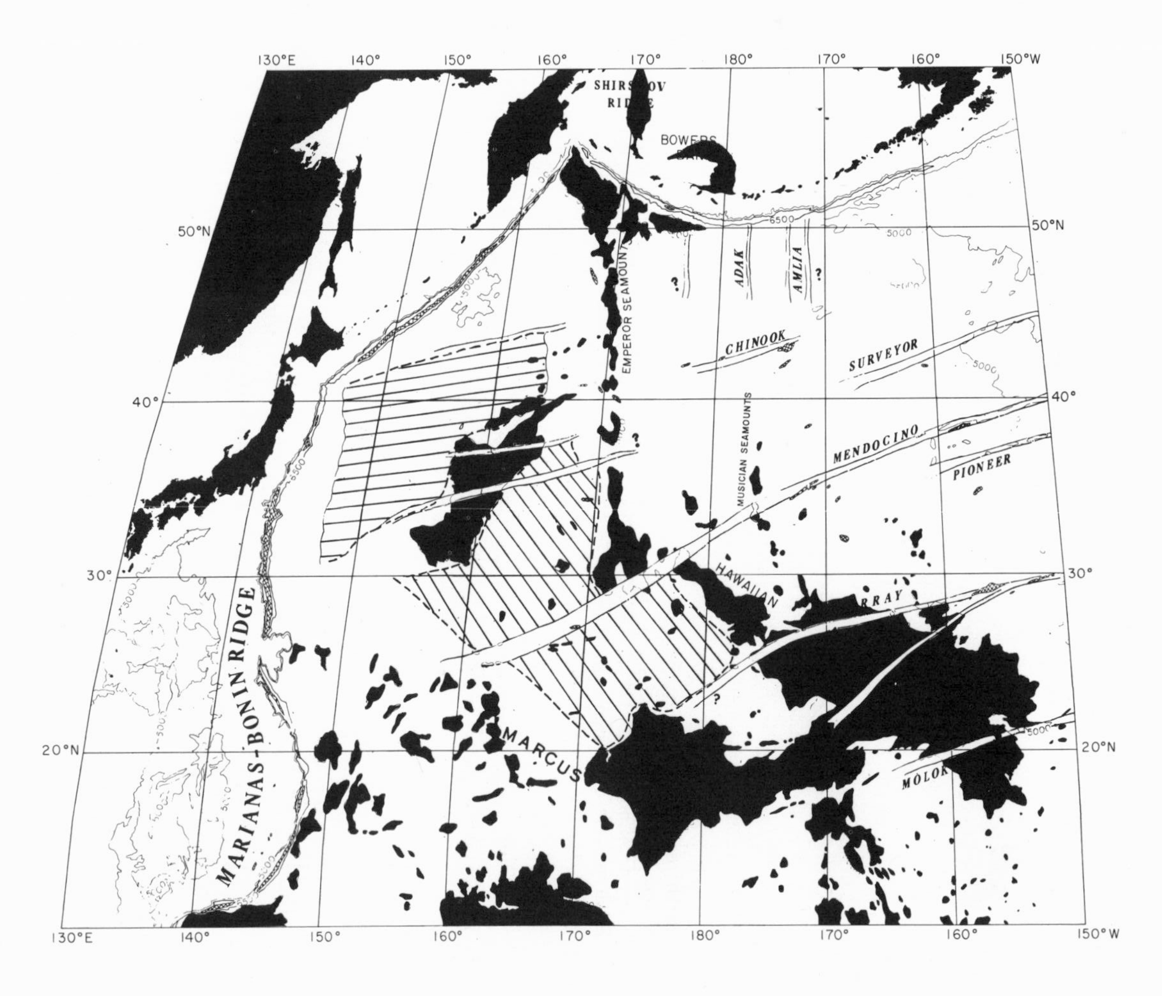

Figure 3. Bathymetric map of the Pacific, modified *from* Udintsev and others (1963), showing the trend and known areal extent of magnetic lineations west of the Hawaiian Arch and Emperor Seamounts. The fracture zones are modified *from* works of Menard (1966), Naugler and Erickson (1968), Malahoff and others (1966), Grim and Erickson (1969), Hayes and Heirtzler (1968), and Peter (1966), and should be regarded as highly schematic. The fracture zones near the Shatsky Rise are speculative. Depths < 5000 m are shaded. The direction of cross-hatching indicates the general trend of magnetic anomaly lineations. Note how the major morphologic features (for example, the Hawaiian Arch) appear to bound the lineations.

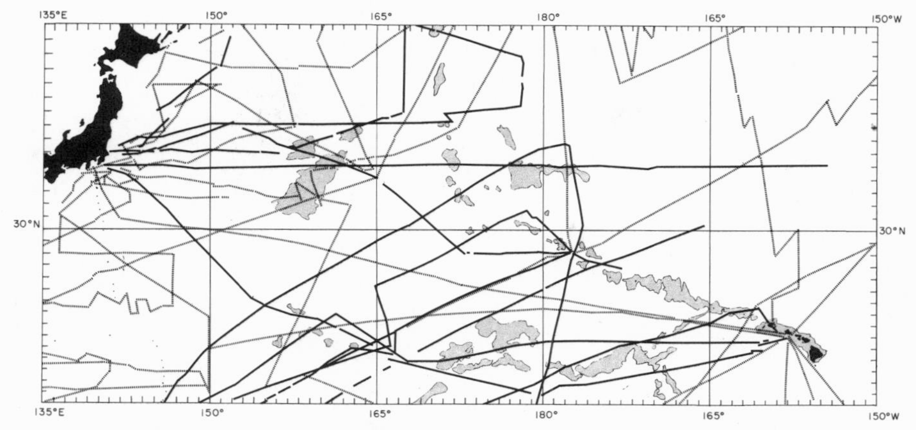

Figure 4. Control lines available for determining magnetic lineations in the northwest Pacific. Dashed lines represent Lamont-Doherty ship tracks of R/V *Vema* and R/V *Robert D. Conrad.* Solid lines represent data collected by the Navy Oceanographic Office, by U.S.C. & G.S., and during Project Magnet flights. Bathymetric data were available along all ship tracks.

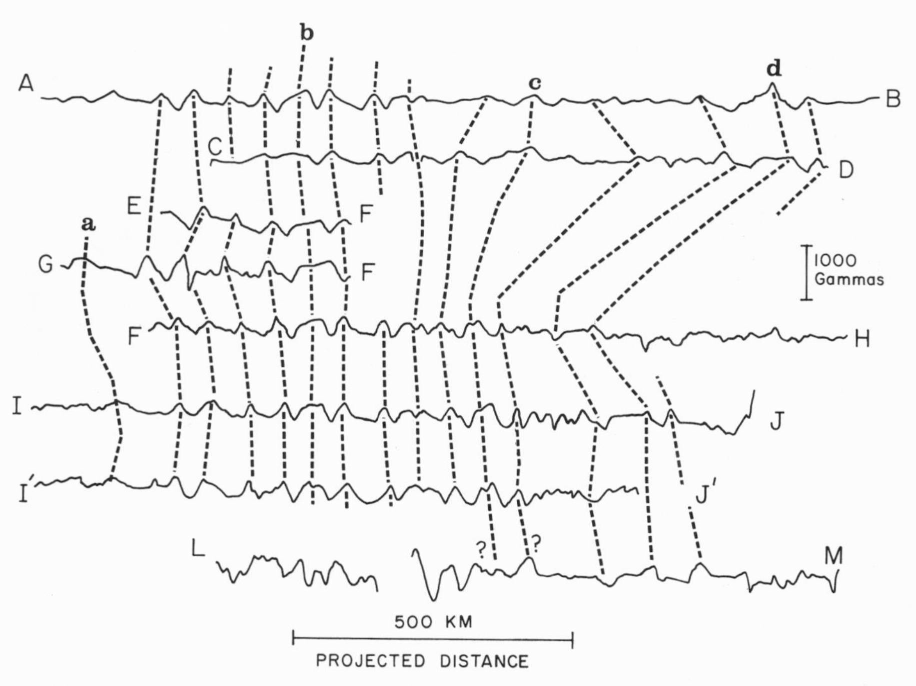

Figure 5. Shows projected profiles of magnetic total intensity anomalies. The profiles have been aligned to illustrate our interpretation of the individual anomaly correlations. Conspicuous seamounts and scarps are present along some tracks and these have been used as control in interpreting the exact position of the fracture zones shown. The original tracks of these profiles are shown in Plate 1, along with selected lineations labeled a, b, c, and d.

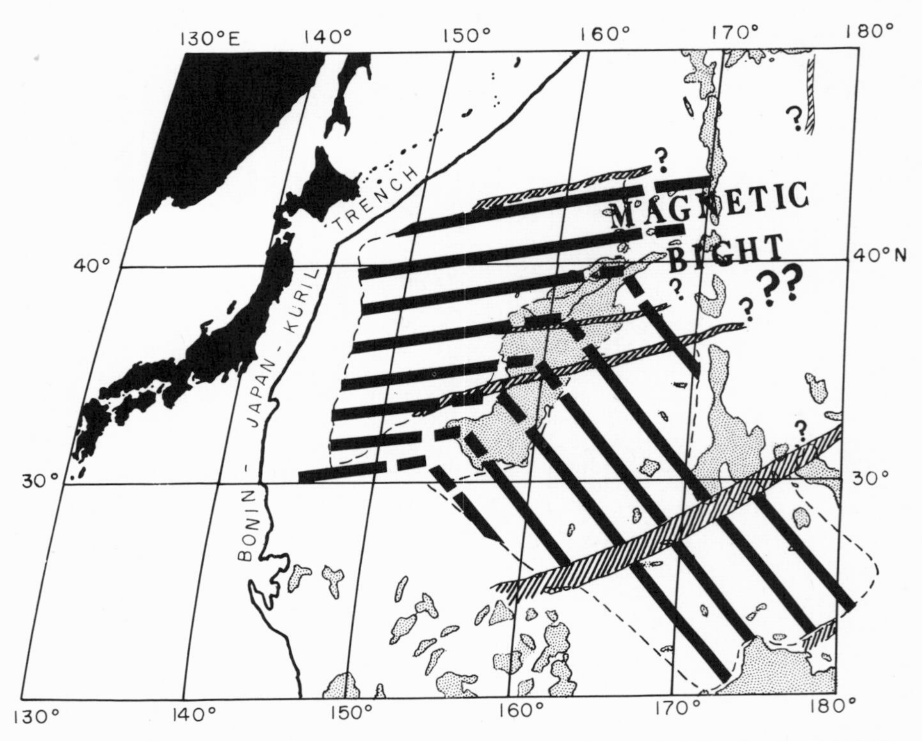

Figure 6. Inferred position of hypothetical magnetic bight in the vicinity of Shatsky Rise. Firm magnetic anomaly correlations around this bight have not yet been established.

The geometry of the two lineation trends and the location of the inferred bight is such that the hypothetical bight could not, without major adjustment of the crustal plates, be an older part of the bight in the northeast Pacific. A simple extension of the hingeline of the Great Magnetic Bight does not pass through the area of the speculative magnetic bight near Shatsky Rise. If the entire magnetic pattern from the western Pacific northeast to the Gulf of Alaska had been generated by one system of migrating ridge axes similar to that suggested by Pitman and Hayes (1968), the evolution of this system must have been much more complicated than depicted in Plate 3. There are no known fracture zones that trend normal to the lineations northwest of the Shatsky Rise. On the contrary, the topography in the area suggests minor fracture zones and offsets subparallel to the magnetic lineations there. There is insufficient evidence to firmly establish the presence of a magnetic bight in the vicinity of the Shatsky Rise. An alternative explanation is that the two lineation patterns may represent two old but independent and non-synchronous patterns of spreading, the younger truncating the pattern generated by the older. All the present data available strongly suggest that this general region of the Pacific is relatively old (Ewing and others, 1968; Langseth and Von Herzen, in press), as compared with other regions of the Pacific, and therefore does not conflict with either of these speculations.

The offset in the lineations west of the Hawaiian Arch is located along a line which probably is the southwest extension of the Mendocino Fracture Zone (Menard, 1966), and which appears to offset the junction of the Emperor Seamounts and the Hawaiian Arch in the same left-lateral sense and by approximately the same amount (330 km). It should be noted, however, that while the inferred offset of the Mendocino Fracture Zone west of California is also left-lateral, the amount of offset is about 1200 km. An example of the topographic expression of the inferred fracture zone and the correlations of offset magnetic anomaly lineations across it is illustrated in Figure 7. There is a subtle change in the regional topographic gradient along profile G-H at the fracture zone. If the fracture zone does in fact represent a continuous extension of the Mendocino Fracture Zone, the contrast in offset across the feature ($\sim$ 300 km on the southwest; 1200 km on the northeast) is difficult to explain. Differential offset along fracture zones is not unique to this area nor does it exclude continuity of the features. The Murray Fracture Zone exhibits differential offset along its eastern extent (Raff, 1966). Christoffel and Ross (in prep.) have found a major fracture zone in the southwest Pacific basin where the offset varies by more than a factor of 2 over a distance of about 700 km.

If we assume the northwest-trending lineations on the west and the northerly trending lineations on the east have been generated by the same segments of migrating ridge, it is interesting to speculate

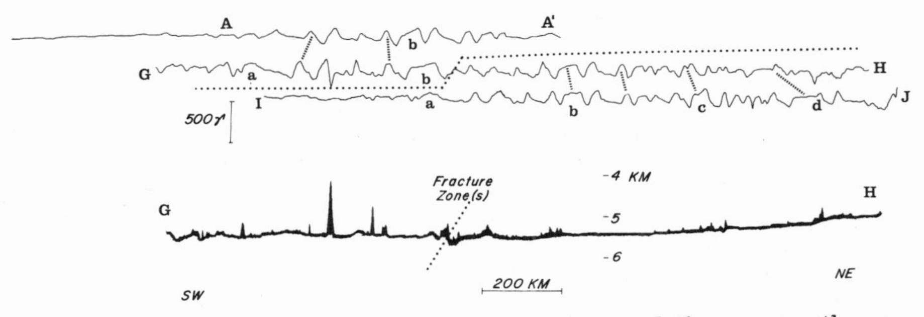

Figure 7. Topography and magnetic anomaly correlations across the proposed southwestward extension of the Mendocino Fracture Zone. Profile locations are indicated in Plate 1. The dotted line separates portions of profile G-H that lie on opposite sides of the fracture zone. Profile A-A' is taken from a Project Magnet flight and anomaly amplitudes are correspondingly attenuated.

on the possible age of the newly discovered lineation pattern. If we extrapolate from the inferred age of anomaly 32 in the east, and assume a constant average spreading rate of 4 cm/yr, the western limit of the new pattern would then be about 180 m.y.B.P. (~ Jurassic-Triassic boundary). We are well aware of the probable error inherent in such an extrapolation.

The southwest edge of the new lineation group is an obvious "quiet zone," as shown in Plate 1 and Figure 7. A magnetic profile from the west coast of the United States to the Marianas Trench and parallel to the Mendocino Fracture Zone would exhibit four zones of distinctive anomalies. This presence of four zones is quite like that found in the North Atlantic (for latitudes ~ 25° to 35° N.) when progressing from the mid-Atlantic Ridge east or west toward an adjacent continent. A possible explanation for the quiet zones adjacent to the continents in the North Atlantic was offered by Heirtzler and Hayes (1967), in which they relate these zones to the long Kiaman magnetic interval of uniform polarity in the late Paleozoic. It is well worth considering that the close correspondence of magnetic zones in the North Atlantic with those of the North Pacific is a possible manifestation of continual spreading that began in these places as early as the late Paleozoic.

Since the Hawaiian Arch and Emperor Seamount Chain appear to disturb or bound the lineation pattern, it seems likely that they postdate the observed magnetic lineations. There appears to be a 300 km left-lateral offset between the northwestern end of the Hawaiian Arch and the southern end of the Emperor Seamount Chain (Fig. 3). The offset is along the extension of the Mendocino Fracture Zone and is in the same sense and of the same magnitude as the offset of the magnetic lineations to the west. We have assumed throughout this paper that the offset of the magnetic lineations is related to transform faults that offset ridge axes (Wilson, 1965). If the Emperor Seamounts and the Hawaiian Arch were once joined, their present offset must be explained by left-lateral transcurrent faulting along the western extension of the Mendocino Fracture Zone. This transcurrent motion could also explain the offset of the magnetic lineations to the southwest. The other possibility, of course, is that these features were formed in their present relative position. If this were the case, the structural significance of the apparent offset remains enigmatic. The highly speculative fracture zones which appear to offset the north-central portion of the Shatsky Rise (*see* Fig. 3) present similar problems of interpretation. The parallel nature of the Hawaiian Arch to one set of lineations, and the near perpendicular configuration of the Emperor Seamount Chain to the other lineation pattern, poses another difficulty in explaining the interrelationship of the morphologic and the magnetic features. The Shatsky Rise clearly affects the two lineation patterns, thus indicating it is younger than either lineation group.

CONCLUDING REMARKS

The magnetic lineations in the northeast Pacific shown in Figure 1 are believed to be well established. They are unquestionably identifiable in terms of the worldwide pattern. Only the details of this magnetic pattern are likely to change as more control lines become available, thus making it possible to recognize minor offsets and irregularities that occur between present track lines. The correctness of Plate 2 is strongly dependent on two major assumptions. These are (1) that the magnetic lineations were formed by a process of sea-floor spreading (moving crustal plates), and (2) that the time scale of the lineations, as proposed by Heirtzler and others (1968) is substantially correct. There have been few published ages obtained by radiometric or paleontological methods from dredge hauls and drill samples in the North Pacific (for example, Carsola and Dietz, 1952; Hamilton, 1956; Menard and others, 1962; Dymond and Windom, 1967; Ozima and others, 1968; Budinger and Enbysk, 1967; Dymond and others, 1968). The only one of the published dates that contradicts the ages of the basement proposed here is the age of 27 m.y. suggested by Buddinger and Enbysk (1967) for the Cobb Seamount. This date was obtained by K-Ar analysis of a basaltic rock from a dredge haul. Dymond and others (1968) subsequently obtained a K-Ar date of approximately 1.6 m.y. from another sample from Cobb Seamount believed to have been *in situ*. One possible explanation for the discrepancy is that the rock used by Budinger and Enbysk was ice-rafted. The tectonic development of the area as depicted in Plate 3 is speculative. We believe it is presently the only plausible explanation of the unique magnetic fabric of the northeast Pacific consistent with the sea-floor spreading hypothesis. The influence of this proposed tectonic evolution on the distribution of terrigenous sediments in the North Pacific during the Cenezoic is profoundly important.

The presence of the newly discovered magnetic lineations southwest of the Hawaiian Arch is well documented, although the exact areal extent of this lineation group is not yet established. The relationship of these lineations to those east of Japan and to the Great Magnetic Bight and lineations far to the northeast is not yet firmly established and is under further study. The apparent continuity of the Mendocino Fracture Zone (and probably the Murray Fracture Zone) through the new magnetic lineation pattern, accompanied by our knowledge of the large areal extent of the pattern, and of the characteristics of the individual anomalies, is strong evidence that the pattern was generated by a process of sea-floor spreading. The age of the new lineations is highly speculative, but because they cannot be correlated with the worldwide pattern, we believe their age to be older than the Late Cretaceous and perhaps as old as Triassic or Jurassic.

ACKNOWLEDGMENTS

The data summarized here are the results of efforts in magnetic data collection and analysis by numerous individuals over many years. Their important contributions are recognized and gratefully acknowledged. Special thanks go to W. Menard, G. Peter, T. Atwater, P. Grim, B. Erickson and S. Uyeda, all of whom provided data prior to publication. Discussion with these individuals and numerous colleagues at Lamont were most valuable. We thank N. Opdyke, J. Hays and R. Wall for critically reviewing the manuscript. This research was supported by the Office of Naval Research Contract N00014-67-A0108-0004, the National Science Foundation, grants NSF GA-1523, NSF GA-10728, NSF GA-1415, and the New York State Science and Technology Foundation SSF (8)-6.

REFERENCES CITED

Anonymous, 1964, Geological Society Phanerozoic time-scale, 1964: Geol. Soc. London Quart. Jour., v. 120, p. 260.

Budinger, T. F., and Enbysk, B. J., 1967, Late Tertiary data from the East Pacific Rise: Jour. Geophys. Research, v. 72, p. 2271-2274.

Burk, C., 1965, Geology of the Alaska peninsula, island arc and continental margin: Geol. Soc. America Mem. 99, 250 p.

Carsola, A. J., and Dietz, R. S., 1952, Submarine geology of two flat-topped northeast Pacific seamounts: Am. Jour. Sci., v. 250, p. 481-497.

Christoffel, D., and Ross, D. A. (in prep.), A fracture zone in the Southwest Pacific Basin south of New Zealand and its implications for sea-floor spreading.

Cox, A., Doell, R. R., and Dalrymple, 1963, Geomagnetic polarity epochs and Pleistocene geochronometry: Nature, v. 198, p. 1049-1051.

Dickson, G. O., Pitman, W. C., and Heirtzler, J. R., 1968, Magnetic anomalies in the South Atlantic and ocean floor spreading: Jour. Geophys. Research v. 73, p. 2087-2100.

Dietz, R. S., 1961, Continent and ocean basin evolution by spreading of the sea-floor: Nature, v. 190, p. 854-857.

Dietz, R. S., 1962, Ocean-basin evolution by sea-floor spreading, *in* Macdonald, G. A., and Kuno, H., *Editors,* The Crust of the Pacific Basin: Am. Geophys. Union, Mono. 6.

Dietz, R. S., 1968, Reply (to paper by Meyerhoff, A. A., and Holmes,

Arthur, originators of spreading ocean floor hypothesis): Jour. Geophys. Research, v. 73, p. 6567.

Doell, R. R., Dalrymple, G. B., and Cox, A., 1966, Geomagnetic polarity epochs: Sierra Nevada data, 3: Jour. Geophys. Research, v. 71, p. 531-541.

Dymond, J. R., and Windom, H. L., 1967, K-Ar ages of rocks dredged southwest of the Hawaiian Islands (abs.): Am. Geophys. Union Trans., v. 48, p. 243.

Dymond, J. R., Watkins, N. D., and Nayudu, Y. R., 1968, Age of Cobb Seamount: Jour. Geophys. Research, v. 73, p. 3977-3979.

Elvers, D. J., Peters, G., and Moses, R., 1967a, Analysis of magnetic lineations in the North Pacific (abs.): Am. Geophys. Union Trans., v. 41, p. 89.

Elvers, D. J., Mathewson, C. C., Kohler, R. E., and Moses, R. L., 1967b, Systematic ocean surveys by the USCGS *Pioneer* 1961-1963: Coast and Geodetic Survey Operational Data Rept. C and CSDR-1, 19 p.

Ewing, J., and Ewing, M., 1967, Sediment distribution on the mid-ocean ridges with respect to spreading of the sea-floor: Science, v. 156, p. 1590-1592.

Ewing, J., Ewing, M., Aitken, T., and Ludwig, W., 1968, North Pacific sediment layers measured by seismic profiling, *in* Knopoff, L., Drake, C., and Hart, P., *Editors.* The crust and upper mantle of the Pacific area: Am. Geophys. Union Mono. 12.

Ewing, M., Saito, T., Ewing, J. L., and Burckle, L. H., 1966, Lower Cretaceous sediments from the northwest Pacific: Science, v. 152, p. 751-755.

Grim, P. J., and Erickson, B. H., 1969, Fracture zones and magnetic anomalies south of the Aleutian Trench: Jour Geophys. Research, v. 74, p. 1488-1494.

Hamilton, E. L., 1956, Sunken islands of the mid-Pacific mountains: Geol. Soc. America Mem. 64, 97 p.

Hamilton, E. L., 1967, Marine geology of abyssal plains in the Gulf of Alaska: Jour. Geophys. Research, v. 72, p. 4189-4214.

Hayes, D. E., and Ewing, M., 1970, Pacific boundary structure, *in* Maxwell, A., *Editor,* The Sea, Vol. IV: Wiley Interscience (in press).

Hayes, D. E., and Heirtzler, J. R., 1968, Magnetic anomalies and their relation to the Aleutian Island Arc: Jour. Geophys. Research, v. 73, p. 4637-4646.

Heirtzler, J. R., Dickson, G. O., Herron, E. M., Pitman, W. C., III, and Le Pichon, X., 1968, Marine magnetic anomalies, geomagnetic field reversals, and motions of the ocean floor and continents: Jour. Geophys. Research, v. 73, p. 2119-2136.

Heirtzler, J. R., and Hayes, D. E., 1967, Magnetic boundaries in the North Atlantic Ocean: Science, v. 157, p. 185-187.

Helsley, C. D., and Steiner, M., 1968, Evidence for long periods of normal magnetic polarity in the Cretaceous period: Geol. Soc. America Spec. Paper 121, p. 133.

Herron, E. M., and Heirtzler, J. R., 1967 Sea-floor spreading near the Galapagos: Science, v. 158, p. 775-780.

Hess, H. H., 1960, Evolution of ocean basins: Report to Office of Naval Research on research supported by ONR Contract No. 1858(10), 38 p.

Hess, H. H., 1962, History of ocean basins, *in* Engel, A. E. J., James, H. L., and Leonard, B. F., *Editors,* Petrologic Studies: A Volume in Honor of A. F. Buddington: Geol. Soc. America, p. 599-620.

Hess, H. H., 1968, Reply (to paper by Meyerhoff, A. A., and Holmes, Arthur, originators of spreading ocean floor hypothesis): Jour. Geophys. Research, v. 73, p. 6569.

Isacks, B., Oliver, J., and Sykes, L., 1968, Seismology and the new global tectonics: Jour. Geophys. Research, v. 73, p. 5855-5899.

Langseth, M., and Von Herzen, R., 1970, Heat flow through the floor of the world oceans, *in* The Sea, Vol. IV: Wiley Interscience (in press).

Larson, R. L., Menard, H. W., and Smith, S. M., 1968, Gulf of California: A result of ocean-floor spreading and transform faulting: Science, v. 161, p. 781-783.

Le Pichon, X., 1968, Sea-floor spreading and continental drift: Jour. Geophys. Research, v. 73, p. 3661-3697.

Le Pichon, X., and Heirtzler, J. R., 1968, Magnetic anomalies in the Indian Ocean and sea-floor spreading: Jour. Geophys. Research, v. 73, p. 2101-2117.

Malahoff, W. E., Strange, W. E., and Woolard, G. P., 1966, Molokai Fracture Zone: Continuation west of the Hawaiian Ridge: Science, v. 153, p. 521-522.

Malahoff, A., and Woollard, G. P., 1968, Magnetic and tectonic trends over the Hawaiian Ridge, *in* Knopoff, L., Drake, C., and Hart, P., *Editors,* The crust and upper mantle of the Pacific area: Am. Geophys. Union Mono. 12.

Mason, R. G., 1958, A magnetic survey off the west coast of the United States between latitudes 32° and 36° N., and longitudes 121° and 128° W.: Royal Astronom. Soc. Geophys. Jour., v. 1, p. 320-329.

Mason, R. G, and Raff, A D., 1961, Magnetic survey off the west coast of North America, 32° N. lat. to 42° N. lat.: Geol. Soc. America Bull., v. 72, p. 1259-1265.

Maxwell, A., 1969, Recent deep-sea drilling results from the South Atlantic (abs.): Am. Geophys. Union Trans., v. 50, p. 113.

McKenzie, D. P., and Morgan, W. J., 1969, The evolution of triple junctions: Nature, v. 224, p. 125-133.

McKenzie, D. P., and Parker, R. L., 1967, The North Pacific: An example of tectonics on a sphere: Nature, v. 216, p. 1276-1280.

Menard, H. W., 1966, Extension of northeastern Pacific fracture zones: Science, v. 155, p. 72-74.

Menard, H. W., Allison, E. C., Durham, J. W., 1962, A drowned Miocene terrace in the Hawaiian Islands: Science, v. 138, p. 896-897.

Menard, H. W., and Atwater, T. A., 1968a, Origin of fracture zone topography: Geol. Soc. America Spec. Paper 121, p. 197.

——1968b, Changes in direction of sea-floor spreading: Nature, v. 219, p. 463-465.

Menard, H. W., and Dietz, R. S., 1951, Submarine geology of the Gulf of Alaska: Geol. Soc. America Bull., v. 62, p. 1263-1285.

Molnar, P., and Sykes, L. R., 1968, Tectonics of the Middle America and Caribbean regions from focal mechanisms and seismicity: Geol. Soc. America Spec. Paper 121, p. 204.

Morgan, W. J., 1968, Rise, trenches, great faults, and crustal blocks: Jour. Geophys. Research, v. 73, p. 1959-1982.

Naugler, F. P., and Erickson, B. H., 1968, Murray Fracture Zone: Westward extensions: Science, v. 161, p. 1142-1145.

Opdyke, N. D., and Foster, J. H., 1970, The paleomagnetism of cores from the North Pacific: Geol. Soc. America Mem. 126, p. 83-119.

Ozima, M., Ozima, M., and Kaneobia, I., 1968, Potassium argon ages and magnetic properties of some dredged submarine basalts and their geophysical implications: Jour. Geophys. Research, v. 73, p. 711-723.

Peter, G., 1966, Magnetic anomalies and fracture patterns in the northeast Pacific Ocean: Jour. Geophys. Research, v. 71, p. 5365-5374.

Peter, G., Erickson, B. H., and Grim, P. J., Magnetic structure of the Aleutian and northeast Pacific basin, *in* Maxwell, A., *Editor*, The Sea, Vol. IV: Wiley Interscience (in press).

Pitman, W. C., III, and Hayes, D. E., 1968, Sea-floor spreading in the Gulf of Alaska: Jour. Geophys. Research, v. 73, p. 6571-6580.

Pitman, W. C., III, and Heirtzler, J. R., 1966, Magnetic anomalies over the Pacific-Antarctic Ridge: Science, v. 154, p. 1164-1171.

Pitman, W. C., III, Herron, E. M., and Heirtzler, J. R., 1968, Magnetic anomalies in the Pacific and sea-floor spreading: Jour. Geophys. Research, v. 73, p. 2069-2085.

Raff, A. D., 1966, Boundaries on an area of very long magnetic anomalies in the northeast Pacific: Jour. Geophys. Research, v. 71, p. 2631-2636.

——1968, Sea-floor spreading—another rift: Jour. Geophys. Research, v. 73, p. 3699-3705.

Raff, A. D., and Mason, R. G., 1961, Magnetic survey off the west coast of North America, 40° N. latitude to 52° N. latitude: Geol. Soc. America Bull., v. 72, p. 1267-1270.

Shor, G. G., 1962, Seismic refraction studies off the coast of Alaska: Seismol. Soc. America Bull., v. 52, p. 37-58.

Sykes, L. R., 1968, Seismological evidence for transform faults, sea-floor spreading and continental drift, *in* Phinney, R. A., *Editor*, The History of the Earth's Crust: Princeton University Press, 244 p.

Talwani, M., Pitman, W. C., III, Heirtzler, J. R., and Mayhew, M., (in prep.) Geologic evolution of the North Atlantic from a study of magnetic anomalies.

Udintsev, G. B., Aqapova, G. V., Beresnev, A. F., Budanova, Ia., Zatonski, L. K., Zenkevich, N. L., Ivanov, A. G., Konaev, V. A., Kucherov, I. P., Farina, N. I., Marova, N. A., Minew, V. A., and Krautskii, E. E., 1963, New bathymetric map of the Pacific Ocean: Oceanological Research No. 9, Acad. Sci. USSR, p. 60-101.

Uyeda, S., and Vacquier, V., 1968, Geothermal and geomagnetic data in and around the island of Japan, in the crust and upper mantle of the Pacific area: Am. Geophys. Union Mono. 12.

Uyeda, S., Vacquier, V., Yasui, M., Sclater, J., Sato, T., Lawson, J., Watanabe, T., Dixon, F., Silver, E., Fukao, Y., Sudo, K., Nishikawa, M., and

Tanaka, T., 1967, Results of geomagnetic survey during the cruise of R/V *Argo* in western Pacific 1966 and the compilation of magnetic charts of the same area: Earthquake Research Inst. Bull., v. 45, p. 799-814.

Vine, F. J., 1966, Spreading of the ocean floor: New evidence: Science, v. 154, p. 1405-1415.

Vine, F. J., 1968, Paleomagnetic evidence for the northward movement of the North Pacific Basin during the past 100 m.y. (abs): Am. Geophys. Union Trans., v. 49, p. 156.

Vine, F. J., and Hess, H. H., 1970 Sea-floor spreading, *in* Maxwell, A., *Editor,* The Sea, Vol. IV: Wiley Interscience (in press).

Vine, F. J., and Matthews, D. H., 1963, Magnetic anomalies over ocean ridges: Nature, v. 199, p. 947-949.

Wilson, J. T., 1965, A new class of faults and their bearing upon continental drift: Nature, v. 207, p. 343-347.

LAMONT-DOHERTY GEOLOGICAL OBSERVATORY CONTRIBUTION NO. 1490
MANUSCRIPT RECEIVED APRIL 10, 1969
REVISED MANUSCRIPT RECEIVED JULY 11, 1969

Author Index

Subject Index